Centrifugation (2nd Edition)

a practical approach

Edited by
D Rickwood

Department of Biology, University of Essex,
Colchester, Essex, England

IRL PRESS
Oxford·Washington DC

W9-BHV-250

IRL Press Limited,
P.O. Box 1,
Eynsham,
Oxford OX8 1JJ,
England

First published April 1984
First reprinting with additions September 1986

British Library Cataloguing in Publication Data

Rickwood, D.
  Centrifugation.—2nd ed.—(The Practical Approach Series)
  1. Centrifugation
  I. Title   II. Series
543′.083   QD54.C4

ISBN 0-904147-55-X

Printed by Information Printing Ltd., Oxford, England

Preface

The centrifuge has become one of the most basic laboratory instruments and as such it is used by a wide range of laboratory personnel. Instruction is usually freely available for the actual operation of all centrifuges from, for example, manufacturers' manuals. However, while there are a number of advanced treatises on centrifugation, these frequently only review various aspects of the subject and as such do not directly relate to the laboratory use of the various centrifugation techniques. This book is designed not only to detail the important criteria for optimising centrifugal separations but also each section includes detailed protocols of experiments designed to illustrate the points made in each section.

While this book has been written primarily for novices, established research workers who already have some experience of centrifugation should find that the text, which emphasises the advantages of using newer types of rotors and gradient media, a useful reference source for their work. In addition, the general appendices at the end of the book provide a great deal of data which are extremely useful for everyone working in the field of centrifugation.

PREFACE TO THE SECOND EDITION

The enormous success of the first edition of this book emphasised the need for a book stressing the practical aspects of centrifugation. This second edition has a similar format to the first in providing extensive experimental details of protocols for all types of centrifugal separations from macromolecules to whole cells. It also describes the applications of centrifuges ranging from simple bench machines to analytical centrifuges. However, the opportunity has been taken to revise the text extensively not only to bring it up to date but also to expand its coverage to make it more comprehensive. The book has been revised and extended not only as a guide to novices but also as a reference source for experienced researchers. Finally, I would like to thank my many colleagues, both those involved in academic research and those associated with centrifuge manufacturers, for their helpful information which has enabled me to assemble such a detailed and comprehensive book.

Contributors

R.J. Barelds
Institute for Experimental Gerontology, TNO, P.O. Box 5815, 2280 HV Rijswijk, The Netherlands.

A. Brouwer
Institute for Experimental Gerontology, TNO, P.O. Box 5815, 2280 HV Rijswijk, The Netherlands.

J.A.A. Chambers
Max-Planck-Institut für Molekulare Genetik, 1000 Berlin 33 (Dahlem), F.R.G.

R. Eason
Department of Biochemistry, University of Glasgow, Glasgow, Scotland, U.K.

T.C. Ford
Department of Biology, University of Essex, Wivenhoe Park, Colchester, Essex CO3 3SQ, U.K.

J. Graham
Department of Biochemistry, St. George's Hospital Medical School, Cranmer Terrace, London SW17 ORE, U.K.

B.D. Hames
Department of Biochemistry, University of Leeds, Leeds LS2 9JT, U.K.

D.L. Knook
Institute for Experimental Gerontology, TNO, P.O. Box 5815, 2280 HV Rijswijk, The Netherlands

D. Rickwood
Department of Biology, University of Essex, Wivenhoe Park, Colchester, Essex CO3 3SQ, U.K.

B.D. Young
Beatson Institute for Cancer Research, Garscube Estate, Bearsden Road, Bearsden, Glasgow, Scotland, U.K.

Abbreviations

BSA	bovine serum albumin
DAB	diaminobenzidine tetrahydrochloride
DFP	di-isopropylfluorophosphate
DMSO	dimethyl sulphoxide
DTT	dithiothreitol
EDTA	ethylenediamine tetraacetic acid
EGTA	ethyleneglycobis(β-aminoethyl)ether tetraacetic acid
GBSS	Gey's balanced salt solution
NP-40	Nonidet P-40
PBS	phosphate-buffered saline
PMSF	phenylmethylsulphonyl fluoride
POPOP	1,4-bis-(5-phenyloxazol-2-yl)benzene
PPO	2,5-diphenyloxazole
PVP	polyvinylpyrrolidone
RCF	relative centrifugal force
mRNP	messenger ribonucleoprotein
SDS	sodium dodecyl sulphate
SV40	simian virus 40
TCA	trichloroacetic acid

Contents

CHAPTER 1

The Theory and Practice of Centrifugation

D. RICKWOOD

1. INTRODUCTION

The aim of this chapter is to introduce the reader to some of the basic concepts in the area of centrifugation. The beginning of this chapter outlines the theoretical bases of centrifugation to introduce the reader to the most important parameters which are likely to be encountered. More rigorous and mathematically detailed treatments of centrifugation theory can be found elsewhere (1,2). The other parts of this chapter deal with the practical aspects of centrifugation in terms of descriptions of centrifuges and rotors, as well as describing the properties of the various gradient media used for centrifugal separations.

2. THEORY OF CENTRIFUGATION

2.1 Sedimentation Theory

In a suspension of particles, the rate at which the particles sediment depends not only on the nature of the particles but also on the nature of the medium in which the particles are suspended as well as the force applied to the particles. Intuitively, one would expect that larger particles should sediment more rapidly than smaller ones. In fact, although biological particles vary enormously in size from relatively small proteins to whole cells, the parameters affecting sedimentation are the same irrespective of size. One important factor affecting the sedimentation of particles is the viscosity of the medium. In 1856, Sir Gabriel Stokes proposed that the frictional force, F, acting on a rigid spherical particle of radius, r_p, was related to the viscosity, η, by the equation:

$$F = 6\pi\eta.r_p.\frac{dr}{dt} \qquad \text{Equation 1}$$

where dr/dt is the velocity of the particle. As shown in *Figure 1* the actual force experienced by particles is determined not only by the gravitational force, g, but also the flotation effects which reflect the differences in the density of the medium (ϱ_m) and the particles (ϱ_p). Thus Equation 1 becomes:

$$(\varrho_p - \varrho_m)\,V.g = 6\pi\eta\,r_p\frac{dr}{dt} \qquad \text{Equation 2}$$

Since the particle is assumed to be spherical, then the volume, V, can be ex-

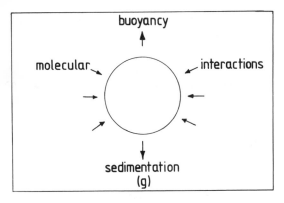

Figure 1. Forces acting on particles in solution.

pressed in terms of the radius of the particle. Thus:

$$\frac{4}{3} \pi r_p^3 (\varrho_p - \varrho_m) g = 6 \pi \eta \, r_p \frac{dr}{dt}$$ Equation 3

In practice, the centrifugal force which moves the particles away from the axis of rotation is very much greater than the Earth's gravitational field and so we can express centrifugal force relative to the Earth's gravitational field by the expression:

$$\text{centrifugal force} = \frac{\omega^2 r}{g}$$ Equation 4

where r is the radial distance of the particle from the axis of rotation and ω is the angular velocity in terms of radians/sec. Substituting this relationship into Equation 4 and simplifying gives the expression in terms of the velocity of particles, namely:

$$\frac{dr}{dt} = \frac{2r_p^2 (\varrho_p - \varrho_m)\omega^2 r}{9\eta}$$ Equation 5

This expression is only true for spherical particles. Non-spherical particles have larger frictional coefficients. In the case of rod-like molecules the frictional coefficient of the molecule (f) can be as much as ten times that of the frictional coefficient of a sphere (f_o). To take this into account, Equation 5 can be modified to give the expression:

$$\frac{dr}{dt} = \frac{2r_p^2 (\varrho_p - \varrho_m)\omega^2 r}{9\eta(f/f_o)}$$ Equation 6

Besides the buoyancy and sedimentation forces, the particles are also subjected to molecular forces of the surrounding medium (see *Figure 1*). If the particles are small then considerable centrifugal force is necessary to counteract these forces and sediment particles.

2.2 Non-ideality of Biological Particles

One factor that can complicate sedimentation studies is that a number of biological particles have a dynamic nature. For example, proteins with a subunit structure (e.g., haemoglobin) may undergo dissociation during centrifugation leading to the formation of multiple peaks. Another problem is that sedimentation down the gradient may alter the properties of the particle. As an example, the rate at which high molecular weight DNA migrates through a gradient depends on the relative centrifugal force, since higher centrifugal forces appear to alter the conformation and hence the sedimentation rate of the DNA (4).

The other feature of centrifugation is that the centrifugal force generates hydrostatic pressure within the solution. The hydrostatic pressure generated can be sufficient to permeabilise membranes to gradient solutes (5), to dissociate nucleoprotein complexes (6) and to disrupt protein complexes (3). Hence, in choosing centrifugation conditions or in interpreting sedimentation patterns, care must be taken to avoid conditions which may lead to the formation of artifacts.

2.3 Sedimentation Coefficients

From Equation 6, it can be seen that it is possible to define a particle in terms of its behaviour in a centrifugal field, that is in terms of its sedimentation coefficient, s, where:

$$s = \frac{dr/dt}{\omega^2 r} \qquad \text{Equation 7}$$

For most biological macromolecules the magnitude of s is about 10^{-13} sec and hence the unit of sedimentation, the Svedberg (S), has been defined as being equal to 10^{-13} sec. The definition of sedimentation coefficients is discussed in greater detail in Chapters 4 and 8. It is also important to realise that not only is the relationship between the sedimentation coefficient of a particle and its molecular weight not linear but also it varies from one type of particle to another.

2.4 Practical Calculations of Centrifugal Force and Centrifugation Times

As shown in Equation 4, the relative centrifugal force (RCF) can be calculated from the expression:

$$RCF = \frac{\omega^2 r}{g} \qquad \text{Equation 4}$$

It is inconvenient to measure the angular velocity, ω, and so it is more convenient to express the RCF in terms of revolutions per minute (r.p.m.), N, and this gives the expression:

$$RCF = 11.18 \times r\left(\frac{N}{1000}\right)^2 \qquad \text{Equation 8}$$

The centrifugal force is usually given in terms of 'g' and is written as such or as 'xg'.

From Equation 8, it can be seen that the centrifugal force acting on the particle is related to the square of the speed and hence doubling the speed increases the centrifugal force by a factor of four. The centrifugal force also increases with the distance from the axis of rotation (r). Hence particles in a homogeneous medium will accelerate as the radial distance increases, although in sucrose gradients, where there is a viscosity gradient, increasing viscosity tends to minimise the effects of increasing the radial distance. The reader should note that centrifugal force can only be calculated if both the speed and radial dimensions of the rotor are known.

Manufacturers usually give the dimensions of rotors in terms of the maximum and minimum radii, r_{max} and r_{min}, respectively (see Appendix III). The r_{max} quoted by manufacturers usually relates to the distance to the bottom of the bucket; especially where thick-walled tubes are being used the r_{max} and hence g_{max} may be significantly less. Throughout this book, unless otherwise stated, the centrifugal force will be given as that at the centre of the solution, that is, the average centrifugal force, g_{av}.

The other parameter of rotors that is usually quoted is the k-factor (see Appendix III). The smaller the k-factor the greater is the pelleting efficiency of the rotor. The k-factor can be calculated for rotors using the expression:

$$k = \frac{2.53 \times 10^{11} \left[\ln\left(\frac{r_{max}}{r_{min}}\right) \right]}{N^2} \qquad \text{Equation 9}$$

If the sedimentation coefficient (s) of particles is known, then the k-factor can be used to calculate the time in hours (t) required to pellet the particles using the relationship:

$$t = \frac{k}{s} \qquad \text{Equation 10}$$

It must be emphasised that the k-factor is related to the speed of the rotor. All k-factors relate to the maximum speed of the rotor; at lower speeds the k-factor is correspondingly increased according to the relationship:

$$k_{actual} = k \left(\frac{N_{max}}{N_{actual}} \right)^2 \qquad \text{Equation 11}$$

Also all k-factors quoted by manufacturers assume that the tubes are full; the k-factors of rotors are smaller if partially filled tubes are used because the shorter pathlength enables the particles to pellet more quickly. The other feature to be remembered is that the k-factor is calculated on the basis that the density and viscosity of the liquid medium in which the particles are suspended are not significantly different from the density and viscosity of water, increasing either of these effectively increases the k-factor.

In the case of sucrose gradients there is a viscosity gradient throughout the tube and hence the k-factor cannot be used. Instead a model system of the time needed to sediment particles to the bottom of a 5 – 20% (w/w) sucrose gradient at 5°C is used to define k' and k* factors (7) which allow one to estimate the sedimentation pattern in a 5 – 20% (w/w) sucrose gradient. A series of k'-values are used depending on the density of the particle (7) while k* factors are calculated on the assumption that the density of the particle in sucrose is

1.3 g/cm³; usually this assumption does not introduce large errors into the calculation (see Appendix III). However, for estimating the sedimentation of particles in sucrose gradients accurately it is often more useful to use computer simulation methods (see Section 4 of Chapter 4).

3. TYPES OF CENTRIFUGAL SEPARATIONS

3.1 **Differential Pelleting**

As might be predicted from Equation 5, centrifugation will first sediment those particles which are largest. In addition, as indicated in Equation 6, very asymmetrical molecules will sediment more slowly than spherical particles of the same mass and density. Increasing either the centrifugation speed or the time of centrifugation will cause smaller particles to pellet also (*Figure 2*). As might be expected from Equation 5, differential centrifugation separates particles not only according to size but also on the basis of density, since particles that are denser (e.g., nuclei) will pellet at a faster rate than less dense particles (e.g., membranes) of the same mass. Hence it is sometimes possible to obtain good separations of particles of similar sizes but different densities by differential pelleting. For particles of similar densities one usually requires about a 10-fold difference in mass to separate one particle from another efficiently by differential pelleting. The major problem with differential pelleting is that, as shown in *Figure 2*, the centrifugal force necessary to pellet the larger particles from the top of the solution is also often sufficient to pellet the smaller particles nearer the bottom of the tube. Hence in a single step it is only possible to obtain a pure preparation of the smallest particles since these will remain in solution after all the other larger particles have pelleted. The yield of such a procedure is, however, likely to be low. An alternative approach is to minimise

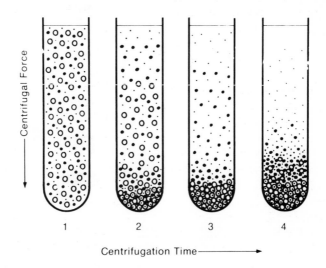

Figure 2. Fractionation of particles by differential pelleting. Reproduced from ref. 7 with permission.

the cross-contamination of the pellet by washing the pelleted material by re-suspension and re-centrifugation; however this does result in losses of the pelleted material. In addition, some more labile particles, for example, cells, may be damaged by repeated re-suspension and re-centrifugation and hence great care is needed. Because of the problems associated with this technique, differential pelleting is usually used for the initial processing of large volumes of heterogeneous mixtures to obtain fractions enriched in the particles of interest prior to further purification. A typical scheme for preparing subcellular fractions and the practical details of this technique are discussed further in Chapter 5.

3.2 Rate-zonal Centrifugation

In rate-zonal centrifugation it is possible to avoid the problem of the co-sedimentation of particles of different sizes by layering the sample as a narrow zone onto the top of a density gradient (*Figure 3*). The primary purposes of the gradient are to facilitate layering of the sample and to minimise convection currents in the liquid column during centrifugation which would otherwise disrupt the particle zones as they move down the tube. An additional advantage in terms of the resolution of sample zones is obtained if the density gradient is also a viscosity gradient; this effect is discussed in detail in Section 3.3 of Chapter 2.

Rate-zonal separations are ideal for particles of defined size (e.g., proteins, RNA and ribosomes). However, particles of the same type are frequently heterogeneous (e.g., membrane fragments, mitochondria and other cell organelles). In this case, rate-zonal separations do not efficiently separate the particles according to type and it is more appropriate to separate particles on

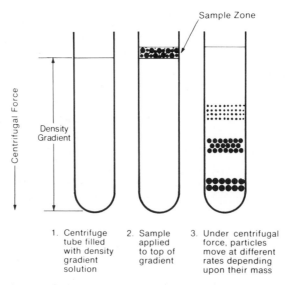

Figure 3. Fractionation of particles by rate-zonal centrifugation. Reproduced from ref. 7 with permission.

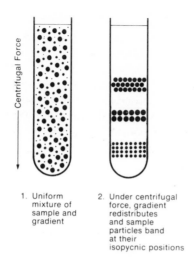

1. Uniform
   mixture of
   sample and
   gradient

2. Under centrifugal
   force, gradient
   redistributes
   and sample
   particles band
   at their
   isopycnic positions

Figure 4. Fractionation of particles by isopycnic centrifugation.

the basis of some other parameter such as density. This can be done by iso-pycnic centrifugation.

3.3 Isopycnic Centrifugation

In isopycnic separations the particles are separated purely on the basis of their density; size only affects the rate at which particles reach their isopycnic positions (*Figure 4*). These separations are based on the centrifugation of particles in a density gradient through which the particles move until their density is the same as that of the surrounding medium (i.e., referring to Equation 5 where ϱ_p equals ϱ_m). The effective density of the particles differs from one medium to another because particles are more hydrated in some media than in others.

The features of isopycnic centrifugation are that, because the separation is an equilibrium one, prolonged centrifugation does not affect the separation as long as the gradient remains stable and the particles are unaffected by centrifugation. For isopycnic separations it is not obligatory to load the sample onto the top of the gradient, indeed by mixing the sample throughout the gradient one removes the possibility of artifactual bands forming as a result of small particles remaining in the sample zone. In addition, some gradient media form gradients when they are centrifuged; these are termed self-forming gradients. Hence it is not always necessary to preform gradients for isopycnic separations. Self-forming gradients are discussed in detail in Sections 4.2 and 5.2 of Chapter 2.

4. CENTRIFUGES AND ASSOCIATED EQUIPMENT

4.1 Types of Centrifuge

Centrifuges can be classified on the basis of a variety of criteria. However, perhaps the most useful approach is to classify preparative centrifuges on the

Table 1. Types of Centrifuges and Their Applications.

	Type of centrifuge		
	low-speed	*high-speed*	*ultracentrifuge*
Speed range (r.p.m. x 10^{-3})	2 − 6	18 − 25	40 − 80
Maximum RCF (g x 10^{-3})	6	60	600
Refrigeration	some	yes	yes
Vacuum system	none	some	yes
Acceleration/braking controls	some	variable	variable
Applications for pelleting:			
cells	yes	yes	yes
nuclei	yes	yes	yes
membranous organelles	some	yes	yes
membrane fractions	some	some	yes
ribosomes/polysomes	−	−	yes
macromolecules	−	−	yes

basis of their maximum speed since this gives an indication of the centrifugal force that can be generated and, in turn, their range of applications. Analytical centrifuges will not be discussed here since they are described in detail in Chapter 8. *Table 1* gives the three main classes of centrifuge, namely, low-speed, high-speed and ultracentrifuge, together with the usual range of applications of each type of machine.

4.1.1 *Low-speed Centrifuges*

These machines range from small bench centrifuges, in which the samples are kept cool by drawing air through the centrifuge bowl, to the large refrigerated floor-standing machines capable of centrifuging up to 6 litres at a time. These machines are used routinely for the initial processing of biological samples. This type of centrifuge can be used for pelleting cells and the faster sedimenting cell organelles such as nuclei and chloroplasts. In addition, they can also be used for the fractionation of cells either on density gradients or by centrifugal elutriation (see Chapter 6).

4.1.2 *High-speed centrifuges*

These centrifuges, usually with maximum speeds of 18 000 − 25 000 r.p.m., are machines that can generate about 60 000 g, they are much cheaper to buy and maintain as compared with ultracentrifuges. High-speed centrifuges are refrigerated and some types also have a vacuum system. However, machines with a vacuum system are usually less convenient to use and more expensive to maintain but they do have the advantage that they have more accurate temperature control systems although, for most practical purposes, this may not be important.

These machines, like the low-speed centrifuges, are used mainly for the preparation of subcellular fractions. The advent of vertical rotors (see Section 4.2.2) has facilitated the use of high-speed centrifuges for gradient separations (8).

4.1.3 *Ultracentrifuges*

These can be subdivided into two types, analytical and preparative ultra-centrifuges; the former are primarily designed to obtain very accurate data on the sedimentation properties of particles and they are described in detail in Chapter 8. Preparative ultracentrifuges are also widely used for quantitative estimations of sedimentation coefficients of particles in sucrose gradients although, using preparative rotors, the data obtained is not as accurate as that obtained using analytical ultracentrifuges. Some types of preparative ultra-centrifuge do have attachments allowing analytical rotors to be used but the quality of the optical system is not usually as good as that of dedicated analytical centrifuges.

The centrifugal force generated by ultracentrifuges can be significantly greater than 600 000 *g* which is sufficient to pellet even quite small proteins.

4.1.4 *Miscellaneous Centrifuges*

Besides the broad classes of preparative centrifuge listed in the previous three subsections, a wide range of other types of centrifuges exist such as the micro-centrifuges, widely used for pelleting small samples, and the specialised continuous flow centrifuges, such as that shown in *Figure 5*, which are used for

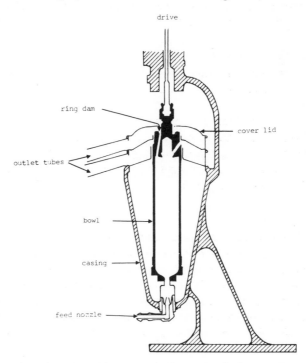

Figure 5. A schematic representation of the construction of a Sharples continuous-flow centrifuge. The liquid is fed into the bottom of the machine, fractionation occurs as the liquid passes up the cylindrical bowl and the fractionated particles are collected by the upper outlets. Different bowls are available for various types of particles and separation.

large volumes of sample. These continuous-flow centrifuges can be used for pelleting, isopycnic separations and phase separations.

4.2 Drive Systems of Centrifuges

This section is not designed to give the reader a detailed technical description of the various drive systems available but rather it surveys the various types of system available to give the reader an insight into the main advantages and disadvantages of each.

4.2.1 Electric Motor Systems

This method is the commonest method used for the manufacture of all types of centrifuge. However, there are a number of very different electric motor systems which vary according to the type of centrifuge. The two commonest forms of motor used are the d.c. brush motor and the induction motor which are used for a wide range of centrifuges. Brushless induction motors tend to require less maintenance although they do require more complex cooling systems, central circuitry and a higher electrical current.

The motors may either drive the rotor directly (direct drive) or indirectly via a belt drive or a gearbox. The direct drive systems tend to be more reliable in that there are fewer components which can fail. On the other hand, direct drives involve greater stresses on the motor assembly, particularly in the case of the high speed and ultracentrifuges where speeds are higher and the precision of the components must be much greater. *Figure 6A – C* illustrates some of the electric motor drives that have been produced.

4.2.2 Turbine Drives

It is of interest to note that, in constructing some of the first ultracentrifuges, Svedberg utilised the potential of the turbine for driving rotors at high speeds for many hours. Although turbines, usually driven by air or steam, continued to be used for specialised machines, such as the type of continuous-flow machines as illustrated in *Figure 5*, they were neglected for a number of years. Then an oil turbine was developed by Sorvall. The advantages of this system (*Figure 6D*) is that only the small turbine assembly and bearing housing are subjected to high stresses (*Figure 7*) and this tends to enhance the overall reliability of the system. It is also claimed that the acceleration control of this system is superior in terms of gentle acceleration at low speeds. However, there is no evidence that, in practice, this system will give better results. The other problem is that the relatively low torque of the turbine limits the accelaration rate of the heavier rotors.

Another interesting centrifuge is the Air-fuge® (Beckman/Spinco) in which the small aluminium rotor is machined to act as a turbine. The air jets are responsible not only for rotating the rotor but also for levitating the rotor. The temperature of the rotor is controlled by controlling the temperature of the air supply. The rotor is small and can be rapidly accelerated to 100 000 r.p.m. (165 000 g). It is especially useful for studying time-dependent reactions.

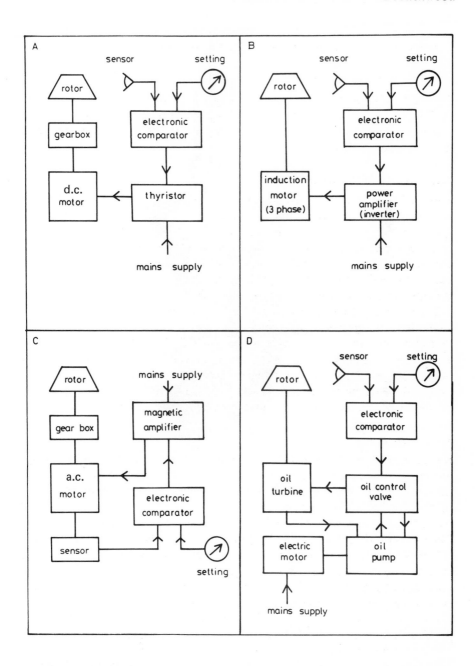

Figure 6. Schematic representations of the various types of drive systems used for ultracentrifuges. **(A)** The typical d.c. motor design; **(B)** induction motor drive system; **(C)** an a.c. motor drive system as used in some older MSE ultracentrifuges; **(D)** oil-turbine drive system as manufactured by Du Pont-Sorvall.

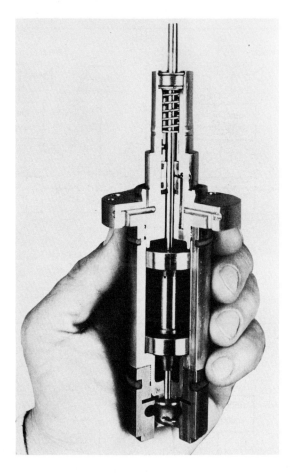

Figure 7. Photograph of the oil turbine drive as used in the Du Pont-Sorvall ultracentrifuge. Reproduced with permission.

However, the necessarily limited volume of the rotor does place restrictions on the range of applications of this type of machine.

4.3 Centrifuge Rotors

Originally rotors were designed in a fairly empirical manner in which, having chosen the capacity of the rotor, a round piece of metal of the appropriate size which was strong enough to reach the required speed without disintegrating was used. However, the use of computer-aided design methods has resulted in the development of rotors with more angular appearances (*Figure 8*). This is because any metal that does not directly strengthen the rotor weakens it because of the centrifugal forces that act on the rotor when it spins; indeed most of the strength of the rotor is required to hold the rotor together rather than to retain the sample tubes within the rotor.

Figure 8. Appearance of a computer-designed fixed-angle ultracentrifuge rotor. Reproduced with the permission of Kontron Instruments.

4.3.1 *Materials Used in the Manufacture of Rotors*

The stress on rotors is related to the square of the speed (see Equation 4) and hence the stress on ultracentrifuge rotors is very much greater than that on low-speed rotors. Low-speed rotors can be made of brass, steel or even moulded plastics. However some modern low-speed rotors are made from aluminium and have computer-optimised designs which combine lightness with high performance. High-speed rotors are usually manufactured from aluminium alloys although some do have titanium buckets. In order to obtain the performance required of ultracentrifuge rotors strong titanium and aluminium alloys have been used but a major innovation has been the introduction of carbon fibre rotors.

All rotors made from aluminium alloys must be treated with extreme care since they are particularly susceptible to corrosion (see Section 4.3.3). Corrosion of aluminium rotors is greatest with acid or alkaline solutions but even solutions containing low concentrations of salts at neutral pH can break down the protective oxide film covering the surface of the aluminium alloy exposing the reactive metal beneath. Aluminium rotors are usually anodised but this does not give a high degree of protection against corrosion. When aluminium rotors are centrifuged at high speed the protective oxide film is disrupted allowing corrosion of the metal within the body of the rotor; this is known as stress corrosion. In addition, aluminium rotors suffer from metal fatigue.

Titanium rotors are much more resistant to corrosion and indeed corrosion is unlikely to present problems using any of the solutions normally used for the separation of biological material since titanium is resistant to salt solutions at both acid and alkaline pH. In addition, titanium rotors do not suffer from metal fatigue and they are also resistant to stress corrosion. In the author's ex-

13

perience no titanium rotor is likely to fail when used within the recommended speed range. The durability of titanium rotors can be judged by the fact that some manufacturers guarantee their titanium rotors indefinitely, irrespective of the number of runs and the total centrifugation time. Other manufacturers only offer relatively short guarantees on titanium rotors before derating is required. Hence, although in terms of price titanium rotors are more expensive, they should last almost indefinitely while aluminium rotors become unusable either as a result of corrosion or metal fatigue.

Carbon fibre composite material, composed as it is of inert carbon fibres in an epoxy resin matrix, is very inert to aqueous and organic solvents and so it should be resistant to corrosion effects. The actual strength of carbon fibre composite depends on the nature of the resin but of even greater importance is the method of manufacture. The strength of carbon fibre material tends to be very directional depending on the orientation of the fibres and so the fibre orientation must be correct; this can be very difficult to achieve. One of the important features of carbon fibre rotors is that they are much lighter than metal rotors and so not only is there less strain on the drive unit but also the times needed for acceleration and braking are much shorter thus a significant reduction in run times is possible.

4.3.2 *Types of Rotor*

Preparative centrifuge rotors can be classified into four main types, namely swing-out (swinging bucket), fixed-angle, vertical and zonal. The last of these is described in detail in Chapter 7 and will not be discussed further in this chapter. In addition, there is a range of analytical rotors and these are described in Chapter 8. The various types of rotor have markedly different characteristics and also, as shown in *Table 2*, different applications. A more detailed discussion of the applications of rotors is given in Chapter 2 (Section 2.2).

Swing-out rotors. In the case of these rotors, the sample solutions in tubes are in individual buckets which move out perpendicular to the axis of rotation as the rotor rotates (*Figure 9*). Hence in these rotors the centrifugal force is exerted along the axis of the tube, however, since the centrifugal force is axial, some particles are sedimented against the wall of the tubes giving rise to 'wall effects' which disrupt the sample zones (see Section 2.2 of Chapter 2).

In the case of most low- and high-speed centrifuge rotors the individual buckets simply slot onto the central part of the rotor. In the case of the ultracentrifuge rotors the necessity for the rotor to withstand very high centrifugal

Table 2. Types of Rotor and Their Applications.

Type of rotor	Type of separation		
	pelleting	rate-zonal	isopycnic
Fixed-angle	excellent	poor	good
Vertical	poor	good	excellent
Swing-out	inefficient	good	adequate
Zonal	poor	excellent	adequate

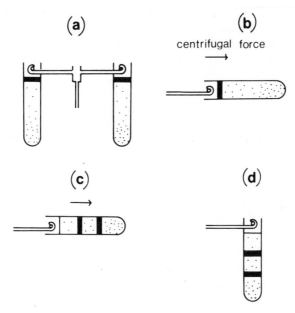

(a)

(b)

centrifugal force

(c)

(d)

Figure 9. Operation of swing-out rotors. **(a)** The gradient is first prepared and the sample is layered on top; centrifuge tubes in buckets attached to swing-out rotor. **(b)** Centrifuge bucket reorients as rotor accelerates to lie perpendicular to the axis of rotation. **(c)** Bands form as the particles sediment. **(d)** Rotor decelerates. Centrifuge bucket comes to rest in its original vertical position.

forces means that much more sophisticated designs need to be employed. The three main systems used for attaching buckets to ultracentrifuge rotors are shown in *Figure 10*. The safest of these three methods is the hinge pin method (*Figure 10a*) in that the bucket is securely attached and cannot be dislodged. However, using this design it is only possible to attach three buckets to the rotor; other designs must be adopted for six-place rotors. The second method of attaching the buckets is the hook-on method in which the bucket hooks on-to the centre of the rotor (*Figure 10b*); the hook may be on the bucket or the yoke of the rotor. Generally, one finds a wide variation in the ease of use of these types of rotor since some hook-on systems are much more secure than others; these variations are found not only between manufacturers but also between different sizes of rotor and depend on the design of the hook. The third method of attaching buckets is the ball and socket method (*Figure 10c*). The ease of use of this type of rotor is dependent on the exact design. In the past some rotors have tended to have rather shallow sockets, in which case a slight disturbance can lead to the displacement of a bucket. A modification of the original design has been introduced by Damon/IEC which allows top loading of the buckets and this facilitates their attachment. In order to ensure that these rotors function efficiently it is important to ensure that the surfaces are free of dirt which otherwise tends to prevent smooth movement of the bucket. Both the hook-on and ball and socket designs can accommodate six

15

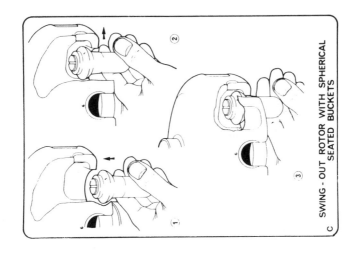

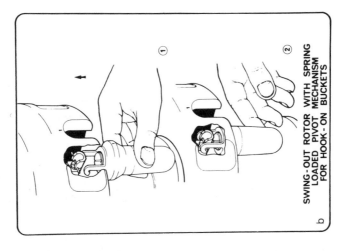

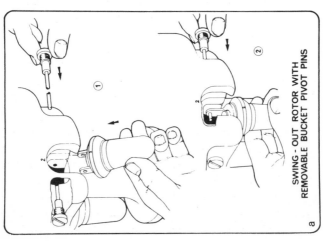

Figure 10. Methods of attaching buckets onto ultracentrifuge swing-out rotors. **(a)** Hinge pin; **(b)** hook-on; **(c)** ball and socket.

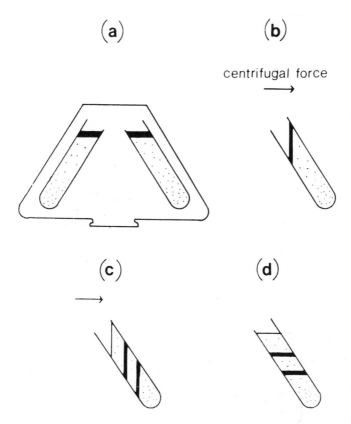

Figure 11. Operation of fixed-angle rotors. **(a)** The gradient is prepared, the sample loaded and the centrifuge tubes placed in the tube holders of fixed-angle rotor. **(b)** Both sample and gradient reorient during acceleration. **(c)** Bands form as particles sediment. **(d)** Bands and gradient both fully reoriented; rotor at rest.

buckets, allowing one to centrifuge two, three, four or six samples each time. It should be noted that rotors should only be run with all six buckets attached and placed in their correct positions to ensure that the dynamic balance of the rotor is maintained. As described in Chapter 2, swing-out rotors are frequently used for rate-zonal separations of particles, they can also be used for isopycnic fractionations although they have a lower capacity than gradients of the same volume in fixed-angle or vertical rotors. Swing-out rotors have relatively high k-factors and hence they are relatively inefficient for differential pelleting.

Fixed-angle rotors. In these rotors, as the name suggests, the tubes are at a fixed angle and when the rotor rotates the solution reorientates in the tubes (*Figure 11*). The angle of the tubes in the rotor can vary from 14° to 40°. Rotors with shallow angles are more efficient at pelleting because the sedimentation pathlength is shorter. Fixed-angle rotors can be designed to withstand very high centrifugal forces (>600 000 *g*) and hence they have relatively low k-factors. In addition, once the particles hit the wall of the tube they slide down and form a pellet at the bottom of the tube (*Figure 12*). Hence fixed-

17

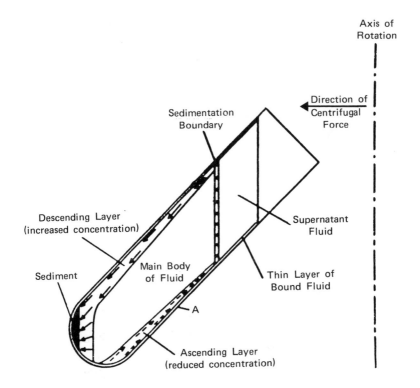

Figure 12. Schematised representation of the boundary sedimentation of particles in a fixed-angle rotor.

angle rotors are especially useful for the differential pelleting of particles. An additional advantage of fixed-angle rotors is that the reorientation of the solution in the tubes enhances the loading capacity of isopycnic gradients (9).

Vertical rotors. These rotors are now available for most types of high-speed and ultracentrifuges. In this rotor, the tubes are held in a vertical position (*Figure 13*) and at first sight one might consider them to be an extreme form of fixed-angle rotor. However, the characteristics of the vertical rotor are sufficiently different for them to merit separate consideration. When the rotor turns the solution begins to reorientate through 90° (*Figure 13*). This reorientation takes place below 1000 r.p.m. and, if the acceleration is sufficiently slow and smooth, the reorientation does not disrupt the gradient or particle zones. The important feature of vertical rotors is their short sedimentation pathlength, namely the diameter of the tube, which endows vertical rotors with their relatively low k-factors. In addition, the design of the rotors enables them to generate very high centrifugal forces. Moreover, the minimum radius and hence minimum centrifugal force is larger than usual and it is thus possible to use high-speed vertical rotors to carry out fractionations which, if other types of rotor are used, would require an ultracentrifuge and rotor (8). Vertical rotors are not suitable for pelleting but they can be used for isopycnic cen-

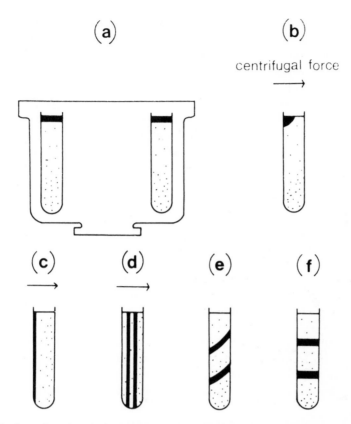

Figure 13. Operation of vertical rotors. **(a)** The gradient is prepared, the sample is layered on top and the centrifuge tubes are placed in the pockets of the vertical rotor. **(b)** Both sample and gradient begin to reorient as rotor accelerates. **(c)** Reorientation of the sample and gradient is now complete. **(d)** Bands form as the particles sediment. **(e)** Bands and gradient reorient as the rotor decelerates. **(f)** Bands and gradient both fully reoriented; rotor at rest.

trifugation. The reorientation of the gradient that occurs means that the capacities of gradients in these rotors are higher than in swing-out and fixed-angle rotors (8). In addition, vertical rotors can be used for rate-zonal separations. It is possible to compare the resolution of swing-out and vertical rotors using computer simulation techniques (10).

4.3.3 *Stability of Rotors*

One of the main considerations in both designing and using rotors is stability during centrifugation. At low speed, even perfectly balanced ultracentrifuge rotors precess, that is the centre of the rotor is not stationary but rather it describes a circular motion. Precession, which is also termed 'synchronous whirl', occurs with all rotors, although the speed at which this occurs and the magnitude of the effect depends upon the design of the rotor and the drive shaft. The degree of precession can be minimised by designing the drive assembly so that the movement of the drive shaft is damped by the bearing housing. An alternative, generally less satisfactory, solution is the incorporation of a stabiliser at the top of the rotor, as used in the Beckman ultracentrifuges prior to the introduction of the L5 model. However, whereas the amplitude of precession may be quite large, there is essentially no evidence to suggest that it spoils separations or affects the resolution of particles.

A more serious problem is that of asynchronous whirl which occurs at higher speeds and which can affect the resolution of rotors. Asynchronous whirl is caused by mechanical imperfections or imbalance of the rotor and drive assembly and such defects are usually detected when the centrifuges and rotors are tested at the factory. However, instability of a rotor resulting in asynchronous whirl can, and does, occur for a number of other reasons. The most likely cause is incorrect balancing of the sample tubes. The magnitude of the effect of a small imbalance depends on the weight, design and rotational speed of the rotor; at high speeds it can be very large indeed. Correct balancing is particularly important when centrifuging density gradients, especially when using swing-out ultracentrifuge rotors; in this case, it is advisable to balance the sample tubes to within 0.1 g. Ultracentrifuge rotors are spun in a high vacuum so that a small leakage of the cap can cause a tube to lose several millilitres of liquid in just a few hours. If the leakage is severe the rotor can become so unstable that the drive shaft is bent; the rotor may even be thrown off the spindle. It must also be remembered that not only should the tubes be balanced as far as the total weight is concerned but also balanced with respect to their centre of gravity. This is particularly important in density-gradient centrifugation; in such experiments the tubes are only correctly balanced if they contain identical gradients. Finally, it must be stressed that the empty buckets and caps together form an integral part of ultracentrifuge swing-out rotors as far as balance is concerned. Consequently, each bucket must be hung at its correct position and in the correct orientation on the yoke of the rotor and, moreover, the tubes and their contents must be balanced by themselves, *not* in the buckets.

A second common cause of asynchronous whirl is running a rotor in a centrifuge with a bent drive shaft. As stated previously, unbalanced rotors can bend the drive shaft, although more usually the damage occurs when rotors are removed from the spindle without due care. Rotors should always be lifted off using the correct extractor key when one is provided, particularly when the rotors are stuck on the spindle. Rotors should always be removed by applying a vertical lifting force and *never* by wrenching or twisting the rotor at an angle to the spindle. If it is suspected that a spindle is damaged it can be checked by placing a rotor in position and spinning it slowly by hand; if the spindle is bent the centre of the rotor will describe a small but discernible circle. Once the spindle is bent it must be replaced before the centrifuge is used again. A bent spindle not only ruins the resolution normally obtainable, but also, in the case of ultracentrifuges, can cause serious (and expensive) damage to the bearing housing and its vacuum seal and the gearbox.

Finally, it should be noted that any object which touches a spinning rotor can also cause asynchronous whirl; consequently, any attempt to restrain the low-speed synchronous whirl movements of a rotor can actually do more harm than good as far as rotor stabilisation is concerned. Also, it is clear that any contact between a spinning rotor and, for example, a temperature sensor will not only damage the sensor but also destabilise the rotor.

4.3.4 *Care of Rotors*

The massive nature of most rotors gives the impression of great strength and it is difficult to imagine circumstances which might result in failure of the rotor. In fact, some rotors can be weakened simply by dropping them onto a solid surface. In addition, although most rotors can be autoclaved at 120°C, repeated autoclaving or heating of the rotor above normal autoclaving temperatures severely weakens the rotor and can lead to failure of the rotor during centrifugation.

However, as described in Section 4.3.1, the major problem with rotors, especially aluminium rotors, is the corrosion. If at all possible, the user should avoid centrifuging concentrated salt solutions in aluminium rotors. When using aluminium rotors particular care must be taken to rinse out the rotor with deionised water after use and allow it to dry. Laboratory detergents should not generally be used since they are usually very alkaline, if it is necessary to use a detergent one should use either one of the proprietary brands of washing-up liquids or household detergents as used for washing clothes. Care must also be taken not to leave aluminium rotors in soak even in distilled water. Some manufacturers sell de-watering sprays for rotors which are similar to the WD 40 spray generally available (e.g., MSE rotor spray). However, the effects of de-watering sprays tend to be fairly temporary and less effective on corroded surfaces. An alternative treatment which has been used very successfully by the author on both new and corroded rotors is to coat the inside of the rotor with Waxoyl (Finnegans Speciality Paints Ltd.); this gives a very durable protective coating which inhibits any further corrosion.

Titanium rotors are essentially resistant to corrosion. However, many so-called titanium rotors also contain aluminium or ferrous components that are susceptible to corrosion. Most ultracentrifuge rotors have a speed disc on the bottom which monitors and controls the speed and these have a fail-safe design so that any damage prevents rotors from reaching their usual maximum speed. Hence, care must be taken not only in washing rotors but it is also important to avoid standing rotors on abrasive surfaces. It is also important that care be taken to ensure that the outer surface of the rotor is not damaged since such damage may lead to inaccurate temperature control by affecting the infra-red pick up of the temperature monitor.

Sometimes rotors cannot be run at their maximum speed, this is known as the derating of rotors. Derating is usually necessary when using solutions with densities greater than 1.2 g/cm^3 since some rotors, especially swing-out rotors, are not designed to reach maximum speed with dense solutions. The maximum amount of derating necessary can be calculated from the equation:

$$N' = N\left(\frac{1.2}{\varrho}\right)^{1/2}$$

where N is the usual maximum speed of the rotor and N' is the maximum speed when using a solution with a density of ϱ g/cm^3.

Rotors must also be derated if they become corroded. Some manufacturers also recommend that rotors should be derated if they have completed a certain number of runs and accumulated a certain number of hours. The exact derating formula varies from one manufacturer to another and each one provides their own formula. Appropriate derating with age is very important for aluminium rotors but much less so for titanium and carbon fibre rotors.

4.4 Centrifuge Tubes, Bottles and Caps

4.4.1 *Materials Used for Tubes and Bottles*

A wide variety of materials have been used for the manufacture of tubes and bottles for centrifuges. It is important to be aware of the limitations of each material since these affect their applications.

Glass. The very inert nature of glass makes it almost ideal as a material for centrifuge tubes. Unfortunately, most types of glass are unable to withstand centrifugal forces in excess of 3000 g. However, exceptional glasses, such as are used for the manufacture of Corex tubes, are much stronger and can withstand stresses in excess of 25 000 g. Even so, if using only small amounts of material, some types of biological molecules can adsorb onto the surface of the glass; this effect can be avoided by siliconising the glass tubes using Repelcote, Siliclad (Hopkin and Williams Ltd. and Clay Adams) or an equivalent compound before use.

Polycarbonate. This plastic has the advantages of being completely transparent, autoclavable and very strong. However, polycarbonate is particularly sensitive to many common solvents (e.g., ethanol). In addition, polycarbonate is attacked by all alkaline solutions (this usually includes most common laboratory detergents), it is sensitive to the nuclease inhibitor diethylpyrocarbonate and also some rotor polishes. The other problem is that because polycarbonate is a brittle plastic some types of thick-walled tubes can be cracked by rapid acceleration or braking.

Polysulfone. This plastic shares many of the desirable properties of polycarbonate in terms of transparency, strength and autoclavability. However, polysulfone is resistant to alkaline solutions and, notably, also to ethanol. However, it is attacked by a number of other common solvents (e.g., phenol).

Polypropylene and polyallomer. These two plastics have similar properties although polyallomer tends to be more transparent than polypropylene and it has marginally better chemical resistance properties. Both plastics can be autoclaved at 120°C for 30 min. These plastics are softer than polycarbonate and thin-walled tubes are readily piercable.

Cellulose nitrate and cellulose acetate butyrate. Cellulose nitrate has the advantage of being transparent, however it is both highly inflammable and explosive! Cellulose nitrate tubes have now been withdrawn and should not be used. Cellulose acetate butyrate tubes have similar desirable properties to cellulose nitrate tubes in terms of transparency, pierceability and hydrophilicity. Tubes of cellulose acetate butyrate are not explosive but neither can they be autoclaved. Both compounds are attacked by caesium trifluoroacetate, strong acids, bases and a range of solvents.

Other materials. A number of other less common materials such as stainless steel are also used for tubes. A full description of their compatibility with various solutions is given in Appendix II. Further details of these materials are given elsewhere (11).

4.4.2 *Sealing Caps for Tubes and Bottles*

Traditionally the most tedious aspect of ultracentrifugation has been the necessity to seal tubes and bottles during centrifugation and indeed perhaps one reason for the popularity of ultracentrifuge swing-out rotors is that it is unnecessary to cap the individual tubes.

In fact it is not always necessary to cap tubes if one uses partially-filled thick-walled tubes; these are available for most fixed-angle rotors and swing-out rotors. However, in the case of ultracentrifuge fixed-angle rotors, it may not be possible to run the rotor at maximum speed when using partially-filled tubes, particularly when using tubes made from softer plastics such as polyallomer and polypropylene; in this case polycarbonate tubes are usually best.

Tubes should always be capped if the sample is radioactive or biohazardous and if it is proposed to run completely full tubes in fixed-angle rotors. When using thin-walled tubes in fixed-angle ultracentrifuge rotors it is always necessary to fill the tube completely and to cap them securely in order to support the tube during centrifugation.

Most simple plastic screw-on or push-on caps are usually not sufficient to prevent leakage from completely full tubes during centrifugation in fixed-angle or vertical rotors. To ensure a leak-proof seal, even for high-speed centrifuges, it is usually necessary to use an 'O' ring sealing system; a typical example of such an assembly is shown in *Figure 14*. Some aluminium screw-on caps are available which are designed to seal under the influence of centrifugal force; usually these caps are only used with thick-walled polycarbonate or cellulose tubes since it is easiest to form a thread on such tubes. It is even easier to use Kontron's self-seal caps (*Figure 15*) which can be used with thin-walled polyallomer tubes. These caps push into the tube and then seal as the centrifugal field forces the cap further into the top of the tube. Beckman have also introduced narrow-necked tubes (Quick-seal® tubes) that can be heat sealed. The heat sealing must be done carefully to ensure that the seal does not fail during centrifugation. Narrow-necked tubes can also be sealed using the Du Pont-Sorvall crimp-sealer which is both simple to use and very efficient.

For vertical rotors where the cap or seal has to withstand large hydrostatic pressures generated by the centrifugal force, there are a number of different

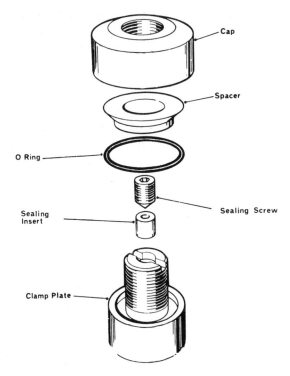

Figure 14. Components of a typical multi-component sealing cap for fixed-angle rotors.

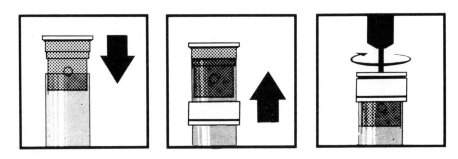

Figure 15. Self-sealing centrifuge caps. The cap is inserted into the top of the tube and the outer titanium collar is slipped up to the top of the tube. The air vent is closed by inserting the screw and the tube is inserted in the rotor. During centrifugation the top of the tube is compressed between the cap and the collar forming a liquid-tight seal. Reproduced with permission of Kontron Instruments.

systems and these have been reviewed elsewhere (8) The Du Pont-Sorvall system of sealing used to use a simple bung system but now the crimp-sealing of tubes is recommended. The Kontron self-seal caps can also be adapted to sealing tubes of vertical rotors. Multicomponent caps have also been devised (8) while Beckman use the same heat-sealed tubes as are used in the fixed-angle rotors.

When using metal caps it is important to remember that they are subjected to the same corrosive forces and the same centrifugal forces as the rest of the rotor and hence caps, especially aluminium ones, should be carefully cleaned after use and checked for cracks and corrosion at regular intervals.

4.4.3 *Choice of Types of Tubes*

The correct choice of tubes is important in order to facilitate centrifugal separations. The choice of tubes depends on several factors including the type of rotor used, nature and volume of the sample as well as the method used to fractionate the solution in the tube after centrifugation. *Figure 16* is a flow-chart devised to aid the reader to make the appropriate choices of tubes for high speed and ultracentrifuge separations.

4.5 **Centrifuge Safety**

The regulations now governing the construction of centrifuges have ensured that, currently, nearly all centrifuges have lid locks to prevent access to the rotor whilst it is moving. Similarly, all centrifuges are now made such that they contain all fragments within the casing should there be a catastrophic failure of the rotor. Should the reader use older centrifuges which are not built to modern specifications, care should be taken not to touch any rotor whilst it is moving since to do so spoils the fractionation obtained and may result in physical injury.

Most of the hazards likely to face the user arise from the centrifuge rotor. Hence, as described in Section 4.3, spillages into the rotor pockets can drastically weaken a rotor but the spilled solution may also contain hazardous compounds. The frequency of such spillages may be increased by inexpert use of tubes and sealing caps. An additional cause of accidents is the failure to balance the rotor. In the case of a swing-out rotor, lack of balance may be because not all of the buckets are attached, or because they are placed at incorrect positions around the rotor. It is of course most important that, in all cases, tubes are balanced to within less than 1% and the balanced tubes are placed symmetrically in the rotor. Failure to balance the rotor correctly will greatly increase the wear on the drive system and may even lead to failure of the drive spindle resulting in extensive damage to the rotor and centrifuge. Whenever possible it is advantageous to use centrifuges with an inbuilt out-of-balance detector since this helps to avoid the accidents that can occur through centrifuging an unbalanced rotor.

Another major hazard arises from the spillage of hazardous samples. There is a particular problem in that in non-refrigerated centrifuges the rotor is cooled by drawing air through the bowl past the rotor. Spillages generally generate aerosols and hence spillages can easily lead to contamination of the whole laboratory. When using hazardous samples in low-speed bench centrifuges it is important to use sealed containers to ensure that there is no possibility of contamination. In the case of centrifuges with vacuum systems, special care must be taken since rotor failure leads to the contamination of the whole vacuum system and will require expert decontamination.

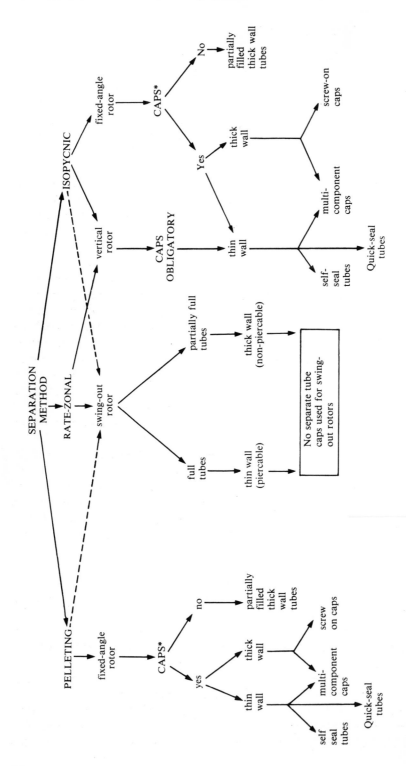

Figure 16. A flow-chart for the selection of tubes.

*caps must always be used if the sample contains hazardous or potentially hazardous material.

5. CENTRIFUGATION MEDIA

5.1 **Introduction**

Many centrifugal separations are preceded by some disruptive process of the starting material. In the case of the separation of cells, it may be necessary first to disaggregate the tissue into single cell suspensions by perfusion (see Chapter 6). In the case of fractionating subcellular particles, cells must be disrupted by homogenisation procedures in media which will ensure that the subcellular components retain their native structure and characteristics; usually isotonic sucrose or salt solutions are used as homogenisation media. A detailed description of homogenisation procedures and homogenisation media are given in Section 2 of Chapter 5. In the case of differential centrifugation, the whole fractionation procedure is carried out in the homogenisation medium. In contrast, when samples are fractionated by rate-zonal or isopycnic centrifugation it is necessary to use gradients whose characteristics are compatible with the samples.

In the case of rate-zonal separations, the primary function of the gradient is simply to stabilise the liquid column in the tube against movement resulting from convection currents, although a secondary role of the gradient is to produce a gradient of viscosity which helps to improve the resolution of gradients (Section 3.3 of Chapter 2). For isopycnic separations, the most important feature of the gradient medium is that the maximum density of its solutions is high enough to be greater than that of the particles. It is important to realise that, as a result of different levels of hydration of particles in different gradient media, the densities of particles vary depending on the particular gradient medium used. Besides these primary requirements for properties of gradient media for rate-zonal and isopycnic separations, a number of other general properties of an ideal density-gradient medium have been defined (12,14) and these can be listed as follows:

(i) the compound should be inert or at least non-toxic to biological material;
(ii) the physico-chemical properties of the solutions of gradient medium should be known and it should be possible to use one or more of these properties (e.g., refractive index) to determine the precise concentration of the gradient medium;
(iii) solutions of the gradient medium should not interfere with monitoring of the zones of fractionated material within the gradient;
(iv) it should be easy to separate the sample material from the gradient medium without loss of the sample or its activity;
(v) the gradient medium should be available as a pure compound and be cheap to use either because it is inexpensive or because it can be recovered after use.

In view of these stringent requirements for an ideal medium, it is not surprising that no single compound is able to meet all of these criteria.

For rate-zonal separations, sucrose meets most of the criteria of an ideal gradient medium, although glycerol and Ficoll gradients are also used occasionally. However, in the case of isopycnic separations no one medium has

Table 3. Applications of the Various Types of Isopycnic Gradient Media[a].

Media	DNA	RNA	Nucleo-proteins	Membranes	Subcellular organelles	Cells	Viruses
Sugars (e.g., sucrose)	−	−	+	+ +	+ +	+	+ +
Polysaccharides (e.g., Ficoll)	−	−	−	+	+	+ + +	+ +
Alkali metal salts (e.g., CsCl)	+ + +	+ +	+	−	−	−	+ +
Colloidal silica (e.g., Percoll)	−	−	−	+	+ +	+ +	+
Nonionic iodinated compounds (e.g., Nycodenz)	+	+	+ + +	+ + +	+ + +	+ +	+ +

[a]The classification is as follows: + + +, good; + +, satisfactory; +, limited applications; −, unsuitable.

proved satisfactory for all types of biological particles. Hence a wide range of gradient media have been used for different types of biological sample and the applications of a selection of these are shown in *Table 3*.

As a general rule, ionic media are used only for viruses, proteins and nucleic acids, though some of the more stable nucleoproteins can be banded in Cs_2SO_4 gradients without prior fixation with formaldehyde. Sucrose gradients are used mainly for organelles and viruses while some of the more recently introduced media such as Nycodenz, metrizamide and Percoll are more versatile and can be used for a wider variety of applications. *Figure 17* summarises the physico-chemical properties of some of the more important gradient media used for rate-zonal and isopycnic separations.

5.2 Properties of Nonionic Gradient Media

5.2.1 Sucrose

Sucrose possesses a number of properties of an ideal gradient medium and thus it has established itself as the premier gradient solute for rate-zonal gradients. In addition, it has also been widely used for the isopycnic fractionation of cell organelles and viruses. The reasons for its popularity lie in its inertness towards biological material, ready availability at low cost and its stable nature. Because of its popularity, the properties of sucrose and sucrose solutions have been fully characterised with respect to the relationships between concentration and viscosity, density and refractive index over a wide range of temperatures (13). In addition, it is usually possible to find in the literature descriptions of suitable gradients and centrifugation conditions to cover the separation of most types of biological samples. However, since new rotors are being developed continuously, it is also necessary to remember that modifications to the gradient and centrifugation conditions can sometimes enhance the degree of resolution obtained as compared with the results obtained using earlier rotors.

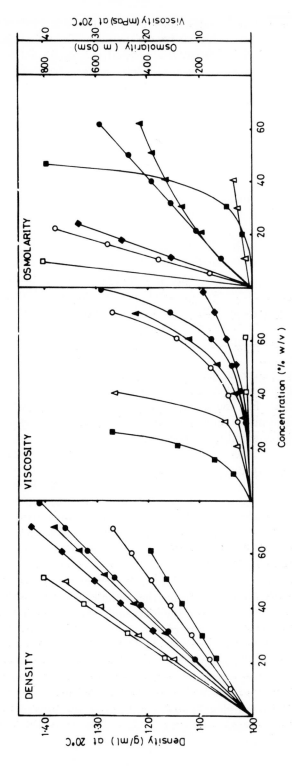

Figure 17. A comparison of physico-chemical properties of gradient media. The relationships between the concentration (% w/v) of gradient solutes and their (A) density (B) viscosity and (C) osmolarity are shown for Nycodenz (●——●), metrizamide (▲——▲), sodium metrizoate (◆——◆), CsCl (□——□), sucrose (○——○), Ficoll (■——■) and Percoll (△——△).

29

The main disadvantages of sucrose are some of its physico-chemical properties (see *Figure 17*). Sucrose solutions have a high osmotic strength and solutions more concentrated than 9% (w/v) are hypertonic; such solutions have a density of only 1.03 g/cm^3 which is rather low to be useful for many isopycnic separations of osmotically-sensitive particles. In addition, concentrated sucrose solutions required for some isopycnic separations are very viscous, thus small particles are unable to reach their isopycnic positions in such gradients. Sucrose is also very susceptible to hydrolysis as the glycosidic bond is labile at pH 3 and below, especially at elevated temperatures. In addition, unless adjusted to pH 5 − 6, concentrated sucrose solutions tend to 'caramelise' when heated to 100°C or above, for example, during sterilisation by autoclaving. This caramelisation gives rise to changes in the properties of the solution and it can be detected by the appearance of a yellow or brownish tinge to the solution. As a general rule, it is much easier to sterilise sucrose solutions by treatment with 0.1% diethylpyrocarbonate; excess diethylpyrocarbonate can be removed by heating the solutions at 60°C for several hours. The other advantage of treating sucrose solutions with diethylpyrocarbonate is that it inactivates most enzymes and as such it is frequently used for removing the ribonuclease activity that almost invariably contaminates most preparations of sucrose. After treatment with diethylpyrocarbonate it is usually necessary to check the pH of solutions since this reagent decomposes to carbon dioxide and ethanol.

There are many grades of sucrose available from a number of different sources. Sucrose from sugar beet usually has lower levels of ribonuclease than that from sugar cane. For some purposes laboratory reagent grade or table sugar are of sufficient purity, however these may contain higher levels of ribonuclease and/or heavy metal ions. The former can be removed with diethylpyrocarbonate and the latter by passing the solution through a small column of chelating resin. Sucrose solutions can be obtained free of u.v.-absorbing material by treatment with activated charcoal (e.g., Norit A). However, it is usually quicker and easier to purchase analytical grade sucrose in order to avoid the problems of purification. A number of companies also sell 'ultra-pure' sucrose which is low in ribonuclease activity and which has a low u.v. absorbance, however such material is expensive and in some cases the quality can vary from one batch to another.

Sucrose gradients must be preformed for rate-zonal and isopycnic fractionations. Gradients can be prepared by using the methods described in Chapter 2. Sucrose gradients can be used for the rate-zonal fractionation of macromolecules and macromolecular complexes, for example, proteins, nucleic acids, ribosomes and polysomes, as described in Chapter 3. Sucrose gradients are used not only for the rate-zonal fractionation of many types of particles, but also for the isopycnic separation of viruses, cell organelles and, where viability is not essential, of cells also; *Table 4* indicates the buoyant densities observed for some such samples. The gradients can be unloaded by any of the standard methods (see Section 7 of Chapter 2) and the shape of the gradient can be readily determined from the refractive indices of the fractions.

Table 4. Buoyant Densities of Biological Particles in Sucrose Solutions[a].

Particles	Medium	Centrifugation conditions	Buoyant density (g/cm^3)
Liver plasma membranes	sucrose	100 000 g for 1.5 h	1.13 – 1.18
Lysosomes	sucrose ⎫		1.21 – 1.22
Mitochondria	sucrose ⎬	59 000 g for 4 h	1.19
Microbodies (peroxisomes)	sucrose ⎭		1.23
Plant microbodies	sucrose ⎫		1.25
Thylakoids	sucrose ⎬	65 000 g for 4 h	1.17
Chloroplasts	sucrose ⎭		1.22
Chromatin	sucrose/glucose	50 000 g for 40 h	1.36
Informosomes	sucrose/D_2O	180 000 g for 20 h	1.29
Mouse sarcoma virus	sucrose	65 000 g for 1 h	1.16
Mouse mammary tumour virus	sucrose	240 000 g for 1 h	1.17
Canine distemper virus	sucrose	88 000 g for 16 h	1.20

[a]Data derived from ref. 20.

5.2.2 *Glycerol*

Glycerol is available as a very pure reagent and its gradients have been used on occasions instead of sucrose gradients for rate-zonal separations. On a weight for weight basis its solutions are less dense and less viscous than the corresponding sucrose solutions. However, glycerol solutions of the same density as sucrose solutions are much more viscous than the equivalent sucrose solutions. The advantages of glycerol are that it helps to preserve the activity of a number of enzymes, it can be removed from the sample under vacuum and it is readily and cheaply available as an analytical grade reagent.

5.2.3 *Polysaccharides*

In order to circumvent the problems that can arise from fractionating osmotically-sensitive particles (e.g., cells and organelles) in the high osmotic strength of sucrose solutions, a number of polysaccharides have been used as density-gradient media. Gradients of glycogen, dextrans and other materials have been described, however the commonest medium used is Ficoll (Pharmacia Fine Chemicals AB). This product is produced by the chemical co-polymerisation of sucrose molecules with epichlorohydrin to give a polysaccharide with a weight average molecular weight of 400 000. Ficoll solutions below 20% (w/v) equivalent to a density of 1.07 g/cm^3 are fairly osmotically inert, though at higher concentrations the osmolarity rises sharply (see *Figure 17*). The other additional major problem with working with these solutions as compared with sucrose solutions is their high viscosity. Even so Ficoll gradients have found a wide application for the fractionation of cells in both rate-zonal and isopycnic separations.

Gradients of Ficoll must be prepared using a gradient mixer since it diffuses

Table 5. Buoyant Densities of Cell Organelles, Cells and Viruses in Ficoll Gradients[a].

Particles	Centrifugation conditions	Buoyant density (g/cm³)
Membranes	100 000 g for 16 h	1.05
Chromatophores	195 000 g for 36 h	1.07
Brain vesicles	21 000 g for 15 min	—
Mitochondria	80 000 g for 2 h	1.136
Hepatocyte cells	6000 g for 2 h	1.10 – 1.15
Fibroblast cells	8000 g for 60 min	1.05
Ehrlich ascites cells	1400 g for 45 min	1.07
Mammary tumour virus	59 000 g for 60 min	1.14

[a]Data derived from ref. 20.

very slowly. However, for the same reason, Ficoll gradients are also more stable than sucrose gradients. Ficoll has a protective effect on some cell organelles and has been used for differential centrifugation as well as for rate-zonal centrifugation, particularly of cells and similar particles. In addition, Ficoll gradients have been widely used for isopycnic fractionations of organelles and cells (*Table 5*). Samples can be recovered from gradient fractions by pelleting them from the pooled fractions diluted as necessary.

5.2.4 *Iodinated Gradient Solutes*

Most iodinated compounds used as gradient media have a structure based on tri-iodobenzoic acid to which hydrophilic groups are attached to increase the solubility of these compounds in water. A detailed description of the various ionic iodinated compounds that have been used has been published previously (14). The structures of most of these ionic compounds are very similar to that of metrizoic acid (*Figure 18*) although others (e.g., ioglycamic acid) have a dimeric type of structure. The important difference seen when comparing ionic iodinated media with metrizamide and Nycodenz (*Figure 18*) is that, in the case of the last two compounds, the carboxyl group is blocked by covalent linkage to a glucosamine and a dihydroxypropylamine group, respectively which makes both of them completely nonionic. These covalent bonds have a dramatic effect on the properties of these compounds in terms of their stability and suitability as gradient media as compared with ionic compounds such as metrizoic acid. (Nycodenz is a product of Nygaard & Co.)

Salts of metrizoic acid, but not free metrizoic acid itself, are soluble in water; metrizoates are usually supplied as solutions. While the sodium salts are very soluble, the presence of calcium or magnesium ions in the gradient can lead to precipitation since these salts have a much lower solubility. At acid pH, that is, less than pH 6, metrizoates are precipitated in the form of metrizoic acid.

In marked contrast to metrizoic acid, both metrizamide and Nycodenz are readily soluble in all aqueous media. In addition, metrizamide, and to a lesser extent Nycodenz, are also soluble in organic solvents such as ethanol, formamide and dimethyl sulphoxide. In marked contrast to the ionic compounds,

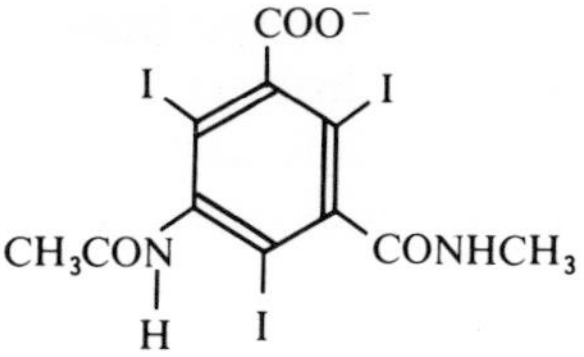

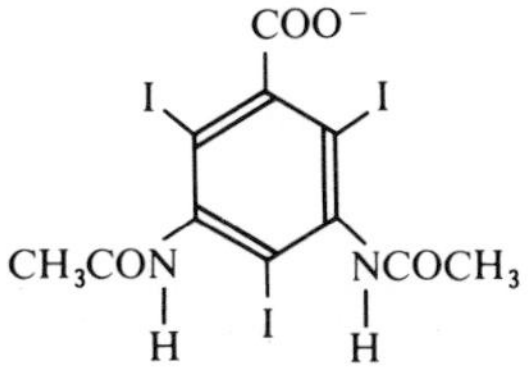

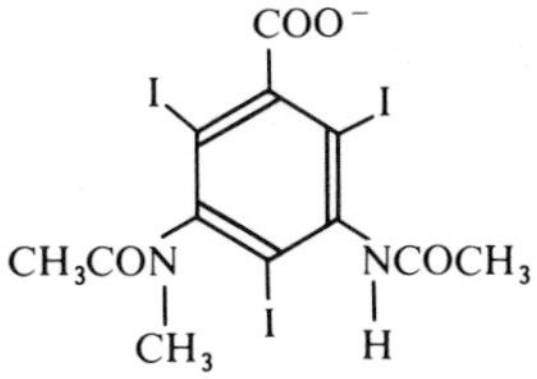

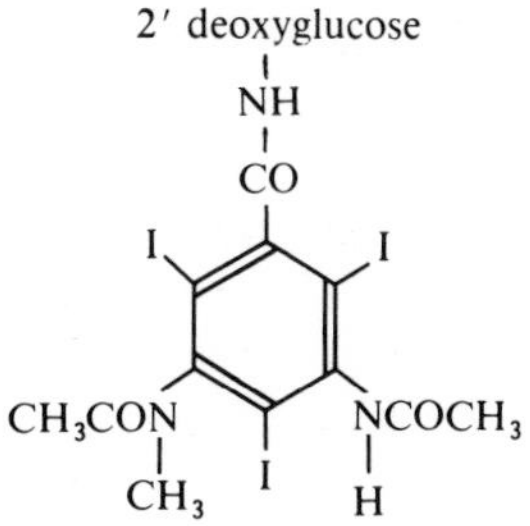

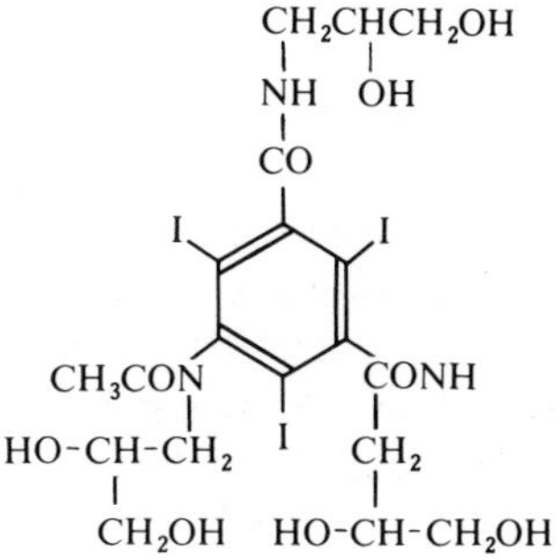

Structure	Mol. wt.	Counter ions	Trivial names	Chemical names
	614	Na$^+$ * MGN$^+$ Na$^+$	Iothalamate Conray Angio-Contrix 48	5-acetamido-2,4,6-tri-iodo-N-methyl isophthalamic acid
	614	MGN$^+$ Na$^+$ Na$^+$;MGN$^+$	Diatrizoates Cardiografin Gastrografin Renografin Hypaque Urografin	3,5-Diacetamido-2,4,6-tri-iodobenzoic acid
	628	Na$^+$	Metrizoates Isopaque Ronpacon Triosil	3-Acetamido-5 (N-methylacetamido)-2,4,6-tri-iodobenzoic acid
	789	None	Metrizamide	2-(3-acetamido-5-N methylacetamido-2,4,6,tri-iodo-(benzamido)-2-deoxy-D-glucose
	821	None	Nycodenz	N,N'-bis(2,3 dihydroxypropyl)-5-N-(2,3 dihydroxypropyl) acetamido-2,4,6-tri-iodo-isophthalamide

*MGN$^+$ = N-methylglucamine

Figure 18. Structures of iodinated compounds used as density gradient media.

the solubilities of metrizamide and Nycodenz are not affected either by the ionic environment or by the pH of the solution. Studies have shown that both metrizamide and Nycodenz are soluble and stable over the range pH 2 to pH 12.5. Solutions of sodium metrizoate and Nycodenz are stable to heat and they can be sterilised by autoclaving. However, the glucosamide group of metrizamide makes it unstable to temperatures above 55°C and so metrizamide solutions must be sterilised by filtration. The presence of a tri-iodinated benzene ring in all of these compounds results in a high absorption

Table 6. Buoyant Densities of Biological Particles in Iodinated Density Gradient Media.

Biological particles	Centrifugation conditions	Buoyant density (g/cm³)		
		metrizoate	metrizamide	Nycodenz
Native DNA	65 000 g for 44 h	1.13	1.11	1.13
Denatured DNA	65 000 g for 44 h	1.14	1.14	1.17
RNA	65 000 g for 44 h	1.23	1.17	1.18
Proteins				
haemoglobin		1.27	–	–
catalase	163 000 g for 72 h	–	1.27	1.29
β-galactosidase		–	–	1.25
serum albumin		–	1.22	–
Polysaccharides				
glycogen		1.48	1.28	1.29
blue dextran		–	1.19	1.19
hyaluronic acid	80 000 g for 48 h	–	1.10	–
chondroitin sulphate		–	1.08	–
proteoglycans		–	1.23	–
Nucleoproteins				
polysomes (Mg²⁺)		–	1.34	1.33
messenger RNP (Mg²⁺)	150 000 g for 68 h	–	1.21	–
80S ribosomes (Mg²⁺)		–	1.33	1.30
80S ribosomes (Na⁺)		–	1.22	–
chromatin	60 000 g for 44 h	1.20	1.16–1.20	1.17–1.18
metaphase chromosomes	16 000 g for 60 min	1.19	1.24	1.29
Organelles				
membranes	100 000 g for 16 h	–	1.14–1.26	1.11–1.19
lysosomes		–	1.13	1.15
mitochondria		1.15	1.16	1.17
peroxisomes	80 000 g for 2 h	–	1.22	1.22
nuclei		–	–	1.23
nucleoli	125 000 g for 2 h	–	1.24	–
Cells				
lymphocytes		1.07	–	1.07
erythrocytes	10 000 g for 30 min	1.15	–	1.11
liver parenchymal		–	1.12	1.14
Viruses				
poliovirus		1.29	1.31	1.30
Coxsackie virus		–	1.18	1.18
Semliki Forest virus	200 000 g for 18 h	–	1.20	1.18
Newcastle disease virus		1.14	–	–
bacteriophage T7		1.27	–	–

in the u.v. region of the spectrum. *Figure 17* shows how the properties of iodinated gradient media compare both with each other and with other types of density-gradient media. For any given concentration, sodium metrizoate solutions are denser and less viscous than equivalent solutions of metrizamide and Nycodenz. However, the osmolarity of sodium metrizoate solutions is much higher than those of the nonionic iodinated media. As can be seen from *Figure 17*, there are no really significant differences between the physico-chemical properties of solutions of metrizamide and Nycodenz.

In comparison with other gradient media, iodinated compounds have a number of advantages. For example, at all densities, iodinated gradient media exhibit much lower osmolarities and viscosities than sucrose (*Figure 17*) while polysaccharides form even more viscous solutions. Although CsCl solutions have high densities and low viscosities their use for many isopycnic separations is precluded because of their high osmolarity and, even more serious, their high ionic strength which can dissociate, denature or otherwise disrupt the integrity of biological samples. Because of their favourable properties, iodinated gradient media have found a wide range of applications from macromolecules to cells, as shown in *Table 6*. Another volume of the Practical Approaches series (19) describes the applications of these gradient media in detail.

Gradients of iodinated media can be preformed and the sample loaded onto the top of the gradient. This method is preferred for the preparation of isotonic gradients and also usually for the separation of large particles (e.g., cell organelles) which need to be centrifuged for only a short period of time (i.e., 2 h or less). Alternatively, for small particles such as macromolecules and nucleoproteins, it is possible to self-form the gradients during centrifugation. In this case, the sample is mixed throughout the gradient solution; this makes it possible to use a large sample volume.

5.2.5 *Colloidal Silica Suspensions*

Colloidal silica suspensions, containing particles in the range 3 − 15 nm in diameter are widely used for various industrial applications and they have also proved useful for centrifugal separations. Such suspensions are not truly nonionic since the particles must be charged in order to stabilise the suspension; neutralisation of these charges destroys the suspension. Originally most experimental work was done using colloidal silica preparations manufactured by Du Pont de Nemours & Co. which are sold under the name of Ludox; *Table 7* details the properties of such preparations. Ludox preparations can contain toxic components, the particles tend to stick to membranes and are unstable in salt solutions. While colloidal silica gradients have been used for a number of years, only one preparation has been specifically developed for centrifugation. This gradient medium, Percoll, is marketed by Pharmacia Fine Chemicals AB. Percoll has a number of advantages over the colloidal silica preparation from which it is derived (15 − 17). The colloidal particles are coated with polyvinyl-pyrrolidone (PVP) which minimises their interaction with biological material and also stabilises the colloid against freezing and thawing and the addition of salts.

Table 7. Properties of Various Colloidal Silica Preparations.

	Percoll	Ludox HS 30	Ludox AM	Ludox SM	Ludox 130M
Density (g/cm³)	1.13	1.21	1.21	1.21	1.23
Silica content (% w/v)	20	30	30	30	26
Particle size (nm)	17	15	15	7	15
Particle charge	negative	negative	negative	negative	positive
Counter ion	Na$^+$	Na$^+$	Na$^+$	Na$^+$	Cl$^-$
pH at 25°C	8.8	10.0	9.0	9.9	4.0
Viscosity (mPas)	10	5	17	6	5
Osmolarity (mOsm/Kg H$_2$O)	20	60	30	–	–
Conductance (mS/cm)	1.0	3.1	2.6	3.1	–

Percoll is a registered trademark of Pharmacia Fine Chemicals AB.
Ludox is available from E.I.Du Pont de Nemours and Co. Inc.; similar preparations are also available from Nalco Chemical Company, Monsanto Chemicals and Nyacol Inc.

The osmotic strength of colloidal silica solutions is also extremely low and it changes little with density. Hence the osmolarity of gradients can be adjusted simply by the addition of an appropriate amount of sucrose or salt solution. For example, for an isotonic solution, one can simply add a ninth volume of 2.5 M sucrose solution to the stock colloidal silica suspension and then the density can be adjusted further by the addition of 0.25 M sucrose solution. Although Percoll solutions as supplied by the manufacturer are stable to auto-claving, they cannot be autoclaved once the tonicity of Percoll has been ad-justed by the addition of NaCl or sucrose. Hence if one wishes to work using sterile Percoll gradients it is necessary to sterilise all the solutions prior to mix-ing. Colloidal silica suspensions are destabilised by conditions which tend to neutralise the charge on the particles, hence, for example, Percoll is precipitated at low pH and high ionic strengths also destabilise the colloidal suspension.

Gradients readily self-form when colloidal silica suspensions are centrifuged even for short times (e.g. 20 000 *g* for 30 min) in fixed-angle rotors; swing-out rotors are not satisfactory for self-forming gradients. As an alternative it is possible to preform gradients using a simple gradient mixer as described in Chapter 2. In spite of the colloidal nature of the gradient, refractive index measurements can be used as a guide to the density profile of gradients. As alternatives the density can also be measured by pycnometry or by measure-ment using a calibrated non-aqueous density gradient. Both of these pro-cedures are rather laborious and instead one usually uses coloured density marker beads manufactured by Pharmacia Fine Chemicals AB which are cen-trifuged in a blank gradient to give an indication of the density profile. Although the buoyant density of the beads is dependent on the ionic composi-tion of the gradients, using interpolation procedures it is usually possible to obtain an estimate of the density of sample bands within individual gradients.

Colloidal silica solutions absorb strongly in the u.v. region of the spectrum, and this absorption is further increased if protective agents such as PVP are

Table 8. Densities, and Separation Conditions for the Isolation of Cells, Viruses and Subcellular Particles in Percoll Gradients.

Particles	Recorded density (g/cm³)	Osmotic balance solution	Starting density Percoll (g/cm³)	Suggested running conditions
Organelles				
plasma membranes	1.02 − 1.03	Sucrose	1.04	
microsomes	1.03 − 1.05	Sucrose	1.05	
peroxisomes	1.05 − 1.07	Sucrose	1.07	63 000 g for 30 min
mitochondria	1.09 − 1.11	Sucrose/Mg²⁺	1.06	50 000 g for 45 min
lysosomes	1.04 − 1.07 1.08 − 1.11	Sucrose } Sucrose }	1.05	50 000 g for 45 min
synaptosomes	1.04 − 1.06	Sucrose/Mg²⁺	1.04	50 000 g for 45 min
nuclei	1.08 − 1.12	Sucrose	1.10	100 000 g for 60 min
chromaffin granules	1.06 − 1.07	Sucrose	Preformed	10 000 g for 30 min
Rat liver cells				
hepatocytes	1.07 − 1.10	Eagle's Media	1.07	30 000 g for 30 min
Kupffer cells	1.05 − 1.06	Eagle's Media or Hepes − NaOH Buffer	1.06	30 000 g for 30 min
Human blood cells				
thrombocytes	1.04 − 1.06 ⎫			See note
lymphocytes	1.06 − 1.08 ⎬ Hepes-NaOH		1.090	See note
granulocytes	1.08 − 1.09 ⎱ Buffer			See note
erythrocytes	1.09 − 1.10 ⎭			See note
Testicular cells				
Leydig cells	1.06	Sucrose	Preformed ⎫	800 g for 20 min
spermatids	1.04	Sucrose	Preformed ⎭	
Bacteria				
E. coli	1.13	PBS	1.10	30 000 g for 20 min
Viruses				
tobacco mosaic virus	1.06	Sucrose	1.06	100 000 g for 45 min
equine abortion virus	1.08	0.01 M Tris-HCl	1.10	40 000 g for 45 min
influenza virus	1.06	0.01 M Tris-HCl	1.05	25 000 g for 25 min
rotavirus	1.08	Sucrose	1.10	50 000 g for 45 min

Note: separation of blood cells is best carried out by preforming the gradient (starting density 1.090 g/cm³) by centrifuging at 20 000 g for 20 min, then layering blood on top of gradient. Then spin at 1000 g for 5 min in a swing-out rotor, leaving the thrombocytes in the serum layer above gradient, which can be removed with a pipette (rate-zonal separation). A further spin for 20 min at 1000 g separates the other cell types, according to at their isopycnic densities.

also present as in the case of Percoll, thus it is not possible to analyse these gradients directly below 280 nm. There is no difficulty in monitoring gradients spectrophotometrically above 350 nm. Percoll does not seriously interfere with the monitoring of radioactivity in fractions and neither does it interfere with the assay of many marker enzymes. However, Percoll does interfere with most types of protein assay and with some types of assay for nucleic acids and polysaccharides. Colloidal silica can be removed from fractions by differential centrifugation to pellet the fractionated material from most of the colloidal silica particles, acidifying solutions to precipitate colloidal silica suspensions and, in addition, some preparations, but not Percoll, can also be precipitated by polyamines. However, even with its associated PVP, Percoll like Ludox does tend to stick to membranes and some types of cells phagocytose the silica particles of Percoll (18). Hence, in these cases, it is impossible to remove all traces of contaminating colloidal silica particles.

Experiments have shown that in the case of combinations of Ludox and polyethylene glycol the presence of polyethylene glycol can modify both the density and even the integrity of some particles. In colloidal silica, in the presence of salt, a single species of particle can, in some instances, form multiple bands. Hence, in some cases, care is necessary in interpreting results.

The use of Percoll gradients is usually restricted to isopycnic separations. The osmotically inert nature of the colloidal suspension makes it easy to adjust the osmotic strength of gradients. These gradients can be used for cells, organelles and even viruses (*Table 8*). The real limitation of these gradients is the particle size of the sample in that it must be significantly larger than the size of the colloidal silica otherwise the particles of silica pellet before the sample bands. Reviews of the applications of Percoll have been published elsewhere (21,22).

5.2.6 *Proteins*

Proteins have a hydrated density of approximately 1.27 g/cm³ and as such it is possible to preform gradients on which it is possible to fractionate biological material either on the basis of size or density. The commonest protein used for such gradients is bovine serum albumin (bovine plasma albumin, BSA) because of its ready availability in a pure form, free of major contaminants and at a reasonable price. Solutions of BSA up to 40% (w/v) can be prepared, however the solutions absorb strongly in the u.v. region of the spectrum and are quite viscous although they are fairly inert osmotically. Gradients can be formed using gradient mixers and they have been used for separating different types of cell.

5.3 **Properties of Ionic Gradient Media**

5.3.1 *Caesium Salts*

Gradients of caesium salts were first used for the isopycnic separation of nucleic acids and as such they have established themselves as some of the most widely used isopycnic gradient media. These media are used almost exclusively for isopycnic separations and as such they are all used as concentrated solutions. All of these solutions are highly ionic and, while they are non-viscous, they all have high osmolarities. There are some differences in the nature of the solutions of these salts with respect to their solubility, maximum density, water activity and steepness of gradients which directly affect the banding of biological materials in gradients of these solutions (see Chapters 2 and 3).

Gradients of solutions of caesium salts can be preformed using any of the

Table 9. Properties of Ionic Gradient Media.

Gradient media	Maximum density (g/cm³)	Slope of gradient[a]	Typical applications	Comments
CsCl	1.91	1.0	Preparation and analysis of DNA; plasmid isolation	RNA pellets
CsHCOO	2.10	0.81	Fractionation of RNA	more viscous than CsCl
Cs_2SO_4	2.01	1.75	DNA, RNA and protein separations; purification of proteoglycans	precipitates scintillation fluids
CsTCA[b]	−	0.97	separation of DNA	absorbs in the u.v.; chaotropic salt
CsTFA[c]	2.6	0.80	DNA, RNA and protein separations; plasmid isolation	absorbs in the u.v.; chaotropic salt
KBr	1.37	0.34	Fractionation of lipoproteins	
KI	1.72	0.66	Fractionation of RNA and DNA	u.v. opaque
NaBr	1.53	0.25	Fractionation of lipoproteins	
NaI	1.90	0.39	Fractionation of RNA and DNA	u.v. opaque
RbCl	1.49	0.51	Analysis of density-labelled proteins	
RbTCA[b]	−	0.86	Fractionation of DNA	absorbs in the u.v.; chaotropic salt

[a]Slope expressed relative to that of an equilibrium CsCl gradient with an initial density of 1.7 g/cm³.
[b]Trichloroacetate salt.
[c]Trifluoroacetate salt. CsTFA is a registered trademark of Pharmacia Fine Chemicals AB.

standard techniques. In addition, gradients can be formed *in situ* by centrifugation. The steepness of the gradients formed depends on the centrifugal field applied and the gradient solute; for example, increasing the centrifugal force steepens the gradient while, for any given centrifugal force, Cs_2SO_4 gradients are steeper than CsCl gradients. The shape of the gradient formed also depends on the type of rotor. It should be noted that preformed gradients redistribute to give the equilibrium gradient shape if they are centrifuged for a sufficient length of time.

The properties and uses of these gradients are shown in *Table 9*. Detailed experimental protocols for these separations are given in Section 6 of Chapter 3. Gradients of CsCl are widely used to separate different species of DNA according to their base composition. Species of DNA which are enriched in the bases cytosine and guanine (GC-rich) band denser. The degree of separation between different species can be optimised by the addition of drugs such as distamycin which selectively bind to regions of the DNA which are enriched in the bases adenine and thymine (AT-rich) and decrease the density of that DNA. In addition, in the presence of ethidium bromide, it is possible to separate DNAs of different conformations (e.g., superhelical and linear species), even if they have the same base composition. Gradients of Cs_2SO_4 can also be used to separate different species of DNA, though the base composition has a smaller effect on the buoyant density of the DNA and the gradients are steeper. One advantage of Cs_2SO_4 gradients, as compared with CsCl gradients, is that they can be used to band RNA. Some species of high molecular weight RNA do tend to aggregate and precipitate in Cs_2SO_4 gradients though this can be minimised by adding 4 M urea to the gradient. Using steep gradients it is possible to band protein, DNA and RNA all on a single Cs_2SO_4 gradient. One problem with Cs_2SO_4 gradients is that the sulphate ions react with and precipitate most scintillation fluids.

It is often not appreciated that, although CsCl and Cs_2SO_4 are the caesium salts most widely used for isopycnic separations of nucleic acids, often better separations can be obtained by using other caesium salts. As examples, the chaotropic nature of both the trichloroacetate and the trifluoroacetate salts have proved useful for obtaining good separations of nucleic acids (24,28). Indeed, as described in Chapter 3, in the case of caesium trifluoroacetate gradients it is possible to obtain a good separation of supercoiled plasmid and linear DNA without the addition of ethidium bromide. Another useful feature of these gradients is that, unlike CsCl and Cs_2SO_4, these salts are soluble in ethanolic solutions and hence it is possible to use ethanol to precipitate nucleic acids from the gradient fractions directly. In addition to these two salts, caesium formate gradients have been used to fractionate density-labelled RNA (25). Some other caesium salts have occasionally been used but generally they do not appear to offer any great advantages over the other caesium salts.

CsCl gradients are often used for preparative purposes and these methods tend to consume significant amounts of CsCl which makes them relatively expensive. It is feasible to recover CsCl after use and the method routinely carried out in the author's laboratory is given in *Table 10*.

Table 10. Recovery of CsCl After Use.

1. Starting with about a litre of used CsCl solution extract it a number of times with an equal volume of acetone (total vol. acetone ~ 7 litres).
2. Filter off the crystals of CsCl and suck dry.
3. Separate the aqueous phase from the acetone in the filtrate.
4. Heat the aqueous phase carefully on a heated magnetic stirrer to drive off the acetone **in a fume cupboard**.
5. Concentrate the solution to ~ 150 ml by heating. Allow the solution to cool.
6. Filter off a second crop of crystals and suck dry. Pool the first and second crystal crops.
7. Heat the crude CsCl in a silica dish to 600°C to melt the CsCl and combust all organic material. Allow to cool.
8. Dissolve the CsCl in a minimal amount of deionised H_2O (~ 500 ml).
9. Filter through a Whatman No. 54 filter on Buchner funnel to remove the carbon.
10. Concentrate the filtrate by heating in a large evaporating dish until a skin of crystals starts to form on the surface of the solution. Allow to cool.
11. Filter off the crystals and dry them at 130°C.
12. Concentrate the filtrate further to get second crop of crystals and filter them off; dry at 130°C.
13. Repeat concentration of the solution and cool in ice.
14. Filter off the crystals and dry them at 130°C; discard the filtrate.

5.3.2 *Sodium and Potassium Salts*

While NaCl and KCl solutions are not dense enough to band most macromolecules, NaBr, NaI, KBr and KI form dense, non-viscous solutions and they have been used for banding lipoproteins, proteins and nucleic acids, the last of which band at lower densities than when banded in CsCl gradients. One advantage of KI and NaI gradients is that, in these media, RNA does not aggregate and precipitate as it does in Cs_2SO_4 gradients. Solutions of these salts also form fairly shallow gradients when centrifuged and thus they are capable of a high degree of resolution. However, the iodide salts in particular do have some disadvantages in that they are prone to oxidation and a reducing agent must be added to prevent the formation of free iodine. The same problem occurs when fractions of gradients are mixed with scintillation fluids, in which case the iodine quenches the ^3H and ^{14}C radioactivity, hence a reducing agent must be added prior to counting. In addition, these media absorb strongly in the u.v. region of the spectrum. Potassium formate and tartrate solutions have also been used for the isopycnic fractionation of viruses.

5.3.3 *Rhubidium Salts*

These have been used for isopycnic fractionations. In particular RbCl gradients have been used instead of CsCl gradients for the isopycnic separation of proteins. While these gradients are shallower than those of CsCl, apart from in the resulting potential improved resolution there are few other advantages in using RbCl. However, as in the case of caesium, gradients of chaotropic salts, notably the trichloroacetate (23), enable one to obtain very good separations not only between different species of DNA but also between native and denatured DNA.

6. CHOICE OF GRADIENT MEDIUM AND CENTRIFUGATION CONDITIONS

Usually fractionations carried out are variations of previously published techniques and the investigator is merely concerned with reproducing the gradient and centrifugation conditions used by others. However, one should always be prepared to consider whether using a different rotor or gradient medium will give better results.

For rate-zonal centrifugation the natural choice for gradients is sucrose although in some instances glycerol is used, particularly for the separation of native enzymes. For some specialised applications other gradient media such as CsCl, Ficoll or metrizamide can be usefully employed for rate-zonal separations.

In contrast, for isopycnic separations, the unsatisfactory nature of some of the more traditional media such as sucrose and CsCl have been recognised and this has led to the introduction of other gradient media (e.g., Ficoll, metrizamide, Nycodenz and Percoll). These other types of media should always be considered since generally they interact less with biological material. In addition, in some cases it is possible to investigate some areas previously closed under the introduction of the new media. An example of this is the use of metrizamide and Nycodenz to study the interactions of low concentrations of anions and cations with DNA and their effect on hydration.

The choice of centrifugal conditions is usually determined by a number of factors. If one is using self-forming gradients for isopycnic separations then the nature of the gradient solute often determines the centrifugation conditions in that too high a centrifugal force will give steep gradients lacking resolution and indeed possibly leading to solute precipitation, a potentially dangerous situation. The centrifugal force generated during the separation may also adversely affect the particles. As examples of this, the metabolic patterns of cells can be disrupted during centrifugal separations (26,27), the characteristics of membranes change when subjected to the high hydrostatic pressures that can be generated in gradients (5) and similar forces appear to be able to disrupt macromolecular complexes (3,6). In the case of DNA the centrifugal force appears to be able to alter the conformation of the DNA and hence the apparent sedimentation coefficient depends on the centrifugation speed (4). Hence it is important not only to use conditions which will minimise the length of the experimental protocol to obtain maximally active preparations but also to choose centrifugal conditions which will not affect the native characteristics of the sample.

7. REFERENCES

1. Spragg,S.P. (1978) in *Centrifugal Separations in Molecular and Cell Biology,* Birnie,G.D. and Rickwood,D. (eds.), Butterworths, London and Boston, p. 7.
2. Sheeler,P. (1982) *Centrifugation in Biology and Medicine,* published by John Wiley, Chichester & New York.
3. Marcum,J.M. and Borisy,G.G. (1978) *J. Biol. Chem.,* **253**, 2852.
4. Clark,R.W. and Lange,C.S. (1980) *Biopolymers,* **19**, 945.
5. Sitiramam,V. and Sarma,M.J.K. (1981) *Proc. Natl. Acad. Sci. USA,* **78**, 3441.
6. Hauge,J.G. (1971) *FEBS Lett.,* **17**, 168.
7. Griffith,O.M. (1979) in *Ultracentrifuge Rotors: A Guide to Their Selection,* Beckman Instruments Inc., Palo Alto, CA.
8. Rickwood,D. (1982) *Anal. Biochem.,* **122**, 33.
9. Flamm,W.G., Bond,H.E. and Burr,H.E. (1966) *Biochim. Biophys. Acta,* **129**, 310.
10. Rickwood,D. and Young,B.D. (1981) *J. Biochem. Biophys. Methods,* **4**, 163.
11. Molloy,J. and Rickwood,D. (1978) in *Centrifugal Separations in Molecular and Cell Biology,* Birnie,G.D. and Rickwood,D. (eds.), Butterworths, London and Boston, p. 289.
12. Cline,G.B. and Ryel,G.B. (1971) *Methods Enzymol.,* **22**, 168.
13. Ridge,D. (1978) in *Centrifugal Separations in Molecular and Cell Biology,* Birnie,G.D. and Rickwood,D. (eds.), Butterworths, London and Boston, p. 33.
14. Hinton,R.H. and Mullock,B.M. (1976) in *Biological Separations in Iodinated Density Gradient Media,* Rickwood,D. (ed.), IRL Press, Oxford and Washington, p. 1.
15. Pertoft,H., Laurent,T.C., Laas,T. and Kagedal,L. (1978) *Anal. Biochem.,* **88**, 271.
16. Pertoft,H., Rubin,K., Kjellen,L. and Klingeborn,B. (1977) *Exp. Cell Res.,* **110**, 449.
17. Pertoft,H. and Laurent,T.C. (1977) in *Methods of Cell Separation,* Vol. **1**, Catsimpoolas,N. (ed.), Plenum Press, New York, p. 25.
18. Wakefield,J.S.J., Gale,J.S., Berridge,M.V., Jordan,T.W. and Ford,H.C. (1982) *Biochem. J.,* **202**, 795.
19. Rickwood,D., ed. (1983) *Iodinated Density Gradient Media: A Practical Approach,* IRL Press, Oxford and Washington.
20. Rickwood,D. (1978) in *Centrifugal Separations in Molecular and Cell Biology,* Birnie,G.D. and Rickwood,D. (eds.), Butterworths, London and Boston, p. 219.

21. Pertoft,H., Laurent,T.C. and Seljelid,R. (1979) in *Separation of Cells and Subcellular Elements,* Peeters,H. (ed.), Pergamon Press, London and New York, p. 67.
22. Pertoft,H., Hirtenstein,M. and Kagedal,L. (1979) in *Cell Populatons Methodological Surveys (B) Biochemistry,* Vol. **9**, Reid,E. (ed.), Ellis Horwood, Chichester, p. 67.
23. Burke,R.L. and Bauer,W.R. (1980) *Nucleic Acids Res.,* **8**, 1145.
24. Burke,R.L., Anderson,P.J. and Bauer,W.R. (1978) *Anal. Biochem.,* **86**, 264.
25. Grainger,R.M. and Wessells,N.K. (1974) *Proc. Natl. Acad. Sci. USA,* **71**, 4747.
26. Walker,G.M. Thompson,J.C., Slaughter,J.C. and Duffus,J.H. (1980) *J. Gen. Microbiol.,* **119**, 543.
27. Allan,I. and Pearce,J.H. (1979) *J. Gen. Microbiol.,* **111**, 87.
28. Andersson,K. and Hjorth,R. (1983) *FEBS Meeting Abstracts,* S01WE022.

CHAPTER 2

Choice of Conditions for Density Gradient Centrifugation

B.D. HAMES

1. INTRODUCTION

In density gradient centrifugation, separation of the components of a sample is achieved by sedimentation through a density gradient, that is, a solution which increases in density with increasing distance down the centrifuge tube. Two distinct approaches are possible using this technique; rate-zonal centrifugation and isopycnic centrifugation.

In rate-zonal centrifugation the sample is loaded as a narrow layer on top of the density gradient. During centrifugation the sample particles separate as a series of bands or zones, each with its characteristic sedimentation rate (*Figure 1a*). Centrifugation is halted before the particles pellet and the separated components are collected by fractionation of the gradient. The rate at which the particles sediment depends on their size, shape and density and the centrifugal force, density and viscosity profile of the gradient (see Chapter 1, Section 2.1). Similar types of biological particles are often similar in shape so that the shape factor is often irrelevant. In addition, their densities fall into a relatively narrow range and the maximum density of the gradient is arranged so as to never exceed the density of the particles. Overall, therefore, separation of similar types of particles by rate-zonal centrifugation can usually be regarded as based mainly upon their size differences.

In contrast to rate-zonal centrifugation, isopycnic centrifugation separates particles solely on the basis of their different densities. The sample can be loaded directly onto a preformed density gradient (*Figure 1b*) and then centrifugation carried out. Alternatively, in some cases it is possible to mix the sample with the gradient medium to give a solution of uniform density and the density gradient actually forms during centrifugation ('self-forming gradients', see *Figure 1c* and Section 4.2). Whereas in rate-zonal centrifugation the density of the gradient must never exceed that of the particles being separated, in isopycnic centrifugation it is a prime condition of the separation that the maximum density of the final gradient must *always* exceed the density of the particles. Each particle will then sediment until it reaches that region of the gradient where the density is equal to that of the particle. Further sedimentation of the particle then ceases. Therefore, in contrast to rate-zonal centrifugation, isopycnic centrifugation is an equilibrium method, yielding a series of zones of particles each at its characteristic 'buoyant density'.

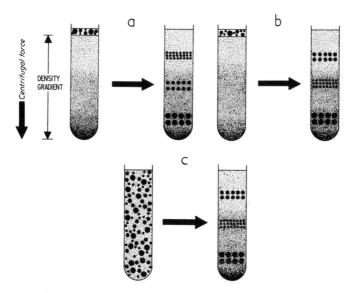

Figure 1. Types of density gradient centrifugation. (**a**) Rate-zonal centrifugation. The sample is loaded onto the top of a preformed density gradient (left) and centrifugation results in a series of zones of particles sedimenting at different rates depending upon the particle sizes (right). (**b**) Isopycnic centrifugation using a preformed density gradient. The sample is loaded on top of the gradient (left) and each sample particle sediments until it reaches a density in the gradient equal to its own density (right). Therefore the final position of each type of particle in the gradient (the isopycnic position) is determined by the particle density. (**c**) Isopycnic centrifugation using a self-forming gradient. The sample is mixed with the gradient medium to give a mixture of uniform density (left). During subsequent centrifugation, the gradient medium re-distributes to form a density gradient and the sample particles then band at their isopycnic positions (right).

Provided that the sample particles are sufficiently different in size or density, rate-zonal or isopycnic centrifugation, respectively, are capable of yielding preparations which are of considerably greater purity than is possible by differential centrifugation. However, maximum resolution of sample components by these techniques requires the selection of optimal conditions for the preparation, centrifugation and analysis of gradients. This chapter describes, in general terms, the practical aspects of rate-zonal and isopycnic centrifugation for analytical and small-scale preparative purposes. Detailed conditions for the separation of macromolecules, subcellular organelles and cells by these techniques are given in Chapters 3, 5 and 6, respectively, whilst large-scale separations using zonal rotors are described in Chapter 7.

2. CHOICE OF ROTOR

2.1 Introduction

As described in detail in Chapter 1, there are basically four types of rotor used for centrifugal separations:

    (i)   swing-out (swinging-bucket) rotor;

    (ii) fixed-angle rotor;

(iii) vertical rotor;

(iv) zonal rotor.

The characteristics of these rotors vary significantly from each other in the context of density-gradient centrifugation. Therefore, in order to achieve optimum results, it is important to choose the correct rotor for each application.

2.2 **Comparison of Rotors for Rate-zonal Centrifugation**

Polyribosomes have been regarded as an ideal test sample for gauging the performance of analytical rate-zonal centrifugation methods (1) in that they consist of a mixture of particles covering a wide range of sizes increasing in discrete steps according to a regular series. Rate-zonal centrifugation of polysomes under ideal conditions produces a series of bands spaced at regular but decreasing intervals in accordance with the increasing number of ribosomes per mRNA molecule. Thus the resolving power[1] of any zonal centrifugation method can be judged by the number of bands that can be observed and the width of those bands. A numerical value cannot be attached to the resolution obtained; only a qualitative comparison between systems can be made. The reader is referred elsewhere (2) for a detailed consideration of this matter and methods whereby resolution can be assessed numerically.

Figure 2 shows the results of rate-zonal separations of identical samples of the same preparation of polysomes from *Dictyostelium discoideum* (3) in swing-out, vertical and fixed-angle rotors. Whereas both the swing-out rotor and the vertical rotor give good polysome profiles with polysomes resolved as far as heptosomes[2], there is a marked lack of large polysomes in the profile obtained using the fixed-angle rotor because these pellet due to the pronounced wall effects in this type of rotor (see Chapter 1, Section 4.3.2). The problem is that, whereas in swing-out rotors the centrifugal force field is almost parallel with the longer axis of the tube, in fixed-angle rotors it is exerted at an angle to the tube wall. Hence in a fixed-angle rotor the particles are forced into the wall of the tube and then sediment to the bottom by sliding down the wall. Even for relatively short times of centrifugation, these wall effects are a serious problem since the larger particles will have pelleted before the smaller particles are optimally resolved. For this reason, fixed-angle rotors are seldom used for rate-zonal centrifugation. During centrifugation in vertical rotors, the particles sediment across the width of the centrifuge tube. While the particles are moving through the first half of the gradient, the walls are divergent and so there are no wall effects but, after particles pass the centre of the tube, the walls converge and wall effects become significant (*Figure 3*). Nevertheless, these effects are considerably less than in the case of fixed-angle rotors.

Although centrifugation in both swing-out and vertical rotors resolves polysomes into a series of discrete peaks (*Figure 2*), polysomes separated using

[1]'Resolution' or 'resolving power' as applied to density gradient centrifugation refers to the volume between the bands or zones of the separated components.

[2]The exact polysome profile observed depends upon the tissue source and method of isolation of the polysomes. In some cases as many as 12 – 16 ribosomes per mRNA molecule can be resolved.

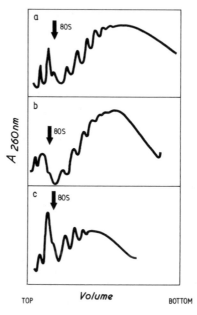

Figure 2. Comparison of different rotors for rate-zonal centrifugation. *Dictyostelium discoideum* polysomes were analysed by rate-zonal centrifugation through 15–30% (w/v) sucrose gradients using the following conditions. (**a**) MSE 6 x 14 ml titanium swing-out rotor, 11.5 ml gradients, 0.2 ml sample volume. Centrifugation was for 2.5 h at 28 000 r.p.m. (102 000 *g*) at 1°C. (**b**) Sorvall SS-90 vertical tube rotor, 35 ml gradients, 0.2 ml sample volume. Centrifugation was for 80 min at 20 000 r.p.m. (32 600 *g*) at 1°C in a centrifuge fitted with a slow acceleration accessory. (**c**) Beckman 75 Ti fixed-angle rotor, 7.5 ml gradients, 0.2 ml sample volume. Centrifugation was for 35 min at 30 000 r.p.m. (60 400 *g*) at 1°C in a centrifuge fitted with a slow acceleration accessory. Each gradient was fractionated by upward displacement and monitored continuously by passage through a spectrophotometer flow cell. The sedimentation position of 80S monosomes is shown.

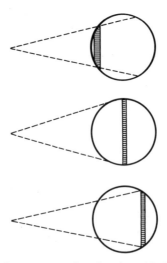

Figure 3. Migration of a sample zone across the tube of a vertical rotor during centrifugation. (Reproduced from ref. 4 with permission.)

a vertical rotor tend to give broader peaks than when separated using a swing-out rotor. Theoretically, this zone broadening, leading to loss of resolution in vertical rotors, can result from two effects. Firstly, the gradient may be disturbed slightly by the reorientation that occurs during the acceleration and deceleration phases, although experiments indicate that this is not a serious problem if acceleration and deceleration are carried out slowly. Secondly, in vertical rotors the pathlength traversed by the particles is short but the band area is large. Therefore more extensive band diffusion may occur during this type of centrifugation than in swing-out rotors where the band area is relatively small. This diffusional effect is most noticeable with low molecular weight particles whose rate of diffusion is high and is of less concern with large particles such as polysomes.

In conclusion, swing-out rotors are the rotors of choice for most analytical and small-scale preparative separations because the relatively long pathlength available to the sedimenting particles and the minimal wall effects optimise the resolution obtained. Many types of swing-out rotors are commercially available. In terms of performance, these differ in the dimensions (both length and width) of the centrifuge tubes and the maximum permissible rotor speed. For optimal resolution in rate-zonal centrifugation, the sample must be loaded to give only a narrow zone on top of the gradient. In practice this means that the sample volume should not exceed $2-3\%$ of the gradient volume (see Section 6.1.1). For an equivalent time period of centrifugation, resolution also increases with increased pathlength available to the sedimenting particles so that the centrifuge tube should be as long as possible. However, if this entails an increase in the time period of centrifugation, the increased diffusion and hence band broadening which occurs may result in no net increase in resolution. Overall, therefore, for small-volume samples, the best resolution is obtained by using narrow, long-tube, high-performance rotors such as the Beckman SW40 or equivalent rotors. Swing-out rotors with wider tubes should be chosen for the analysis of larger volume samples so that the sample can still be loaded as a narrow zone.

Vertical rotors are also useful for rate-zonal separations, especially for large particles such as cells or subcellular organelles. For macromolecules with molecular weights less than 2×10^5, there is significant broadening of the bands due to the enhanced diffusion which occurs in vertical rotors. Nevertheless, rate-zonal separations of macromolecules can successfully be carried out using this type of rotor. The main advantage of vertical rotors is the considerably shorter time needed to achieve any given separation compared with other rotor types. This occurs for two reasons. Firstly, the particles only need to sediment a short distance across the width of the tube rather than down its length. Secondly, the tubes in a vertical rotor are located near the edge of the rotor so that the minimum centrifugal force is higher than in comparable swing-out or fixed-angle rotors. As a result, it is often possible for particles to sediment across the tube in as little as one-fifth of the time taken to sediment from the top to the bottom of a tube in a swing-out rotor run at equivalent speed. Rickwood (4) has recently assessed the performance of vertical rotors. Fixed-angle rotors are not normally used for rate-zonal centrifugation. Large-scale rate-zonal separations are best carried out using a zonal rotor (Chapter 7).

2.3 Comparison of Rotors for Isopycnic Centrifugation

Unlike rate-zonal centrifugation, wall effects are much less important when considering isopycnic centrifugation; whether the sample particles sediment uninterrupted through the gradient or sediment down the walls of the centrifuge tube, they will cease sedimenting when they reach their buoyant density. Therefore swing-out, fixed-angle and vertical rotors may all be used for isopycnic centrifugation. In fact, fixed-angle and vertical rotors have the following advantages for isopycnic centrifugation when compared with swing-out rotors.

(i) *Decreased time of centrifugation.* Because the effective pathlength is shorter for fixed-angle rotors than swing-out rotors, a considerably shorter time period of centrifugation is necessary for equilibrium banding of the sample particles. Since vertical rotors have the shortest gradient pathlength, they are the rotor type of choice for rapid isopycnic separations.

(ii) *Increased resolution.* For self-forming density gradients, the use of a fixed-angle rotor or vertical rotor instead of a swing-out rotor leads to greater resolution at the same rotor speed (see ref. 5 and Section 4.2.2).

(iii) *Increased capacity.* The larger band area, as a result of reorientation of the gradient, and the increased number of tubes in fixed-angle and vertical rotors as compared with swing-out rotors means that the sample capacity per rotor is very much greater.

3. CHOICE OF DENSITY GRADIENT FOR RATE-ZONAL CENTRIFUGATION

3.1 Introduction

A density gradient is essential for rate-zonal centrifugation to support the zones of particles as they sediment. If centrifugation is carried out in a medium of uniform density, the sample zones tend to fall through the liquid column simply because the combined density of sample particles plus medium in the zone is greater than the density of the medium itself. Additional advantages of using a density gradient for rate-zonal centrifugation are:

(i) The higher density of the gradient compared with that of the sample supports the sample as it is layered onto the gradient. This enables the sample to be loaded as a narrow zone which maximises resolution during the subsequent fractionation.

(ii) The increasing density from the top to the bottom of the density gradient stabilises the liquid column against mixing by convection or mechanical disturbances.

(iii) The presence of a gradient of increasing viscosity serves to sharpen the sample zones during centrifugation and so improves the degree of resolution attainable (see Section 3.3.1).

It is important to realise that, since rate-zonal centrifugation is a rate separation, the particles should be able, in principle, to sediment down the whole length of the tube. Thus the maximum density of the gradient must *never* exceed that of the particles. With this proviso, the properties of all density gradients used for rate-zonal centrifugation are determined by the choice of gradient solute and solvent and the density profile of the gradient during centrifugation.

3.2 Choice of Gradient Solute and Solvent

A number of authors (e.g., refs. 6, 7) have described the properties of an ideal density gradient solute. These include cheapness, inertness, high solubility, low osmolality, stability in solution and low optical absorbance (see Chapter 1, Section 5.1). Unfortunately there is no ideal density-gradient solute for all separations although some are much more suitable than others for particular applications of rate-zonal centrifugation. Sucrose is suitable for most rate-zonal fractionations although it has the disadvantage of being rather viscous at densities greater than about 1.13 g/cm^3 ($\sim 30\%$ w/w sucrose). Since sucrose also exerts very high osmotic effects even at low concentrations, the separation of osmotically-sensitive particles may require the use of other solutes, for example Ficoll (Pharmacia Fine Chemicals AB) which is a large copolymer of sucrose and epichlorhydrin.

In most cases the gradient solvent used is an aqueous solution buffered near physiological (neutral) pH. Rate-zonal centrifugation of nucleic acids can pose particular problems. Although conventional aqueous sucrose gradients may be used for the sedimentation analysis of double-stranded molecules, analytical rate-zonal centrifugation of single-stranded DNA and RNA is complicated by the fact that these molecules can form intermolecular and intramolecular hydrogen bonds which are stable in aqueous solutions containing appreciable concentrations of salt. This may lead to erroneous estimates of the true size of the component single-stranded molecules. Therefore, for these applications, rate-zonal centrifugation is often carried out using denaturing, non-aqueous solvents (DMSO or formamide) as described in Chapter 3.

Clearly, the choice of solute and solvent depends upon the nature of the particles to be fractionated and so the reader is referred to the relevant chapter (Chapters 3, 5 and 6) for detailed consideration of this topic. Since most applications at the present time employ aqueous sucrose gradients for rate-zonal centrifugation, the following sections will concentrate on the preparation and use of this type of gradient. However, the basic techniques of rate-zonal centrifugation are the same irrespective of the solute and solvent chosen.

3.3 Choice of Gradient Shape

The shape of a gradient refers to its concentration profile, that is, the variation of concentration of the gradient medium along the tube. The density gradients used for rate-zonal centrifugation are usually continuous gradients, that is the density increases smoothly with increasing distance from the axis of rotation

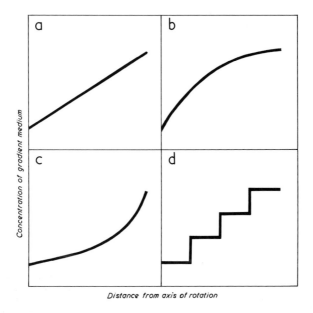

Figure 4. Types of gradient shape. (a) Linear gradient; (b) convex gradient; (c) concave gradient; (d) discontinuous (step) gradient.

(*Figure 4a − c*). Discontinuous (step) gradients (*Figure 4d*), which are used for some types of isopycnic centrifugation (Section 4.2.3), are not usually used for rate-zonal separations since they tend to generate artifactual bands at the solution interfaces. Therefore, the following discussion on the choice of gradient shape for rate-zonal centrifugation considers only continuous gradients.

3.3.1 *Linear Gradients*

A linear gradient, in which the density increases linearly with increasing distance from the centre of rotation (*Figure 4a*), is sufficient for most rate-zonal separations in swing-out or vertical rotors. For swing-out rotors, a linear gradient can also be defined as one where the density increases at a constant rate with volume. Note that this is not so for zonal rotors (see Chapter 7).

When designing a linear gradient for swing-out or vertical rotors, the density at the top of the gradient must be sufficient to support the sample whilst the density at the bottom of the gradient must not exceed the density of the particles to be separated. In general, the greater the slope of the gradient, the better the resolution obtained. This is because viscous drag rises enormously as the sucrose concentration increases (8). Therefore as a zone sediments down a gradient its sedimentation rate slows, particles at the front edge of the zone move slower than those at the trailing edge, and the zone becomes narrower. The greater the increase in viscosity down the length of the tube the greater this zone-sharpening effect becomes (*Figure 5*). Therefore, where possible, a steep 5 − 30% or 10 − 40% sucrose gradient should be employed.

The slope of the gradient is also important from the point of view of loading

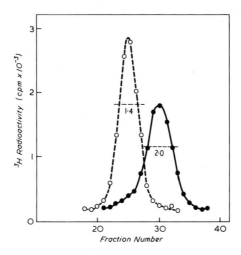

Figure 5. Gradient-induced zone sharpening. Tritiated λ phage DNA was analysed by rate-zonal centrifugation in a $5-20\%$ linear sucrose gradient (●——●) and in a $5-30\%$ linear sucrose gradient (○-----○) followed by collection of fractions from the bottom and counting in a liquid scintillation spectrometer. Measurement of the zone widths at 60% of their height (indicated by the dotted lines) shows that the band in the $5-30\%$ gradient is much narrower than in the $5-20\%$ gradient, that is, the steeper gradient has induced a substantial degree of zone sharpening compared with the less steep gradient (adapted from ref. 9).

capacity, as *Figure 6* shows. A given slope of gradient can only tolerate a certain amount of sample in a zone before gradient inversion occurs, this makes the zone unstable and it begins to spread. Increasing the slope of the gradient could theoretically allow larger loads but then viscosity and hence band sharpening increases, increasing the effective concentration in the zone still further. Poor resolution during rate-zonal centrifugation is almost always the result of overloading, irrespective of the nature of the particles being fractionated. The loading of samples is discussed in more detail later (Section 6.1.2).

3.3.2 *Isokinetic Gradients*

An isokinetic gradient (11) is one in which constant rotor speed will cause a given particle to sediment at a constant velocity throughout the gradient. This will occur if the increase in density and viscosity of the gradient solution balances the increase in centrifugal force along the length of the centrifuge tube. The distance travelled by a particle in an isokinetic gradient is proportional to the time elapsed since the start of centrifugation, the centrifugal force applied and the sedimentation coefficient of the particle. Therefore, if the sedimentation coefficient of a single particle is known, this can be used to calculate the sedimentation coefficients of other particles in the same gradient, *provided* that the density of the sample particles is the same as that of the calibration marker particle. When using an isokinetic gradient in a swing-out rotor, a plot of distance from the centre of rotation (or of volume of gradient) against sedimentation coefficient will give a straight line.

To construct an isokinetic gradient it is necessary to consider the specific

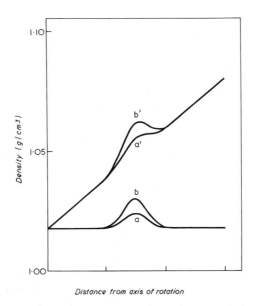

Figure 6. Sample capacity of a gradient. If the density of the sedimentation medium were constant (lower curves) its local perturbaton by a sample zone, (**a**) or (**b**), would always give rise to a negative density gradient on the leading edge of the zone. This would be an unstable situation so that the zone would spread out until the negative gradient had disappeared. If instead, a positive density gradient is incorporated in the sedimentation medium (upper curves), this positive density gradient will compensate the negative gradient due to the zone (**a′**). However, beyond a certain amount of macromolecules, the resultant density gradient on the leading edge of the zone again becomes negative (**b′**) and the zone will spread. Thus a given density gradient has a defined sample capacity which must not be exceeded if band broadening is to be minimised (from ref. 10).

rotor and tubes to be used and the concentration, density and viscosity of the 'gradient medium at the temperature of centrifugation. In the case of swing-out rotors, linear 5 − 20% (w/v) sucrose gradients at 5°C are isokinetic for the separation of proteins (12) and nucleic acids (13), as are linear 10 − 30% glycerol gradients. However, in most cases, isokinetic gradients have convex exponential shapes. Various workers (14 − 16) have devised such gradients. Most of these are for sucrose although there is no reason why isokinetic gradients cannot be constructed for any gradient solute provided that the relationship between viscosity and concentration is known.

3.3.3 Convex and Concave Gradients

For some isokinetic gradients and some other applications it may be necessary to generate non-linear gradients. A gradient which needs to be steep near the top to support the sample zone need not be so steep further down where the zones have separated from each other. In this situation one could make a case for using a convex gradient (*Figure 4b*). Alternatively, provided the light end of the gradient is sufficiently dense to support the sample layer, a concave gradient (*Figure 4c*) might be preferred since it will have a rapid increase in viscosity near the bottom which can be used to slow the sedimentation of particles over a given size while the slower zones continue to separate.

3.3.4 *Complex Gradient Shapes*

If a programmable gradient generator is available, the investigator can generate gradients which have both steeper and shallower regions to sharpen or resolve particular zones at will (see ref. 17 and Section 5.1.5) although sharp discontinuities in any gradient tend to flatten quite quickly by diffusion. Usually the use of these gradient makers is restricted to zonal rotors.

4. CHOICE OF DENSITY GRADIENT FOR ISOPYCNIC CENTRIFUGATION

4.1 **Choice of Gradient Solute and Solvent**

Almost all isopycnic centrifugation analyses have been carried out using a suitable aqueous buffer as the solvent. The buffer composition and pH depend upon the nature of the biological sample being fractionated and are chosen to optimise the recovery of viable cells, active subcellular organelles or intact macromolecules. Similarly, the choice of gradient solute depends upon the nature of the biological sample and the aim of the fractionation. A description of the more commonly-used gradient solutes is given in Section 5.1 of Chapter 1 and detailed experimental protocols for their use are given in subsequent chapters. The following sections discuss the general properties of isopycnic density gradients.

4.2 **Choice of Gradient Shape**

4.2.1 *Introduction*

When a solution is centrifuged, all of the components in it are subject to the resulting force field. Thus, the density gradient solute molecules tend to sediment towards the bottom of the tube. If their rate of sedimentation is initially greater than their rate of diffusion, a continuous density gradient is formed. In practice these 'self-forming' gradients (*Figure 1c*) are used by mixing the sample with the gradient medium to give a solution of uniform density. The mixture is then centrifuged. The solute molecules sediment to form the gradient and then the sample molecules band at their isopycnic points. The alternative approach is to prepare the gradient prior to centrifugation (preformed gradient; *Figure 1b*) then layer the sample on top and carry out centrifugation until the sample particles have reached equilibrium.

Preformed gradients used for isopycnic centrifugation can be either discontinuous (step) gradients or continuous gradients. Step gradients are most suitable for separating whole cells or subcellular organelles from plant or animal tissue homogenates (see Chapter 5) and for purifying some viruses (Chapter 7). Continuous gradients are preferred for the fractionation of complex mixtures of biological particles since the gradual change in density along the gradient yields much higher resolution of the sample components.

The choice as to self-forming or preformed continuous gradients depends on several factors, including the nature of the gradient solute itself. Suspensions of colloidal silica, such as Ludox and Percoll, generate self-forming gradients in less than 30 min at 10 000–20 000 g. Because gradient formation is so rapid there is often no need to preform the gradients although this may be necessary in some cases to generate the required gradient shape. However, most gradient solutes will produce self-forming gradients only slowly. Thus CsCl and metrizamide require several hours at 50 000–100 000 g. In these cases, the choice of self-forming or preformed gradients depends upon both the gradient shape desired (Sections 4.2.2 and 4.2.3) and the nature of the sample being fractionated. The major advantage of preformed gradients is the considerably shorter time needed for centrifugation since time is not needed for the density gradient to form but only for the sample components to band at their isopycnic points. Therefore, preformed gradients are especially useful when the biological particles to be separated are large (>100S) or labile. However, one potential source of artifacts is that the sample is loaded onto the top of the gradient in a relatively small volume (compared with the gradient). In some cases, biological samples lose activity during concentration or, especially in the case of cells and viruses, may form aggregates of particles. Self-forming gradients avoid this problem since it is possible to use large sample volumes and adjust these to the correct starting density by the addition of solid gradient solute or a concentrated stock solution. Furthermore, top-loaded samples on a preformed gradient reach their isopycnic positions in an exponential manner so that when separating low molecular weight particles, such as proteins, it is important to prove that the position of the particle in the density gradient after centrifugation represents the true density of that particle. In self-forming gradients, since particles are initially distributed evenly throughout the gradient, bands can only be produced by particles reaching their isopycnic points. This also avoids the possibility of artifactual bands which can occur with preformed gradients as a result of sample components remaining in the loading zone during centrifugation.

4.2.2 *Self-forming Gradients*

As described above, many gradient media give useful self-forming gradients when centrifuged. However, the nature of the gradient generated depends upon the gradient medium. Ludox and Percoll gradients form rapidly but the colloidal silica particles tend to pellet to the bottom of the centrifuge tube. In contrast, continued centrifugation of gradient media that form true solutions (such as CsCl, Cs_2SO_4 and metrizamide) eventually produces a stable equilibrium gradient where the sedimentation of the gradient solute is balanced by diffusion of solute molecules in the direction opposite to that of sedimentation. The exact shape of the equilibrium gradient produced depends upon not only the particular gradient solute used but also the exact conditions of centrifugation.

Ideally the aim should be to choose conditions which generate a gradient whose density range exceeds that of the particles to be fractionated, so that all of the sample particles band, but which is sufficiently shallow to allow maximum resolution of the individual sample components. Within any given band of particles separated by isopycnic centrifugation, the sample components are often heterogeneous in size or density or both so that the band is quite broad. In such cases, beyond a certain point, the use of a shallower gradient will result in bands which are further apart than in a steeper gradient but the bands will also be broader so that there may be no real increase in resolution.

Fortunately, the principles which govern the shape of self-forming gradients are the same for all gradient media that form true solutions although, for historical reasons, most quantitative analyses have been done for CsCl gradients. Using the following equations it is possible to predict the profile of any self-forming gradient and hence to design a suitable gradient for use (see also ref. 18).

Slope of the gradient. The final slope of the density gradient $\left(\dfrac{d\varrho}{dr}\right)$ can be predicted from the following equation:

$$\frac{d\varrho}{dr} = \frac{\omega^2 r}{\beta^\circ} = \frac{1.1 \times 10^{-2} N^2 r}{\beta^\circ} \qquad \text{Equation 1}$$

where ω is the angular velocity (rad/sec), r is the distance (cm) from the axis of rotation, β° is the density gradient proportionality constant for the gradient solute in the gradient solvent and N is the rotor speed (r.p.m.).

In practice, this equation can predict the slope of the gradient for only about the middle three-quarters of the gradient. *Table 1* lists the β° values that have been evaluated for a number of gradient solutes. The method of calculation of β° values for other gradient solutes is published in ref. 19. From Equation 1 it

Table 1. Values of β° for Ionic Gradient Media.

Gradient solute	β° values $(\times 10^{-9})$ for solutions with initial densities (g/cm^3) of							
	1.1	1.2	1.3	1.4	1.5	1.6	1.7	1.8
CsCl	–	2.04	1.55	1.33	1.22	1.17	1.14	1.12
Cs_2SO_4	–	1.06	0.76	0.67	0.64	0.66	0.69	0.74
CsTCA[a]	–	–	–	1.18	–	–	–	–
CsTFA[b]	–	–	–	–	–	1.29	–	–
KBr	6.20	3.80	3.05	–	–	–	–	–
KI	4.28	2.55	1.96	1.73	–	–	–	–
NaBr	7.70	5.20	–	–	–	–	–	–
NaI	5.10	3.19	2.82	–	–	–	–	–
RbBr	–	2.15	1.56	1.34	1.22	–	–	–
RbCl	–	3.42	2.76	2.25	–	–	–	–
RbTCA[a]	–	–	–	–	1.33	1.38	–	–

[a]TCA is the trichloroacetate salt.
[b]TFA is the trifluoroacetate salt; CsTFA is a registered trademark of Pharmacia Fine Chemicals AB.

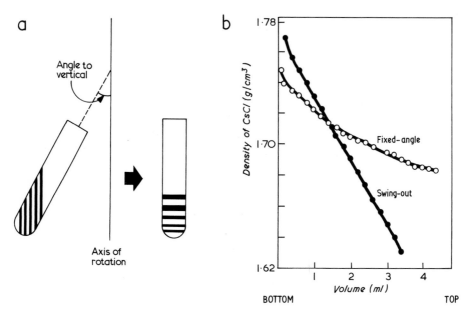

Figure 7. Density gradients in swing-out and fixed-angle rotors. (**a**) Diagram of a gradient in a tube of a fixed-angle rotor during centrifugation (left) and at rest after reorientation to the vertical position (right). (**b**) Profiles of density gradients formed by centrifugation of a solution of CsCl (initial density = 1.72 g/cm³) in a swing-out and a fixed-angle rotor (from ref. 5).

is clear that the slope of the density gradient becomes steeper with the square of the rotor speed, increasing distance from the axis of rotation and decreasing β° values. In addition, the slope is usually steeper with increasing initial concentration of the gradient solute.

Density range of a gradient. This can be calculated for an equlibrium gradient using the following equation:

$$\varrho_2 - \varrho_1 = \frac{\omega^2}{2\beta^\circ}(r_2^2 - r_1^2) = \frac{1.1 \times 10^{-2} \times N^2}{2\beta^\circ}(r_2^2 - r_1^2) \qquad \text{Equation 2}$$

where ϱ_1 and ϱ_2 are the densities (g/cm³) at points r_1 and r_2 (cm) from the axis of rotation.

Effect of rotor choice on gradient slope and range. It is important to note that the density range and slope of an equilibrium gradient will depend upon the type of rotor used for isopycnic centrifugation. This is because the density range of the gradient depends upon the horizontal distance between the gradient top and bottom *during centrifugation.* This distance is unchanged during centrifugation in a swing-out rotor because the rotor buckets swing out to a horizontal position during acceleration and then back to the vertical during deceleration (see Chapter 1, Section 4.3.2). However, during acceleration and deceleration of vertical and fixed-angle rotors, the pockets retaining the centrifuge tubes are fixed and it is the gradients which reorientate. Therefore, using either of the latter rotors, the distance between the top and bottom of each

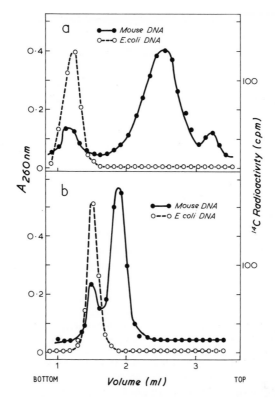

Figure 8. Fractionation of a mixture of unlabelled mouse liver DNA and [^{14}C]DNA from *Escherichia coli* by isopycnic centrifugation using a CsCl gradient, (**a**) in a fixed-angle rotor and (**b**) in a swing-out rotor. The DNA was centrifuged to equilibrium and the gradients then fractionated. The positions of the mouse and *E. coli* DNA were determined by measurement of the absorbance at 260 nm (● —— ●) and ^{14}C radioactivity (○— —○) respectively (from ref. 5).

gradient increases during reorientation. The gradient is forced to extend over a large distance and so it becomes shallower (*Figure 7*) with a consequent increase in resolution of the sample bands. An example of this phenomenon is shown in *Figure 8*. The degree of extension of the gradient during deceleration, and hence the potential increase in resolution, depends upon the angle of the centrifuge tube to the direction of the centrifugal fields; the larger the angle the greater is the effect. Therefore, vertical rotors, where the tubes are at right angles to the centrifugal force field, give greater resolution than fixed-angle rotors.

Maximum and minimum densities in a gradient. These values can be calculated provided that a reference point of known density is available. This is provided by the isoconcentration point (r_c) which is that point in a gradient where the density is the same as in the original solution. For swing-out rotors,

$$r_c = [1/3 \, (r_t^2 + r_t r_b + r_b^2)]^{1/2} \qquad \text{Equation 3}$$

where r_t and r_b are the distances (cm) of the gradient top and bottom from the axis of rotation.

59

No mathematical analysis for the calculation of r_c for fixed-angle and vertical rotors, which takes account of gradient reorientation during acceleration and deceleration, has yet been published. However, empirically it has been found that for practical purposes Equation 3 is also valid for these types of rotor.

Once r_c is known, the minimum and maximum densities at the top (ϱ_t) and bottom (ϱ_b) of a gradient centrifuged at N r.p.m. can be calculated as:

$$\varrho_t = \varrho_i - \frac{1.1 \times 10^{-2} \times N^2}{2\beta^\circ} (r_c^2 - r_t^2) \qquad \text{Equation 4}$$

$$\varrho_b = \varrho_i + \frac{1.1 \times 10^{-2} \times N^2}{2\beta^\circ} (r_b^2 - r_c^2) \qquad \text{Equation 5}$$

where ϱ_i is the density of the initial solution. Alternatively, knowing the desired density range of the gradient, it is possible to use Equations 4 and 5 to calculate the rotor speed necessary to produce that gradient. Note that these equations should always be used prior to isopycnic centrifugation using new conditions since this allows one to calculate whether, at the chosen rotor speed, the concentration of solute at the bottom of the final gradient will exceed that which leads to crystallisation or precipitation of the solute. For example, if the density at the bottom of a CsCl gradient exceeds 1.9 g/cm^3, then crystals of CsCl will form. This is potentially a very dangerous situation in that the dense CsCl crystals (4 g/cm^3) can overstress the rotor and lead to catastrophic failure during centrifugation. Because of the importance of predetermining the shape of the gradient for sensible and safe use of isopycnic centrifugation, the Appendix to this chapter lists a microcomputer program which can be used to carry out the necessary calculations.

Finally, one should be aware that for some applications, for example when using metrizamide, the equilibrium shape may not be optimal for the separation required. In these cases it may be possible to choose a time period of centrifugation which will yield a non-equilibrium gradient but which has the required shape.

4.2.3 *Preformed Gradients*

Discontinuous (step) gradients. These are prepared by successively layering solutions of different density in a centrifuge tube (Section 5.1.2). The sample is then loaded on top and centrifugation is started immediately. Each particle of the sample sediments until it encounters a layer of greater density than itself. Therefore the sample is separated into a series of bands, each band being located at the interface between two density layers (*Figure 9*). Usually only a few density layers or 'steps' are used, perhaps three or four. Therefore only a crude fractionation occurs to give families of particles each containing particles of different densities but falling within a specific density range. Clearly, gradient shape in this context refers to the number of steps and the density for each step used. The choice of these conditions will vary according to the complexity of the sample mixture and the densities of the sample particles which

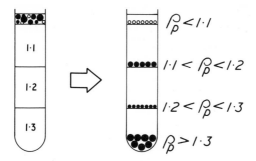

Figure 9. Use of discontinuous (step) gradients for isopycnic centrifugation. (**a**) The step gradient is prepared by overlayering or underlayering solutions of the gradient medium of different density (see the text). In the example shown, the gradient consists of three layers with densities of 1.1, 1.2 and 1.3 g/cm³. The sample is then loaded on top of the gradient. (**b**) After centrifugation, the sample has been fractionated into families of particles, each family consisting of particles with densities (ϱ_p) falling within the density range defined by two gradient layers (from ref. 24).

are to be fractionated (see Chapter 5).

Continuous gradients. When using continuous density gradients for isopycnic centrifugation, the choice of gradient shape will depend on the range of densities of the particles to be separated. Linear gradients are usually used if the densities of the particles are fairly evenly distributed. It is important to note that, unless the preformed gradient shape is identical to the predicted equilibrium gradient under the centrifugation conditions used, the preformed gradient will be unstable because the gradient solute will redistribute during centrifugation. The time t (in hours) for which the initial gradient slope at the midpoint of the gradient is stable is given by:

$$t = k' \, (r_b - r_t)^2 \qquad\qquad \text{Equation 6}$$

In the case of CsCl, the value of k' is 0.3 (20).

In practice, the gradient shape in the middle third of the gradient is maintained for about half as long ($t/2$ hours). Clearly, from Equation 6, the greater the physical length of the preformed gradient, the longer it is maintained. Therefore for the isopycnic fractionation of sample particles by this method, the preformed gradient should be as long as possible. The time taken for the particles to reach their isopycnic positions in a gradient of this kind can be determined as described in Section 6.2.4.

5. FORMATION OF GRADIENTS

5.1 Preparation of Gradients for Rate-zonal Centrifugation

5.1.1 *Introduction*

Gradients can be produced by freezing and then thawing a homogeneous solution of the gradient medium in a centrifuge tube one or more times (e.g., ref. 21). Although this simplest of methods has been used successfully with various gradient media to prepare gradients for certain fractionations (e.g., ref. 22), it is not a commonly-used technique because of the difficulty in predicting the

Table 2. Dilution of Stock Solutions of Sucrose.

Desired final concentration % (w/w)	Volume (ml) of stock solution to be diluted to 1 litre[a]					Density (g/cm³)	Refractive index of the final solution at 20°C[b]
	60% (w/w)	66% (w/w)	70% (w/v)	80% (w/v)	2.0 M		
5	66	58	73	64	75	1.0179	1.3403
10	134	119	149	130	152	1.0381	1.3479
15	205	182	228	199	233	1.0592	1.3557
20	279	247	310	271	317	1.0810	1.3639
25	357	315	396	346	405	1.1036	1.3723
30	437	387	485	424	496	1.1270	1.3811
35	521	461	578	506	591	1.1513	1.3902
40	609	539	676	591	691	1.1764	1.3997
45	701	620	777	680	795	1.2025	1.4096
50	796	704	883	773	903	1.2296	1.4200
55	896	792	994	870	–	1.2575	1.4307
60	1000	884	–	971	–	1.2865	1.4418

[a]All values quoted are at 4°C.
[b]Refractive index of water at 20°C is 1.3330. Note that the presence of solutes other than sucrose will increase the observed refractive index.

final shape of the gradient and the final distribution of buffer ions. Gradients can also be produced by layering solutions of decreasing densities one upon the other followed by diffusion to produce a smooth gradient. However, it is more usual to prepare the gradients using a gradient maker. This allows the production of gradients in a much shorter time than the diffusion method, that is, in minutes rather than hours. Gradient makers range from the simplest equipment capable of generating linear or exponential gradients to sophisticated commercial apparatus capable of being programmed to yield any desired shape of gradient. Whichever method is chosen, it is important to make up the gradient solutions accurately. Traditionally, sucrose solutions for rate-zonal centrifugation are usually expressed in terms of percentage weight/weight (w/w), that is, by mixing weighed amounts of sucrose and buffer. In practice, the simplest way to prepare gradient solutions is by dilution of concentrated stock solutions. This can be readily carried out using the values given in *Table 2*. The reader should note that diluting a 60% (w/w) solution to twice its volume does *not* give a final concentration of 30% (w/w). In all cases the concentrations of the sucrose gradient solutions should be checked by refractometry before use.

5.1.2 *Diffusion Method*

Layers of different concentrations of gradient solute in the same buffer are arranged in the centrifuge tube with the most dense layer at the tube bottom and

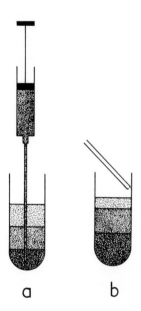

Figure 10. Gradient preparation using the diffusion method; (**a**) underlayering, (**b**) overlayering. The techniques are described in the text. In each case the denser solutions are depicted by the degree of stippling.

the least dense at the top. One procedure uses a calibrated syringe fitted with a blunted, large-bore needle to introduce a known volume of the least dense solution into the centrifuge tube, and then this is underlayered slowly with a more dense solution (*Figure 10a*). The process is repeated to yield a series of layers increasing in concentration toward the tube bottom. Alternatively, the most concentrated solution is placed in the centrifuge tube first and then overlayered with the next most concentrated (*Figure 10b*). The procedure is repeated with each of the solutions to form the desired step gradient. Either technique may be used to prepare the gradients although it is somewhat easier to avoid undue mixing and to prepare gradients in a reasonable period of time using the underlayering procedure. Once formed, the step gradients are left to equilibrate at the temperature of centrifugation to allow continuous gradients to form by diffusion. This is carried out either with the gradients held vertically or sealed (for example, with a rubber bung) and laid horizontally. The time required to obtain a continuous gradient depends on a number of factors which relate to the diffusion rate of the gradient solute. These include the concentration range and viscosity of the gradient, the thickness of each layer, temperature, and most importantly, the molecular weight of the solute. As a general rule, gradients may be preformed by diffusion only if the molecular weight of the gradient medium is less than 1000 daltons, since gradients prepared from high molecular weight gradient media diffuse only very slowly. As an example, for a 10 ml sucrose gradient prepared from five layers of sucrose each about 2 cm deep, the minimum period of time required is approximately 12 h at 4°C if the tube is held vertically or about 60 min at 4°C if the tube is laid horizontally. It is important to realise that the longer the gradients are left, the less steep the gradient will become as diffusion proceeds.

5.1.3 *Simple Linear Gradient Maker*

The simplest and most widely used gradient maker is that for producing linear gradients. It consists of two vessels of equal cross-sectional area joined by a connecting channel which can be opened and closed by means of a stopcock. One chamber has an exit which leads to the centrifuge tube via a length of flexible tubing. Such devices are marketed by several companies including a number of centrifuge manufacturers and they can be purchased in a series of chamber sizes to suit most centrifuge tube capacities, including large-scale versions for use with zonal rotors. Some commercial gradient makers of this kind have multiple outlets with the purpose of allowing identical gradients to be produced in several different centrifuge tubes (usually three) at the same time (*Figure 11*). In practice, it is reasonably easy to produce a satisfactory home-made device with single or multiple outlets provided care is taken to ensure that the two chambers have equal cross-sections. If this is not the case then the gradient will not be linear.

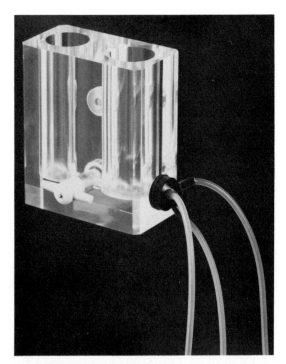

Figure 11. A simple two-chamber linear gradient maker with three outlets from the mixing chamber (courtesy of MSE Scientific Instruments).

There are two methods for preparing linear gradients using a gradient maker of this kind which will be referred to here as dense end first and light end first. When forming gradients dense end first, the outlet tube from the gradient maker is allowed to rest against the top of the centrifuge tube and the gradient liquid flows down the wall of the tube until the tube is filled (*Figure 12a*). This method is readily applicable to centrifuge tubes made of hydrophilic materials such as cellulose nitrate and cellulose acetate butyrate. However, some centrifuge tubes, for example Ultra-Clear™ (Beckman), polyallomer and polycarbonate tubes, are hydrophobic and it is not feasible to prepare gradients this way; instead of the desired gentle flow of liquid down the tube wall,

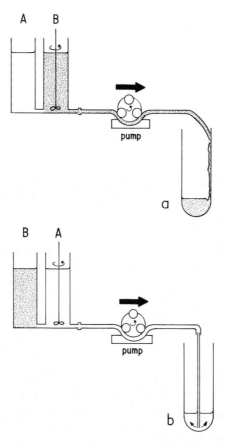

Figure 12. Gradient preparation using a simple linear gradient maker. (**a**) Formation of gradients dense end first: the reservoir contains the less dense solution A and the mixing chamber contains the denser solution B. The gradient is poured with the outlet touching the wall of the centrifuge tube at the top. (**b**) Formation of gradients light end first: the reservoir contains the denser solution B and the mixing chamber contains the less dense solution A. The gradient is poured with the outlet touching the bottom of the centrifuge tube. Although the gradient medium can be allowed to enter the tubes by gravity, more usually a peristaltic pump is employed as shown. Linear gradients will be produced only if the cross-sectional areas of the reservoir and mixing chamber are identical (see *Figure 13*).

large droplets form which cascade down the wall and disrupt the gradient. This problem can be overcome by pre-treating polyallomer tubes with chromic acid for about 30 min (23). Alternatively, one can place a narrow strip of dialysis tubing or a thin glass rod down one side of the tube down which the gradient can be poured. However, a much easier remedy is to form the gradient light end first. In this method the outlet tube of the gradient maker is led to the bottom of the centrifuge tube (*Figure 12b*) so that the gradient does not need to flow down the tube wall. Therefore, in contrast to the formation of gradients dense end first, it is possible to form gradients light end first using any type of centrifuge tube. Furthermore, the latter method is faster than the former since in the former method the flow rate cannot be allowed to exceed about 1 ml/min or the excessive flow down the side of the tube stirs up the gradient. When the gradient is prepared light end first, the rates of flow can be allowed to reach at least 2 ml/min without disturbance.

Formation of a gradient dense end first. This is carried out as follows.

(i) Clean the gradient maker and sufficient small-bore, flexible tubing (e.g., Tygon tubing) thoroughly before use. Allow them to air-dry. The length of the tubing used (and its bore) should be kept small to reduce laminar mixing of the gradient as it flows from the gradient maker to the centrifuge tube. For rate-zonal analysis of nucleic acids, particularly RNA, it is important that the apparatus and all tubing should be free of nuclease contamination. To do this, rinse the apparatus and tubing with 0.1% diethyl pyrocarbonate in water and then with autoclaved distilled water.

(ii) Set up the required number of centrifuge tubes held firmly in a test-tube rack or clips, preferably at the temperature required during centrifugation.

(iii) The next stage is to connect the outlet(s) from the gradient maker to the centrifuge tube(s) using the tubing. Although the density gradients may be delivered to the centrifuge tubes by gravity, greater control over the flow rate is possible if it is regulated by a peristaltic pump (*Figure 12a*). Neither impeller pumps nor piston pumps are suitable for pouring gradients because of their sizeable 'dead volume'. An additional advantage of peristaltic pumps is that the liquid is always confined to the small-bore tubing and does not contact the pump itself. Make sure the tubing is arranged so that the gradient flows continually downwards, otherwise mixing of the gradient will occur in the tubing as a result of gradient inversion. Finally, either allow the outlet end of the tubing to rest against the tube wall near the top or clip it in place there.

(iv) Make sure that the outlet of the gradient maker and the channel connecting the two chambers of the apparatus are closed. Pipette the required volume (depending on the size of the gradient required) of the *less* dense solution into the reservoir and the denser solution into the mixing chamber (*Figure 12a*). It is usually easiest to allow the gradient maker to empty itself when preparing gradients but a small volume of liquid is always retained by the gradient maker and outlet tubing and so it is necessary first to carry out a preliminary experiment to determine this retention volume. The volume placed in each chamber in subsequent experiments can then be adjusted to allow the gradient maker to deliver reproducibly the desired volume to the centrifuge tube. For the liquid levels in each chamber to be at equilibrium initially, the volumes (assuming the cross-sectional areas of the two chambers are identical) must be in inverse proportion to their density so that equal *weights* of each solution are present. Otherwise if equal *volumes* are used, once the connecting channel is opened the denser solution will immediately flow into and contaminate the less dense solution.

(v)   Stir the solution in the mixing chamber using a small magnetic stirring bar inserted in this chamber with the gradient maker supported above a magnetic stirrer, or by use of a helix of stainless steel wire driven by an overhead motor. The stirring should be efficient but not too vigorous (to prevent excessive vortexing).

(vi) Open the outlet and set the peristaltic pump at a speed pre-determined to give a slow steady flow rate (about 1 ml/min with 5 – 14 ml gradients) and then open the channel connecting the two chambers of the gradient maker. The dense solution will flow out of the mixing chamber and be replaced by less dense solution from the reservoir to re-establish hydrostatic equilibrium. The solution flowing into the centrifuge tube becomes progressively more dilute, resulting in the formation of a density gradient in the tube. The concentration of the gradient solute leaving the mixing chamber at any time is given by:

$$C_t = C_M + (C_R - C_M) \frac{V_t}{2V_0} \qquad \text{Equation 7}$$

where C_t is the concentration of gradient being delivered at time t, C_M is the initial concentration of solute in the mixing chamber, C_R is the initial concentration of solute in the reservoir, V_t is the volume of gradient poured at time t and V_0 is the original volume of liquid in each chamber. Sheeler (24) has described examples of the use of this equation.

(vii) Allow the gradient maker to deliver the required volume of solution to the centrifuge tube but prevent the very last part of the gradient from entering the tube since this usually causes bubbles to form on the top of the gradient. Provided the correct volumes of solutions have been used, the centrifuge tube should now be full to within about 5 mm of the tube top.

(viii) Remove the tubing from the centrifuge tube and store the gradient at the temperature at which it is to be used until all the gradients have been prepared and the samples are ready for loading. Gradients prepared from high molecular weight solutes are more stable than those prepared from low molecular weight solutes which diffuse more rapidly. In general, use the gradients as soon as possible after preparation and certainly within an hour or so.

Formation of a gradient light end first. This is carried out essentially in the same manner as for the previous method except that:

(i) The outlet tubing is connected to a long, blunted syringe needle which is placed with its free tip touching the bottom of the centrifuge tube (*Figure 12b*).

(ii) The denser solution is now placed in the reservoir of the gradient maker and the less dense solution in the mixing chamber. Using this method, less dense solution enters the centrifuge tube first and is displaced by the solution of increasing density which follows.

It is worth noting that, in the absence of even a simple gradient maker, linear gradients can theoretically be generated using a three-channel peristaltic pump, all channels of which have the same flow rate. The more dense solution is placed in an appropriate sized mixing chamber and the less dense solution is fed into it through one channel. The solution is taken out of the mixing chamber into the centrifuge tubes, two at a time, using the other two channels. The linearity of the gradients formed depends on the similarity of flow in the channels. In practice it is very difficult, if not impossible, to obtain identical flow rates in all channels using a standard peristaltic pump. Finally, Corless (25) has described a procedure for the preparation of small-volume density gradients (<1 ml) by using a combination of a standard gradient maker and one or two peristaltic pumps.

5.1.4 *Exponential Gradient Makers*

Although linear gradients are the type most commonly used, concave and convex gradients can also be easily prepared. One approach is to use a simple two-chamber gradient maker where the chambers have different cross-sectional areas (*Figure 13*; see also ref. 8). Alternatively, for generating true exponential gradients, one can use an exponential gradient maker. In this type of apparatus the gradient liquid is withdrawn from a mixing chamber, the volume of which is kept constant, and this in turn is replenished from a reservoir. If the mixing chamber initially contains the less dense solution, the resulting gradient profile is convex exponential, whereas if the mixing chamber contains the more dense solution the resulting profile is concave exponential. The degree of non-linearity depends upon the relative volume of the mixing chamber compared with the final volume of the gradient according to the equation:

$$C_t = C_R - (C_R - C_M).e^{-\dfrac{V_t}{V_M}} \qquad \text{Equation 8}$$

where C_t is the concentration of gradient solute in the liquid leaving the mixing chamber at time t, C_R is the concentration of gradient solute in the reservoir liquid, C_M is the initial concentration of gradient solute in the mixing chamber, V_t is the volume of gradient liquid already withdrawn from the mixing chamber at time t and V_M is the volume of the mixing chamber. Note that the highest concentration of gradient solute in the final gradient never reaches the concentration of gradient solute in the reservoir.

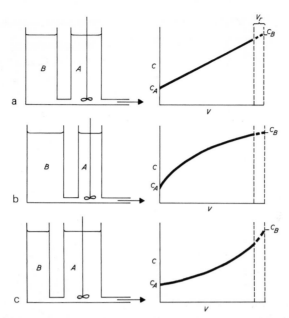

Figure 13. Preparation of convex and concave gradients using a simple gradient maker. (a) Linear gradients are produced when the cross-sectional area of the gradient maker reservoir and mixing chamber are identical, (b) convex gradients are produced when the cross-sectional area of the reservoir exceeds that of the mixing chamber, (c) concave gradients are produced if the cross-sectional area of the reservoir is less than that of the mixing chamber. In each case the mixing chamber contains the less dense solution, A, with concentration C_A, and the reservoir contains the denser solution, B, with concentration C_B. C and V refer to the concentration of the gradient medium and the volume of the gradient leaving the gradient maker, respectively, and V_r is the residual volume of the gradient maker (from ref. 8).

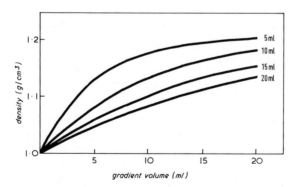

Figure 14. Dependence of exponential gradient shape upon mixing volume. The four curves illustrate the effect of changing the ratio of the volume of liquid in the mixing chamber to the total gradient volume. The liquid present initially in the mixing chamber had a density of 1.0 g/cm³ and that in the reservoir had a density of 1.2 g/cm³. In each case, the total volume of the gradient formed was 20 ml. As the mixing chamber volume increases from 5 ml to 20 ml, so the curvature and range of the final gradient decreases (adapted from ref. 10).

Figure 14 shows a typical family of convex gradient profiles which can be generated by varying the mixing chamber (fixed) volume. Clearly, since the exact shape of the gradient produced depends on the volume of the mixing

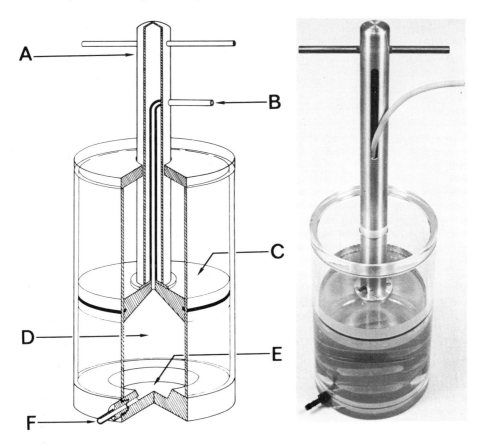

Figure 15. An exponential gradient maker with a variable-volume mixing chamber. **A**, piston handle; **B**, inlet from reservoir; **C**, tapered piston; **D**, variable-volume mixing chamber; **E**, recess for magnetic stirring bar; **F**, outlet to centrifuge tube. The desired fixed volume is placed in the mixing chamber and the piston is lowered onto the liquid surface to expel air from both the chamber and the inlet line. The inlet is then linked to the liquid reservoir and gradient formation started (from ref. 24 with permission). A commercial version of this apparatus is available from Bio-Sep (USA).

chamber relative to the total gradient volume, it is advantageous to use an exponential gradient maker in which the mixing chamber volume is adjustable. An apparatus of this type is shown in *Figure 15*. An even simpler type of apparatus has been described by Noll (26). Examples of the use of Equation 8 to produce concave or convex gradients with specific profiles are described by Sheeler (24).

5.1.5 *Programmable Gradient Makers*

For some purposes it may be desirable to construct gradients which are more complex than those described above, for example containing both linear and exponential regions. This can be achieved using one of the more sophisticated, programmable gradient makers which are commercially available, although these are mainly used in centrifugation studies with zonal rotors. These machines make use of two reservoirs from which the solutions are pumped either by piston feed, peristaltic pumps, or diaphragm pumps and variations in gradient shape are achieved by varying the relative pumping rates from these two reservoirs. The programme is selected either by a cut-out template which is scanned by the gradient maker or by pre-setting a number of regulator dials.

5.1.6 *Production of Multiple Gradients*

Several swing-out rotors in use for rate-zonal centrifugation employ six buckets and some vertical rotors have eight to sixteen tube holders. The production of these numbers of gradients individually can be a fairly time-consuming process. Therefore a number of types of apparatus have been devised for the production of several density gradients simultaneously.

Although multiple discontinuous density gradients can be produced manually, the procedures of overlayering or underlayering (Section 5.1.2) are time-consuming and tedious. McRee (27) has described a simple, inexpensive hydraulic system for preparing multiple discontinuous gradients of almost any desired volume. With regard to continuous gradients, the simplest multiple gradient maker is a standard gradient maker fitted with multiple outlets (usually three) from a single mixing chamber (*Figure 11*). To form identical gradients using such equipment, it is obviously essential that the flow rate to each centrifuge tube must be identical. This is often attempted by passing the multiple outlet tubes through a multichannel peristaltic pump set to deliver equal flow rates in all channels. However, in practice, it is rarely possible to obtain exactly identical gradients by this method since flow rates in the different channels inevitably vary because of differences in tubing length, wall uniformity, pump roller wear, etc. Despite this, the gradients produced are usually sufficiently similar to be useful for the multiple analysis of samples which contain internal marker components. Nevertheless, it is necessary to use more complex equipment to obtain truly identical gradients for accurate quantitative work. One example of this type of equipment is the multiple density gradient maker marketed by Beckman Instruments Inc. which can produce three identical gradients simultaneously with each gradient consisting of up to 57 ml volume. Home-made devices have also been designed for this purpose (24,28).

5.2 **Preparation of Gradients for Isopycnic Centrifugation**

5.2.1 *Self-forming Gradients*

The sample, gradient solute and buffer are mixed in proportions so as to give a solution with a density which will generate an equilibrium gradient of the appropriate density range (Section 4.2.2). This can either be done by mixing the sample with a concentrated stock solution of the gradient medium or by the addition of gradient solute directly to the sample. The latter approach is particularly useful when the sample volume is large. Whichever method is used, thorough mixing is essential to ensure a uniform density throughout the liquid prior to centrifugation. *Table 3* gives the properties of aqueous solutions of CsCl, the medium most commonly used for self-forming gradients. Since

Table 3. Properties of Aqueous CsCl Solutions at 20°C[a]

Concentration		Density (g/cm³)	Refractive index at 20°C
Percent w/w	*Percent w/v*		
20.0	23.51	1.1756	1.3507
22.0	26.33	1.1967	1.3528
24.0	29.24	1.2185	1.3550
26.0	32.27	1.2411	1.3572
28.0	35.40	1.2644	1.3594
30.0	38.66	1.2885	1.3617
32.0	42.03	1.3135	1.3641
34.0	45.54	1.3393	1.3666
36.0	49.18	1.3661	1.3691
38.0	52.96	1.3938	1.3717
40.0	56.90	1.4226	1.3744
42.0	61.00	1.4525	1.3771
44.0	65.27	1.4835	1.3800
46.0	69.73	1.5158	1.3829
48.0	74.37	1.5495	1.3860
50.0	79.23	1.5846	1.3892
52.0	84.30	1.6212	1.3925
54.0	89.62	1.6596	1.3960
56.0	95.19	1.6999	1.3996
58.0	101.05	1.7422	1.4035
60.0	107.21	1.7868	1.4076
62.0	113.71	1.8340	1.4120
64.0	120.59	1.8842	1.4167

[a]*Data derived from ref. 32.*

most self-forming gradients are relatively shallow, it is important that the density of the initial mixture is exactly that required. The density should be checked either by refractometry or pycnometry (see Section 8.1) and the density adjusted as necessary before centrifugation. Ideally the initial density should be slightly higher than the banding density of the most dense sample component to avoid the possibility of the latter pelleting before the density gradient forms.

5.2.2 Preformed Gradients

Discontinuous gradients for isopycnic centrifugation can be prepared as described in Sections 5.1.2 and 5.1.6. Unlike their use for rate-zonal centrifugation, they must be used immediately after preparation to prevent diffusion smoothing out the step interfaces. Continuous density gradients for isopycnic centrifugation can be prepared by diffusion of step gradients (Section 5.1.2) but they are more routinely prepared using gradient makers as described earlier for rate-zonal centrifugation (Sections 5.1.3 – 5.1.6). The densities of the initial solutions should be checked either by refractometry or pycnometry (see Section 8.1) before the gradients are prepared. If necessary, fine adjustments to the densities can be made by the addition of more gradient solute or solvent.

6. CENTRIFUGATION OF GRADIENTS

6.1 Rate-zonal Centrifugation

6.1.1 *Composition of the Sample*

Once the density gradients have been formed, the tubes must be handled extremely carefully since vibrations or sudden temperature variations will disturb the gradients. The sample should be ready for loading *before* the gradients are prepared so that the minimum amount of time elapses between the preparation of gradients and their use. For many preparations, the sample must be kept cold to preserve the native structure and activity of the biological components and thus the sample (and the gradient) should be prepared and maintained at 4°C prior to and during centrifugation.

The composition of the sample will depend upon the type of gradient medium to be used and the aim of the fractionation. However, several aspects of sample composition are similar irrespective of the type of particles being fractionated.

(i) The sample solution must have a density less than that of the top of the gradient which is to support it or the sample will fall through the gradient until it reaches its isopycnic position or the bottom of the tube. The sample is usually prepared in the same buffer as the gradient. When working with large macromolecules or nucleoprotein complexes (e.g., ribosomes), it is often advantageous to omit sucrose from the sample since the particles then sediment considerably faster through the sample zone than through the top of the gradient and so concentrate at the interface at the top of the gradient before fractionation. This zone-sharpening of the sample potentially increases the resolution obtainable.

(ii) Care must be taken to avoid the phenomenon of 'droplet sedimentation' (also called streaming). This type of instability occurs because the sample is necessarily less dense than the gradient and so contains a lower concentration of gradient solute than the top of the gradient (29). Hence there will be a concentration gradient resulting in the solute diffusing into the sample zone. On the other hand, in rate-zonal centrifugation the sample particles will necessarily be larger than the gradient solute molecules and therefore will diffuse much more slowly. The result is an overall increase in density in the sample zone immediately adjacent to the top of the gradient so that this begins to sink into the gradient. The periphery of the sample zone tends to sink fastest and causes the formation of droplets which fall through the gradient. Disturbance of the interface between the sample and gradient by vibration or minor convective mixing may also result in droplet sedimentation (10). The effect of droplet sedimentation is a slow streaming forward of the leading edge of the sample zone. This can affect the final resolution achieved. In practice, droplet sedimentation tends to be a problem when using low centrifugal fields and low molecular weight solutes. In high centrifugal fields the droplets seldom become large enough or dense enough to affect resolution seriously. Nevertheless, droplet sedimentation can occur prior to centrifugation if the period of time between loading of the sample and centrifugation is excessive. Therefore, gradients should be centrifuged as soon as possible after the sample has been loaded. The use of high molecular weight solutes (for example, Ficoll) with low centrifugal fields minimises droplet sedimentation since the large solute molecules diffuse only slowly. In serious cases of droplet sedimentation, the sample zone may be stabilised by loading the sample as an inverse gradient (8).

(iii) For optimal resolution, the amount of sample loaded must not exceed the capacity of the particular gradient to be used. The volume of the sample which should be loaded is a function of the cross-sectional area of the centrifuge tube. For centrifuge tubes $1.0 - 1.6$ cm diameter, the sample volume should be about $0.2 - 0.5$ ml whereas for tubes 2.5 cm in diameter the volume loaded can be as large as $1.0 - 2.0$ ml. The overriding criterion of sample volume loaded is the thickness of the initial sample zone; in general, the thinner it is the better the resolution that can be attained. The maximum mass of sample material which can be loaded (and hence the maximum concentration of the sample) depends upon the slope of the gradient and needs to be determined empirically. The minimum amount of sample that can be loaded is usually defined by the sensitivity of the detection method to be employed. For nucleic acids and proteins, the load will depend upon whether these are to be detected spectrophotometrically after centrifugation (in which case the amount loaded should be sufficient to observe absorbance peaks in excess of 0.1) or by determining radioactivity (in which case less material can be loaded, depending on the specific radioactivity of the sample). Clearly, the minimum loading for enzymes and subcellular components which are to be detected by enzyme assay will depend upon the sensitivity of the particular assay used. Another consideration is that sample losses by adsorption to the centrifuge tube wall can become significant when separating microgram amounts of sample. This is usually most problematical when analysing macromolecules by density gradient centrifugation, for example, losses of nucleic acid by adsorption to cellulose nitrate or polypropylene tubes, the latter mainly at high pH. In this particular case, the problem can usually be avoided by using polyallomer tubes. A more generally applicable remedy is to include carrier macromolecules or millimolar concentrations of suitable detergents (e.g., Triton X-100) in the sample.

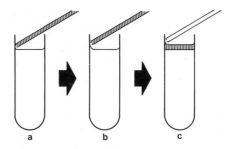

Figure 16. Technique for loading a sample onto a gradient. (**a**) The pipette containing the sample is touched against the meniscus at the tube wall. (**b**) For wettable tubes, the sample is loaded more gently if the pipette tip is now moved slightly upwards to leave a trail of wetted tube. (**c**) The sample is allowed to run slowly out of the pipette.

6.1.2 *Loading the Sample*

Loading of the sample onto the density gradient is one of the most crucial steps in rate-zonal centrifugation. The resolution achieved in the fractionation is greater the narrower the initial sample zone relative to the length of the gradient. Therefore the sample must be layered onto the gradient extremely gently to minimise mixing with the top of the gradient. Individual researchers prefer different types of dispenser for loading samples. The simplest method is to use a narrow-bore, low-volume glass pipette or a fine-bore Pasteur pipette for intermediate-size volumes and an automatic micropipettor (e.g., Gilson Pipetman) for smaller volumes, although successful use of the last of these requires more practice. Using the first method, the sample is sucked up into the pipette and then the pipette tip touched gently against the meniscus of the gradient at the tube wall at an angle of about 45° to the wall (*Figure 16*). In the case of hydrophilic tubes the sample can be loaded more easily if the pipette is then moved 2 − 3 mm up the tube wall leaving a trail of wetted tube connecting the pipette tip to the meniscus. Sample is then allowed to flow *slowly* out of the pipette and should creep across the surface of the gradient to form a sharply-defined sample layer.

6.1.3 *Centrifugation*

Before beginning the centrifuge run, check that the centrifuge tubes are dry on the outside and almost completely filled (to within ~3 mm of the tube rim) to prevent collapse of the tubes during centrifugation. Check that opposing tubes are identical in weight as they should be if identical gradient volumes and identical samples volumes have been used.

The temperature of the centrifuge run will depend upon the nature of the particles to be separated. As mentioned earlier, many biological samples are temperature-labile and hence, as a general rule, centrifugation is best done at low temperatures, often 4°C. Because of their large heat capacity, rotors and their gradients change temperature only slowly. Therefore it is important that the rotor is pre-equilibrated at the temperature of the centrifugation and the gradients either formed at the desired temperature or pre-incubated for 2 – 3 h prior to the loading of the samples. The centrifuge chamber should also be pre-equilibrated at this temperature. In general, centrifugation needs to be carried out at the maximum speed compatible with sample stability and the nature of the gradient medium so that the centrifugation time is minimal. This ensures that loss of sample activity due to lability and loss of resolution due to zone broadening by diffusion are also minimal. Since the sedimentation velocity of particles is proportional to the square of the rotor speed (Chapter 1, Section 2.1), even a modest increase in rotor speed can shorten the time of centrifugation appreciably. The rotor should be accelerated slowly at the beginning of the centrifugation run since the greatest angular acceleration occurs during the initial acceleration of the rotor from rest and is responsible for most of the mixing of the sample zone which occurs. These swirling forces are known as Coriolis' forces and their magnitude increases with increasing diameter of the centrifuge tube. Swirling during deceleration should be minimised by decelerating the rotor without use of the brake. However, since most swirling occurs at low rotor speeds just before the rotor comes to rest, it is sometimes more convenient to decelerate with the brake on until the rotor has decelerated to about 2000 r.p.m. then to allow the rotor to coast to rest with the brake off. A number of preparative ultracentrifuges have pre-programmed slow acceleration and deceleration modes so that manual adjustment is unnecessary.

6.1.4 *Predicting the Duration of Centrifugation*

Isokinetic gradients. As described earlier (Section 3.3.2), a given particle will sediment at constant velocity through an isokinetic gradient. Furthermore, the distance sedimented at any time is directly proportional to the sedimentation coefficient of the particle. If the distance sedimented (l_1) by a particle of sedimentation coefficient s_1 is known at the chosen speed, then the distance sedimented (l_2) by the sample particle of sedimentation coefficient s_2 can be calculated simply as:

$$l_2 = \frac{s_2 \cdot l_1}{s_1} \qquad \text{Equation 9}$$

The reference particle with a sedimentation coefficient of s_1 *must* have the same density as the sample particle. Whilst this is difficult to achieve precisely, it would be sufficient for most purposes to use particles of similar type and size to the sample particles for calibration. For example, rRNA markers might be used to standardise isokinetic gradients for the separation of other types of RNA molecules.

Non-isokinetic gradients. For rate-zonal centrifugation in non-isokinetic gradients, the duration of the centrifugation run required for optimal resolution of the sample components will depend on the speed and temperature of centrifugation, the shape, viscosity and density range of the gradient and the sedimentation coefficients of the sample particles. The obvious approach is trial and error, that is, varying the duration of centrifugation until the desired results are obtained with that particular set of sample components and that particular type of gradient. An alternative approach is to use the data on sedimentation characteristics which are available from centrifuge manufacturers. Several types of data are available and therefore not all can be discussed here. One method involves the k* factor. This is a rotor-specific factor (see Appendix III) which allows one to predict the time (t) necessary to sediment a particle with a density of 1.3 g/cm³ and a sedimentation coefficient, s, through a linear 5 − 20% (w/w) sucrose gradient at 5°C; the chosen density used in this calculation corresponds to the densities of DNA and proteins in sucrose gradients. Thus the time in hours is given by:

$$t = \frac{k*}{s}$$
Equation 10

In practice, of course, it is unlikely that one would want to sediment the particle completely through the gradient, that is to pellet it. Usually the aim is to determine how long centrifugation should be continued for the particle to sediment a given distance down the gradient. Therefore, as an approximation, halving the calculated time required for pelleting will give an estimate of the time needed for particles to reach about the middle of the gradient and so on.

A more accurate method for estimating run times is to use the $s\omega^2t$ charts produced by Beckman Instruments Ltd. These refer to 5 − 20% (w/w) or 15 − 30% (w/w) linear sucrose gradients at 4°C or 20°C where the sample/gradient interface is 3 mm below the top of the tube. Each chart contains curves only for particles of density 1.4 and 1.8 g/cm³ but interpolation around the curves give usable results for particles of other densities. A typical chart is shown in *Figure 17*. To determine the time of centrifugation required for the

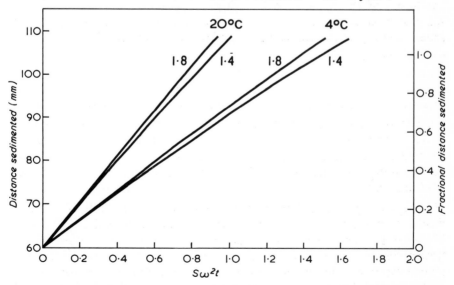

Figure 17. The $s\omega^2t$ chart for 5 − 20% (w/w) sucrose gradients centrifuged in the SW50.1 rotor (courtesy of Beckman Instruments Ltd.).

sample particles to sediment a given distance at the chosen speed, draw a line from the ordinate to the curve and then, from the point of intersection, draw a line to the abscissa. From the values of $s\omega^2 t$ indicated, and knowing s and ω, the time, t (in sec), can be readily calculated.

Unfortunately both these types of calculation are only applicable to particles of certain density centrifuged in a few types of density gradient at a few temperatures. However, considerably more useful are the computer programs given in Chapter 4. These can be applied to a wider range of rate-zonal centrifugation conditions since one can simulate sedimentation in sucrose gradients of any shape and in any rotor over a wide range of temperatures. Hence such simulation procedures can be used to optimise rate-zonal separations.

Finally, one should mention the $\omega^2 t$ integrator which is available on some preparative ultracentrifuges. This device automatically records the total centrifugal force experienced by the sample components during the centrifuge run including that during acceleration and deceleration. Having attained the optimal conditions for a particular rate-zonal separation and recorded the $\omega^2 t$ value at the end of the run, one can set the integrator for future runs at this value. The future runs will then be controlled automatically to duplicate the total centrifugal field experienced during the original centrifugation run.

6.2 Isopycnic Centrifugation

6.2.1 *Sample Composition*

As with rate-zonal centrifugation, it is important that gradients used for isopycnic centrifugation should not be overloaded with sample or this will result in poor separation of the sample components. The capacity of a gradient used for isopycnic centrifugation (that is the mass of sample particles which can be present in a band without distortion) depends upon both the slope of the gradient and the difference in density between the individual components to be separated. There is no mathematical procedure for calculating the actual gradient capacity at the present time and hence this must be determined empirically.

The volume of sample which can be loaded onto the gradient will depend upon whether the gradient has been preformed or will be allowed to self-form during centrifugation and is discussed in the next section.

6.2.2 *Loading the Sample*

Self-forming gradients. The sample is either mixed with a concentrated solution of the gradient solute or solid gradient solute is added to give the correct initial density. Large volumes of sample can therefore be used. It is important that mixing of the sample and gradient medium should be carried out thoroughly so that the mixture has a uniform density throughout. The density is then usually checked by refractometry (see Section 8.1) and adjusted to the correct value, if necessary, by the addition of extra gradient solute or buffer. The appropriate volume of the mixture is then placed into each centrifuge tube. For thin-wall tubes, this should fill the tube to within about 3 mm of the top or the tube will collapse during centrifugation. Sometimes it is desirable to use a gradient which only partially fills the tube to reduce the centrifugation time required (Section 6.2.4). In these cases the gradient mixture must be overlayered with liquid, for example liquid paraffin, to fill the tube (6). Alternatively, one can also use thick-walled polyallomer and polycarbonate tubes which can be centrifuged only partially filled without the tubes collapsing during centrifugation. The minimum volume required for such tubes and the maximum speed depends on the rotor used and so the reader should consult the manufacturer's instructions.

Opposing tubes must be accurately balanced both in terms of weight and the centre of gravity. The best procedure is to ensure that opposing tubes receive the same volume of the identical gradient mixture and are then capped (if necessary) and balanced. If the gradient mixture is to be overlayered, the tubes should be balanced and then the overlay added. The tubes are then capped and re-balanced. If a blank gradient is needed, it should be identical to the sample gradient except that it should contain buffer instead of sample. Finally, the tubes should be dried on the outside and loaded into the rotor.

Preformed gradients. When using preformed gradients, the tubes must be handled very carefully prior to centrifugation since vibration or temperature variations will disrupt the gradients. To minimise diffusion of the gradients, the samples should be ready for loading before the gradients are prepared so that the minimum time elapses between gradient formation and use. Each sample, which must have a density less than that of the top of the gradient, is gently layered onto the gradient following the procedure described in Section 6.1.2 for rate-zonal centrifugation. To minimise changes in the density profile at the top of the gradient during centrifugation, the sample volume should preferably be small in relation to the gradient volume. If this is not possible, the sample density should be arranged to be only slightly less than that of the top of the gradient. Balancing of the gradients and overlayering of samples, if necessary, are carried out as for self-forming gradients.

Flotation. This is a variation of the more usual method of using preformed gradients of isopycnic centrifugation. The sample is mixed with a dense solution and layered under the gradient. During centrifugation, the sample particles rise or 'float' to their isopycnic points. Flotation is slower than sedimentation but is useful when the densities of the particles to be separated are inversely proportional to their sizes, that is the smaller the particle the denser it is. As described in Chapters 3, 5 and 6, this technique has been used for separations of lipoproteins, nucleoproteins, membrane fractions and cells.

6.2.3 Centrifugation

As in the case of rate-zonal centrifugation, the rotor must be pre-equilibrated to the temperature chosen for fractionation for several hours before centrifugation. Prior to centrifugation, the researcher should check that the rotor speed to be used will not result in the precipitation or crystallisation of gradient solute at the bottom of the tube with the attendant risk of rotor failure (see Section 4.2.2). In addition, when the average density of the gradient exceeds 1.2 g/cm³ (for example when using CsCl or Cs_2SO_4 as solute), the maximum speed of the rotor should usually be reduced or rotor failure may occur. The most that any rotor should be 'derated' can be calculated as:

$$N' = N \left(\frac{1.2}{\varrho}\right)^{1/2}$$
Equation 11

where N' is the maximum speed (r.p.m.) for a solution with a density of ϱ g/cm³ and N is the normal maximum speed of the rotor.

Assuming that the rotor speed selected does not exceed either of the two limitations described, the rotor is accelerated to the chosen speed at the required temperature. After the allotted time period of centrifugation has elapsed, the rotor is allowed to decelerate without use of the brake. The tubes are then carefully removed from the rotor with minimum disturbance and maintained at the temperature of the fractionation whilst awaiting analysis.

6.2.4 Predicting the Duration of Centrifugation

Since isopycnic centrifugation is an equilibrium method, the actual length of run is much less critical than in the case of rate-zonal separations as long as prolonged centrifugation does not adversely affect the sample. The following mathematical methods of predicting centrifugation times to band all the sample particles during isopycnic centrifugation apply mainly to the analysis of macromolecules with known sedimentation coefficients and densities. The time required to band much larger particles, such as cells, is usually best determined empirically.

Self-forming gradients. The time period required for isopycnic banding using self-forming gradients is a summation of the time needed for the gradient to form plus the time needed for the sample particles to band at their isopycnic points. The time to form an equilibrium gradient is given by:

$$t = k'' (r_b - r_t)^2$$
Equation 12

where t is the time required (in hours), r_b is the distance (cm) of the bottom of the gradient from the axis of rotation, r_t is the distance (cm) of the top of the gradient from the axis of rotation and k'' is a constant which is inversely proportional to the diffusion coefficient of the gradient solute. For CsCl at 20°C, k'' has a value of 5.6.

From Equation 12, the time needed to obtain an equilibrium gradient depends upon the size of the gradient solute and the length of the density gradient but is independent of the rotor speed. Clearly, the major way to reduce centrifugation time with any given gradient solute is to keep the effective length of the gradient as short as possible. When using a swing-out rotor, this can be achieved by only partially filling the centrifuge tube with the gradient. For thick-wall tubes this is all that is required. However, for the centrifugation of short-column salt gradients in thin-wall tubes, the rest of the tube should be filled with liquid paraffin (Section 6.2.2.). An alternative approach is to use a fixed-angle or vertical rotor since the reorientation of the gradient during acceleration considerably reduces the effective length of the gradient. In fact, the effective pathlength decreases as the angle of the tube to the vertical decreases. Therefore the shortest pathlength, and hence the fastest time for attainment of an equilibrium gradient, is obtained using vertical rotors where the effective length of the gradient is only the diameter of the tube.

The time taken for a sample particle to band depends upon its size. Larger particles sediment faster than smaller particles and so will reach their isopycnic points faster. Therefore the minimum time necessary to band all the components of a sample is the time taken for the smallest particle to reach its isopycnic point. The time required is calculated as:

$$t = \frac{9.83 \times 10^{13}.\beta° \, (\varrho_p - 1)}{N^4 r_p^2 s_{20,w}} \qquad \text{Equation 13}$$

where t is the time required (in hours), ϱ_p is the buoyant density of the sample particle (g/cm³) in that gradient medium, N is the rotor speed (r.p.m.), r_p is the distance (cm) of the particle at equilibrium from the axis of rotation and $s_{20,w}$ is the sedimentation coefficient of the particle at 20°C in water. Note that for gradient media which have viscosities significantly greater than water, an appropriate correction to $s_{20,w}$ must be made.

In contrast to the time necessary to form the equilibrium gradient (Equation 12), the length of the gradient has little effect upon the time needed for the sample particles to reach equilibrium (Equation 13). However, whereas the time needed to form an equilibrium gradient is independent of rotor speed (Equation 12), the time taken for particles to reach their isopycnic points is inversely proportional to the fourth power of the rotor speed (Equation 13). Advantage has been taken of these facts in the gradient relaxation technique of Anet and Strayer (30) using fixed-angle rotors. Centrifugation is first carried out at relatively high speed to band the particles and then at a lower speed calculated to give the desired gradient shape (Section 4.2.2). As the equilibrium gradient is formed, the particle bands now re-distribute to their final positions. This approach can reduce the overall centrifugation time by more than half with no loss of resolution.

Preformed gradients. The duration of centrifugation is reduced dramatically when using preformed gradients since time is required only to band the sample particles and not also to form the gradient. When the sample is layered onto the top of the preformed gradient, the time (in hours) required for centrifugation is given by:

$$t = \frac{2.53 \times 10^{11} \, (\varrho_p - 1)}{N^2 r_p \, s_{20,w} \times \dfrac{d\varrho}{dr}} \left[\ln \left(\frac{r_p - r_t}{r_t} \right) + 4.61 \right] \qquad \text{Equation 14}$$

where r_t is the distance (cm) from the top of the gradient to the axis of rotation and $\dfrac{d\varrho}{dr}$ is the slope of the density gradient. The other symbols are as defined for Equation 13.

7. FRACTIONATION OF GRADIENTS

7.1 **Introduction**

Great care must be taken in fractionating gradients after centrifugation since resolution is easily lost at this stage. The ideal gradient fractionation scheme should incorporate the following features:

(i) All operations should be designed to minimise disturbance of the gradient. During removal of tubes from the rotor and during the fractionation process, the tubes must be handled extremely carefully to avoid mechanical disruption of the gradients. Moreover they should be kept at the same temperature as during centrifugation otherwise convection currents may disrupt the gradient profile.

(ii) During fractionation, density inversion of the fractionated gradient should be avoided since this leads to mixing and loss of resolution.

(iii) Whenever possible, the gradient should be monitored in a continuous fashion during the fractionation process.

(iv) The volume of the tubing from the gradient to the monitoring apparatus and hence to the collection apparatus should be kept as small as possible.

(v) When the liquid leaves the centrifuge tube it must not be contaminated by any material which is not within the range of the gradient, that is, pelleted material.

(vi) The gradient should be fractionated at a slow flow-rate since viscous zones of, for example, high molecular weight DNA, will tend to be retarded by viscous drag on the tube walls. If the rate of gradient collection is too high, these zones can be displaced from their original positions after centrifugation.

Several methods which have been devised for fractionating density gradients are described here. Suitable apparatus is available from a number of manufacturers. Alternatively their manufacture is usually within the capabilities of many departmental workshops.

7.2 **Unloading Gradients from the Bottom**

The original method that was widely used was to clamp the tube in a vertical position, simply pierce the tube with a sharp needle and collect a defined number of drops per fraction (*Figure 18a*). However, if the hole made is too small the gradient drips out too slowly whereas too large a hole allows air bubbles to enter and pass up the gradient. To some extent these problems can be overcome by first sealing the top of the centrifuge tube with a rubber bung through which a needle passes and is linked to a large-volume syringe. The flow-rate is then controlled by the rate at which air is allowed to enter the top of the tube from the syringe. However, the procedure is still tedious and cumbersome. A considerable improvement is the use of commercial tube-piercing apparatus. This type of apparatus holds the centrifuge tube containing the gradient firmly in a vertical position and then a hollow needle is screw-

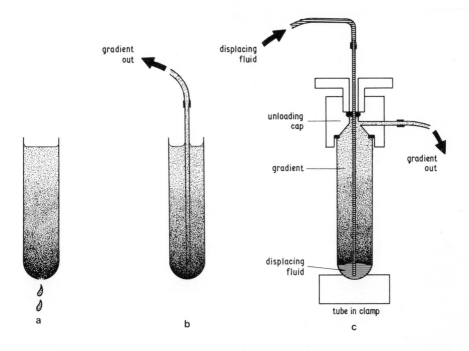

Figure 18. Methods for collecting density gradients after centrifugation. (**a**) The bottom of the tube is pierced and fractions collected as the gradient drips out. (**b**) The gradient is pumped out via a narrow tube inserted through the gradient to the tube bottom. (**c**) Collection by upward displacement. The gradient is displaced upwards through an unloading cone by delivering a dense liquid to the bottom of the centrifuge tube via a narrow cannula.

ed up into the bottom of the tube. The hollow needle remains in place throughout the subsequent fractionation. This outlet can be led to a fraction collector which incorporates a drop counter previously programmed to collect a fixed number of drops per fraction. The whole fractionation procedure then becomes automated once the tube has been pierced. However, there are also several disadvantages of this method of fractionation. Firstly, drop size can vary not only as a function of the outer diameter of the outlet tubing but also with the solute concentration of the gradient solution so that the fraction size varies along the gradient. Theoretically this can be overcome by using a peristaltic pump between the gradient and the fraction collector and collecting on a constant time basis with the pump set at a constant flow-rate. However, although this yields fractions of equal volume, it is not a method to be recommended since a significant loss of resolution occurs as the gradient traverses the pump. Secondly, the presence of a pellet in the centrifuge tube can result in contamination of the first few fractions or even preclude this type of fractionation altogether. Moreover, some tubes, for example thick-wall polypropylene and polycarbonate tubes, cannot be easily pierced.

7.3 Unloading Gradients from the Top

7.3.1 *Direct Unloading*

The simplest procedure for unloading gradients from the top is to pass a fine-bore stainless steel or glass tube through the gradient to the tube bottom and then pump the gradient out via a peristaltic pump (*Figure 18b*). This is often the instinctive method used by novices. However, it is not to be recommended. Not only is resolution lost as the gradient passes through the pump tubing but the gradient is forced to invert as it passes up the narrow tube, causing yet more mixing.

7.3.2 *Upward Displacement*

This method is undoubtedly the one of choice for most applications, giving the best resolution whilst being convenient to use. The technique is to displace the gradient upwards into an unloading cone using either a denser solution of the gradient medium or a denser inert organic liquid as the displacing medium. If the gradient is displaced with a dense aqueous solution it must contain the same solute as the gradient solution otherwise diffusion rapidly occurs at the interface of the displacing medium and gradient causing disruption of the bottom of the gradient. Early studies made use of dense ($>55\%$ w/w) sucrose solutions to displace sucrose gradients but these are difficult to use because of their very high viscosity. Alternatively, investigators have used an organic liquid such as Fluorochemical 43 (FC43), which is marketed as 'Maxidens' by Nyegaard & Co. Maxidens is dense (1.9 g/cm^3), inert, non-viscous, immiscible with aqueous gradients and relatively cheap. Furthermore, if Maxidens does become contaminated with gradient solution it can easily be decontaminated by extraction with a suitable aqueous phase.

The displacing medium can be introduced either via a thin stainless steel tube which is inserted into the gradient so as to reach the bottom (*Figure 18c*) or via a hollow needle pierced through the bottom of the tube. The latter technique is theoretically superior as there should be less disturbance of the gradient but it is not suitable for thick-wall tubes or if a large pellet is present. Whichever method is used, it is important to ensure that the tube or hollow needle which is introducing the displacing medium is completely full of this medium before use, otherwise air bubbles will be expelled into the gradient ahead of the displacing fluid and will disrupt the gradient. Using either method, the centrifuge tube contents are displaced upwards towards a funnel cap and drops are collected from the outlet either manually or using a fraction collector. Unlike the top or bottom unloading methods, collection of fractions of known volume is easy to achieve; using a fraction collector programmed to collect fractions on a constant time basis, the displacing fluid is introduced using a peristaltic pump set to operate at a constant flow rate. This gives fractions of a constant volume and, since the gradient solution has not itself encountered the pump, undue mixing of the gradient is avoided. However, Noll (26) has pointed out that peristaltic pumps, by virtue of their pulsating mode of pumping, will tend to cause some turbulence and loss of resolution and hence he recommended the use of a syringe pump where possible. In fact it is possible to obtain very good resolution and constant volume fractions by the upward displacement method without the use of either a pump or a fraction collector. The technique is to introduce the displacing fluid from a calibrated burette with the meniscus of the displacing fluid arranged to be above that of the gradient. A known volume of displacing fluid allowed to enter the centrifuge tube under gravity from the burette will displace an equivalent volume of the gradient and fractions can be collected in the usual way.

In summary, the technique of upward displacement is widely applicable to many types of gradient. It has the advantages of avoiding the problems associated with the presence of a pellet, gives the best resolution of the methods currently available, and need not require piercing of the tube thus allowing the method to be used with all types of centrifuge tube. Furthermore,

Figure 19. A commercial gradient unloader for collecting gradients by upward displacement using the method shown in *Figure 18c*. The apparatus is shown with a centrifuge tube in position. The unloading cones are available in several sizes. One cone is in place on the apparatus and two others are shown alongside (courtesy of MSE Scientific Instruments).

since the tubes need not be pierced for gradient retrieval they can often be washed and re-used. A typical apparatus for the collection of gradients by upward displacement without tube piercing is shown in *Figure 19*.

7.4 **Direct Recovery of Bands**

If, after centrifugation, the sample bands are visible and sufficiently well-spaced apart, each can be recovered directly from the gradient from the top using a Pasteur pipette with its tip bent at right angles to the stem. The bent tip of the pipette is placed in or just below the band which is then recovered slowly with a to-and-fro sweeping motion of the tip. Alternatively, in the case of thin-wall tubes, a single band may be recovered through the side of the centrifuge tube using a hypodermic syringe and needle. This is the preferred method for recovery of a single sample band when recovery via the top of the tube using a

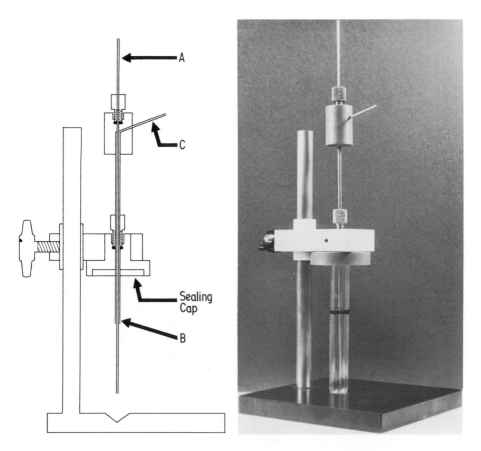

Figure 20. Commercial gradient unloader for the collection of entire gradients by upward displacement or for the collection of individual bands in the gradient. The apparatus is shown on the right in the format used for direct collection of bands from a gradient. A diagrammatic representation is shown on the left. **A**, inner tube; **B**, outer tube; **C**, outlet tube (courtesy of Nyegaard & Co.).

Pasteur pipette could lead to unwanted contamination with other component bands through which the pipette tip would need to pass initially.

A commercial gradient unloader is also available which can be used for either fractionation of the entire gradient by upward displacement (Section 7.3.2) or the collection of individual bands in the gradient (*Figure 20*). Using the apparatus in this latter mode, the centrifuge tube is clamped firmly in a vertical position with its base seated in a depression in the apparatus stand and its top sealed by the sealing cap. The inner tube (*A*) is connected to a reservoir of displacing fluid via plastic tubing and before use it is completely filled with displacing fluid to exclude air bubbles. The outlet tube (*C*) is clamped shut. The outer tube (*B*) is then moved gently into the gradient until its lower end is just above the band to be recovered and is then clamped in position. Next, the inner tube (*A*) is moved gently through the gradient to the bottom of the centrifuge tube. The outlet tube is then unclamped. The displacing fluid is passed through the inner tube (*A*) to the bottom of the centrifuge tube and that part of the gradient containing the band passes up through the space between the inner (*A*) and outer (*B*) tubes by upward displacement.

The direct recovery of bands from density gradients can be used only for preparative work and only when the band required is clearly visible. For analytical work the entire gradient should be fractionated by one of the procedures described in Sections 7.2 and 7.3.

8. ANALYSIS OF GRADIENTS

8.1 Determination of the Gradient Profile

After collection of the gradient, the density of each fraction should be measured accurately so that the density profile of the gradient can be determined. The most direct method is to weigh known volumes of liquid accurately using a pycnometer but this method is extremely time-consuming: It is more convenient to measure the refractive index using an Abbé refractometer and then determine the concentration of gradient solute and density by reference to appropriate tables (e.g. *Tables 2* and *3* for sucrose and CsCl, respectively). The added advantage of this method is that it requires as little as 25 μl of sample to determine the refractive index, thus preserving the rest of the fraction for further analysis. Essentially all density gradient solutes show a linear relationship between the refractive index (η) and the density (ϱ) and concentration. In each case the relationship can be expressed in the form of a simple linear equation:

$$\varrho = a\eta - b \qquad \text{Equation 15}$$

The values of *a* and *b* for a number of density gradient solutes are given in Appendix IV. Before applying this equation it is important to adjust the value of the refractive index to correct for the effect of other solutes other than the gradient solute on the refractive index.

When using Percoll gradients, the density profile of the gradient can be determined by the alternative procedure of using density marker beads (Pharmacia Ltd.). These comprise a set of colour-coded beads, each of known density in Percoll, which are centrifuged in a 'blank' gradient at the same time as the centrifugation of those gradients containing sample. After centrifugation, the distance of each type of bead from the top of the gradient is measured and used to plot the density profile for that set of gradients. Care must be taken in that the actual densities of the beads vary depending on the solutes present in the gradient solution.

8.2 Analysis of Samples

Two basic types of analysis are possible. Either the gradient can be collected directly in a series of fractions which are then examined individually for the biological components of interest, or the gradient is allowed to flow through a spectrophotometer flow cell for optical analysis (continuous monitoring). For many purposes it is advantageous to employ both methods, that is to pass the gradient through a spectrophotometer flow cell and then collect it into fractions for additional analysis (*Figure 21*).

8.2.1. *Continuous Monitoring*

As one would expect, the collection of fractions results in far lower resolution than can be achieved by continuous optical monitoring using a flow cell such that the monitoring of closely-spaced components may be possible only by the flow cell method. The distribution of particulate matter throughout the gra-

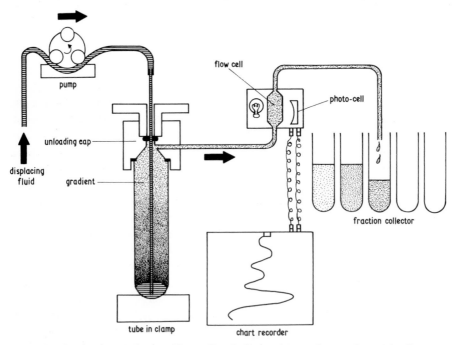

Figure 21. Continuous monitoring. The gradient is displaced upwards towards an unloading cap by introduction of a dense displacing fluid at the tube bottom either using a peristaltic pump (as shown) or by gravity using a burette reservoir (see the text). The displaced gradient is monitored optically by passage through a spectrophotometer flow cell before collection as a series of fractions.

dient can be determined by light-scattering measurements at $500-600$ nm, proteins and nucleic acids can usually be monitored at 280 nm and 260 nm respectively, whilst proteins with certain prosthetic groups (such as FAD, NAD^+, haem, etc.) can be monitored at the corresponding absorption maxima of these groups in order to determine the distribution of these proteins or the organelles that contain them. The flow rate and design of the flow cell are very important if turbulence is to be minimised and maximum resolution obtained. Allington *et al.* (2) have shown that a bulk flow cell gives superior resolution compared with a laminar flow cell and that resolution increases as the illuminated volume of the flow cell decreases. Several commercial models of flow cell are available with different illuminated volumes depending on the application. Throughout monitoring, one should take precautions to minimise mixing of the gradient. In addition to using a correctly designed flow cell this involves keeping flow lines as short and narrow as possible and avoiding gradient inversions. Thus if the gradient is being fractionated by upward displacement, the flow lines should run upwards all the way to the flow cell, and feed the gradient in at the bottom of the cell and out at the top. If fractions are to be collected, the flow lines should then continue upwards to the fraction collector. Alternatively, if the gradient is to be fractionated by unloading from the bottom, the flow through the system should be downwards. However, this method of unloading is less recommended for use with a flow cell since bubbles are easily trapped, causing turbulence and hence high background 'noise' and loss of resolution.

8.2.2 *Analysis of Fractions*

It is sometimes not possible to monitor the component of interest using the continuous optical scanning method, either because there is too little present for optical detection or lack of a suitable wavelength. In these cases, detection of the sample components relies upon the analysis of gradient fractions.

Radioactive analysis. Frequently the distribution of the sample throughout the gradient can be determined by using radioactively labelled samples and measuring the radioactivity in each fraction using a liquid scintillation spectrometer. In this case, if the whole fraction is to be used for analysis, it is usually convenient to collect fractions directly into scintillation vials. The experimenter then need only add a suitable water-miscible, liquid scintillation cocktail before counting.

One problem associated with such analyses is that some scintillation fluids, for example Triton X-100/toluene scintillant, cause the gradient solute to precipitate out or to form a two-phase mixture with resultant quenching. In the case of sucrose this can be avoided by using Bray's scintillation fluid (31) which can tolerate much higher concentrations of sucrose without quenching although the overall efficiency of counting is lower than with Triton X-100/toluene scintillant. Sample components radiolabelled with γ-emitters or strong β-emitters can be detected and quantitated without mixing them with scintillation fluid by using a gamma counter or by Cerenkov counting, respectively.

Enzymic or chemical analysis. In some cases, the distribution of the sample may be determined by enzymic or chemical assays. The use of organelle marker enzymes is described in Chapter 5 and protocols for assays are given in Appendix V. It may be necessary to remove the gradient solute prior to analysis. Whilst low molecular weight molecules can be readily removed by dialysis or gel filtration, it is more difficult to recover fractions containing high molecular weight or colloidal gradient compounds. One general method that can be used for recovering samples is to dilute the samples such that the mean density of the medium is less than that of the sample which can then be recovered by differential centrifugation. This method works only if the particle size of the sample is much greater than that of the gradient medium. When using colloidal gradient media such as Percoll, it is sometimes possible to pellet the silica particles through a cushion of sucrose whose density is greater than that of the sample. Nucleic acids can be recovered from CsCl or Cs_2SO_4 by precipitation with ethanol after appropriate dilution of the salt; caesium trifluoroacetate is ethanol soluble and hence dilution is unnecessary.

9. REFERENCES

1. Noll,H. (1969) *Anal. Biochem.*, **27**, 130.
2. Allington,R.W., Brakke,M.K., Nelson,J.W., Aron,C.G. and Larkins,B.A. (1976) *Anal. Biochem.*, **73**, 78.
3. Cordingley,J.S. and Hames,B.D. (1977) *FEBS Lett.*, **82**, 263.
4. Rickwood,D. (1982) *Anal. Biochem.*, **122**, 33.
5. Flamm,W.G., Birnsteil,M.C. and Walker,P.M.B. (1972) in *Subcellular Components: Preparation and Fractionation*, Birnie,G.D. (ed.), Butterworths, London, p. 279.
6. Griffiths,O.M. (1975) *Techniques of Preparative Zonal and Continuous Flow Ultracentrifugation*, published by Beckman Instruments Inc., Palo Alto, California.
7. Hinton,R.H., Mullock,B.M. and Gilhuus-Moe,C.D. (1974) in *Methodological Developments in Biochemistry*, Vol. 4, Reid,E. (ed.), Longmans, London, p. 103.
8. Ridge,D. (1978) in *Centrifugal Separations in Molecular and Cell Biology*, Birnie,G.D. and Rickwood,D. (eds.), Butterworths, London, p. 49.
9. Fritsch,A. (1975) *Preparative Density Gradient Centrifugation*, published by Beckman Instruments International S.A., Geneva.
10. Hinton,R. and Dobrota,M. (1976) *Density Gradient Centrifugation*, published by North Holland Publishing Co., Amsterdam, New York and Oxford.
11. Noll,H. (1967) *Nature,* **215**, 360.
12. Martin,R.G. and Ames,B.N. (1961) *J. Biol. Chem.*, **236**, 1372.
13. Burgi,E. and Hershey,A.D. (1963) *Biophys. J.*, **3**, 309.
14. McCarty,K.S., Stafford,D. and Brown,O. (1968) *Anal. Biochem.*, **24**, 314.
15. Spragg,S.P., Morrod,R.S. and Rankin,C.T. (1969) *Separation Sci.*, **4**, 467.
16. McCarty,K.S.,Jr., Vollmer,R.T. and McCarty,K.S. (1974) *Anal. Biochem.*, **61**, 165.
17. Graham,J.M. (1973) in *Methodological Developments in Biochemistry*, Vol. 3, Reid,E. (ed.), Longmans, London, p. 205.
18. Birnie,G.D. (1980) *Trans. Biochem. Soc.*, **8**, 513.
19. Ifft,J.B., Martin,W.R. and Kinzie,K. (1970) *Biopolymers*, **9**, 597.
20. Baldwin,R.C. and Shooter,E.M. (1983) in *Ultracentrifugal Analysis in Theory and Experiment*, Williams,J.W. (ed.), Academic Press, New York and London, p. 143.
21. Davis,P.B. and Pearson,C.K. (1978) *Anal. Biochem.*, **91**, 343.
22. Stoker,N.G., Pratt,J.M. and Holland,I.B. (1984) in *Transcription and Translation — A Practical Approach*, Hames,B.D. and Higgins,S.J. (eds.), IRL Press Ltd., Oxford and Washington DC.
23. Wallace,H. (1969) *Anal. Biochem.*, **32**, 334.
24. Sheeler,P. (1981) *Centrifugation in Biology and Medical Science,* published by Wiley-Interscience.

25. Corless,J.M. (1978) *Anal. Biochem.,* **84**, 251.
26. Noll,H. (1969) in *Techniques in Protein Biosynthesis*, Vol. **2**, Campbell,P.N. and Sargent. S.R. (eds.), Academic Press, New York, p. 112.
27. McRee,D. (1978) *Anal. Biochem.,* **87**, 648.
28. Sheeler,P., Doolittle,M.H. and White,H.R. (1979) *Anal. Biochem.,* **87**, 612.
29. Remenyik,C.J., Dombi,G.W. and Halsall,H.B. (1980) *Arch. Biochem. Biophys.,* **201**, 500.
30. Anet,R. and Strayer,D.R. (1969) *Biochem. Biophys. Res. Commun.,* **34**, 328.
31. Bray,G.A. (1960) *Anal. Biochem.,* **1**, 279.
32. Weast,R.C., ed. (1974) *Handbook of Chemistry and Physics, 55th edition,* published by The Chemical Rubber Co., Cleveland, Ohio, p. D199.

APPENDIX

A Computer Program for the Calculation of Equilibration Times and Profiles of Self-Forming Gradients

D. RICKWOOD

This program is written in PET-BASIC and will run on an 8k machine. It will calculate the time required for the gradient and sample to equilibrate. It also calculates the density at any point within the gradient. Input the data as requested; to escape from the program press 'RETURN'.

```
READY.


READY.
 150 PRINTTAB(12)" CALCULATION OF "
 160 PRINTTAB(11)"▓▓_____"
 170 PRINTTAB(11)" GRADIENT PROFILE "
 180 PRINTTAB(8)"▓▓▓▓▓▓▓DATA PROCESSING PROGRAM"
 200 PRINT"▓▓▓▓":GOSUB 5000
 210 PRINT"▓"
 230 PRINTTAB(6)"▓ PLEASE INPUT THE FOLLOWING ":PRINT""
 240 FORI=1TO19:PRINT"♦":NEXT
 250 PRINT"▓":PRINT"▓▓"
 260 GOSUB5100
 270 INPUT"▓▓MAXIMUM RADIUS (CENTIMETRES)=";RX
 280 GOSUB5100
 290 INPUT"▓▓MINIMUM RADIUS (CENTIMETRES)=";RN
 300 IFRX>RNGOTO360
 310 PRINTTAB(3)"▓▓▓MINIMUM RADIUS MUST BE LESS THAN"
 320 PRINTTAB(13)"▓MAXIMUM RADIUS"
 330 FORI=1TO3000:NEXT
 340 GOTO210
 360 RC=SQR((RN↑2+RN*RX+RX↑2)/3)
 370 GOSUB5100
 380 PRINT"▓▓▓(LESS THAN 1.8G/CC)"
 390 INPUT"▓▓▓DENSITY OF MEDIUM (G/CC)=";DO
 400 IFDO<=1.8THENGOTO430
 410 PRINTTAB(3)"▓▓▓DENSITY MUST BE LESS THAN 1.8G/CC"
 420 GOTO330
 430 GOSUB5100
```

```
440 INPUT"█▌ROTOR SPEED (RPM)              =";RP
450 GOSUB5100
460 PRINT"█▌ARE THE ABOVE CORRECT(YES/NO)?"
470 GETBB$:IFBB$=""GOTO470
480 IFBB$="N"GOTO210
490 W=2*RP*π/60
500 PRINT"⏷"
510 GOTO4000
520 PRINT"█▌TYPES OF MEDIA ARE AS FOLLOWS:              CODE        MEDIUM"
530 PRINT"█      120---CAESIUM CHLORIDE"
540 PRINT"       130---CAESIUM SULPHATE"
550 PRINT"       140---SODIUM IODIDE"
560 PRINT"       150---RHUBIDIUM CHLORIDE"
570 INPUT"█ ENTER CODE FOR MEDIUM USED....";M
590 H=10*DO
600 IFM=120THEN B=CS(2,H):GOTO700
610 IFM=130THEN B=C4(3,H):GOTO700
620 IFM=140THEN B=NA(4,H):GOTO700
630 IFM=150THEN B=RB(5,H):GOTO700
640 GOTO570
700 BO=B*1E9
702 GOTO 2010
710 PRINT"█THE ISOCONCENTRATION POINT (I.E. WHERE THE"
720 PRINT"⏷DENSITY ="DO;"G/CC)IS AT A RADIUS OF"
730 PRINTTAB(12)"█"INT(RC*100)/100;"█CM"
740 PRINT"█▌▌PLEASE INPUT THE RADIUS FOR WHICH A    DENSITY IS REQUIRED"
750 PRINT""
800 INPUT"RADIUS (CM).....";RI
810 IFRI<RNTHEN PRINT"█VALUE SMALLER THAN MINIMUM RADIUS":GOTO750
820 IFRI>RXTHENPRINT"█VALUE LARGER THAN MAXIMUM RADIUS":GOTO750
830 DI=DO+(W↑2/(2*BO))*(RI↑2-RC↑2)
840 PRINT"DENSITY AT "RI"CM ...."
845 PRINTTAB(24)"⏷█"DI
850 GOTO750
2010 PRINT"█▌▌▌▌▌DO YOU WISH TO CALCULATE THE "
2020 PRINT"APPROXIMATE TIME REQUIRED FOR GRADIENT"
2030 PRINT"█▌FORMATION AND SAMPLE EQUILIBRATION"
2040 PRINT"█▌▌▌▌▌▌▌▌▌▌▌▌▌ YES/NO █▌▌▌▌?"
2045 GETGG$:IFGG$=""GOTO2045
2050 IFGG$="N"THEN GOTO 710
2060 IFGG$<>"Y"THENGOTO2010
2070 PRINT"█▌▌▌▌▌▌▌▌▌▌▌USE AN ESTIMATE IF YOU DO NOT KNOW THE   CORRECT VALUE
2080 INPUT"▐▐▐▐▐▐▐PLEASE INPUT THE S20,W OF YOUR SAMPLE       S=.........";SW
2085 INPUT"█....AND THE BUOYANT DENSITY              G/CC=......";BD
2100 PRINT"█THE TIME TAKEN FOR THE GRADIENT TO FORM"
2105 TG=INT(5.6*((RX-RN)↑2)*100)/100
2110 PRINT"WILL BE APPROXIMATELY"
2115 PRINTTAB(10)".....";"█"TG;"█ HOURS"
2120 DT=DO+(W↑2/(2*BO))*(RN↑2-RC↑2)
2130 RQ=SQR(RN↑2+((2*BO)/W↑2)*(BD-DT))
2140 SQ=(9.83E13*BO*(BD-1))/(RP↑4*RQ↑2*SW)
2150 PRINT"█▌THE TIME REQUIRED FOR THE SAMPLE TO"
2160 PRINT"REACH EQUILIBRIUM WILL BE APPROXIMATELY"
2170 PRINTTAB(10)".....";"█"INT(SQ*100)/100;"█ HOURS"
2180 PRINT"█▌AT EQUILIBRIUM THE SAMPLE WILL BE AT "
2190 PRINT"AN APPROXIMATE RADIUS OF"
2200 PRINTTAB(10)".....";"█"INT(RQ*100)/100;"█CM"
2300 PRINT"█▌THE ABOVE FIGURES ARE ONLY A GUIDE AND"
2330 PRINT"█TO CALCULATE THE EXACT GRADIENT PROFILE"
2340 PRINT"           PRESS ANY KEY"
2350 GETZZ$:IFZZ$=""GOTO2350
2360 GOTO 710
3999 END
4000 DIMD(1,20),CS(2,20),C4(3,20),NA(4,20),RB(5,20)
4010 DATA0.00,2.04,1.55,1.33,1.22,1.17,1.14,1.12
4020 DATA0.00,1.06,0.76,0.67,0.64,0.66,0.69,0.74
4030 DATA5.10,3.19,2.82,0.00,0.00,0.00,0.00,0.00
4040 DATA0.00,3.42,2.76,2.25,0.00,0.00,0.00,0.00
4050 READCS(2,11),CS(2,12),CS(2,13),CS(2,14),CS(2,15),CS(2,16),CS(2,17)
4055 READCS(2,18)
4060 READC4(3,11),C4(3,12),C4(3,13),C4(3,14),C4(3,15),C4(3,16),C4(3,17)
4065 READC4(3,18)
4070 READNA(4,11),NA(4,12),NA(4,13),NA(4,14),NA(4,15),NA(4,16),NA(4,17)
```

```
4075 READNA(4,18)
4080 READRB(5,11),RB(5,12),RB(5,13),RB(5,14),RB(5,15),RB(5,16),RB(5,17)
4085 READRB(5,18)
4090 PRINTSPC(2)"DENSITY   GRADIENT   PROPORTIONALITY"
4100 PRINT"CONSTANTS   (BETA-VALUES)   FOR SOME"
4110 PRINT"COMMONLY USED DENSITY GRADIENT MEDIA"
4120 PRINT"█DENSITY","████BETA-VALUE *10E9"
4130 PRINT"█R█T"
4140 PRINT"█25'C","███🄲CSCL";"🄳███CS2SO4";"████NAI";"🄳███RBCL"
4150 FORI=1TO39:PRINTTAB(I)"█▊█":NEXT
4160 DATA1.1,1.2,1.3,1.4,1.5,1.6,1.7,1.8
4165 PRINT""
4170 READD(1,11),D(1,12),D(1,13),D(1,14),D(1,15),D(1,16),D(1,17),D(1,18)
4190 FORI=11TO18
4200 PRINT"█"D(1,I):PRINT"██████🄳"CS(2,I):PRINTTAB(14)"🄳"C4(3,I)
4210 PRINTTAB(24)"🄳"NA(4,I):PRINTTAB(33)"🄳"RB(5,I):NEXT I
4220 FORI=1TO39:PRINTTAB(I)"█▊█":NEXTI:GOTO 520
4999 END
5000 PRINTTAB(6)"PRESS ANY KEY TO CONTINUE"
5010 GETAA$:IFAA$=""GOTO5010
5020 RETURN
5100 PRINT"█◆◆◆◆◆◆◆◆◆◆◆◆◆◆◆◆◆◆◆◆◆◆◆◆◆◆◆◆◆◆◆◆◆◆◆◆◆◆◆◆◆":RETURN
```

CHAPTER 3

Centrifugal Methods for Characterising Macromolecules and their Interactions

D. RICKWOOD AND J.A.A. CHAMBERS

1. INTRODUCTION

Macromolecules are the basic building blocks of all cellular structures. Hence a great deal of effort has been spent on methods to isolate, fractionate and characterise the various types of macromolecules and to study their interactions. Of the various types of techniques that have been used for fractionating macromolecules, centrifugation has been the most pre-eminent in terms of both versatility of applications and the quality of separations obtainable. Macromolecules can be separated either according to their sedimentation rate or on the basis of their buoyant density in the various gradient media. By using either one or a combination of these two methods it is possible to obtain extremely good separations of macromolecules not possible using other techniques. In addition, it is possible to use both rate-zonal and isopycnic centrifugation to study the interaction between macromolecules.

2. PRECAUTIONS REQUIRED WHEN FRACTIONATING MACRO-MOLECULES

The main problem of working with macromolecules is the avoidance of degradation of the molecules during experimental manipulations since a single break can partially or completely alter the characteristics of macromolecules depending on the type of macromolecule and the position of the break along the macromolecule.

One of the most frequent causes of degradation of macromolecules is the presence of hydrolytic enzymes either inherently present in the sample or introduced as a result of contamination of the sample during its preparation. Enzymes in the sample can often be inactivated by specific inhibitors, for example, serine proteases can be inhibited by phenylmethylsulphonyl fluoride (PMSF) or di-isopropylfluorophosphate (DFP). However, usually it is much more convenient to prepare pure samples free of hydrolytic enzymes and to ensure that there is no contamination of the sample. Because of the extreme damage that even a single scission can cause, the reader is advised to take the most stringent precautions against enzymic degradation especially when working with nucleic acids which can have molecular weights well in excess of 10^6 daltons. To remove all nuclease activity it is necessary to autoclave all solutions or to treat them with 0.1% diethylpyrocarbonate and all containers, pipettes, pipettor tips etc., should be sterilised also. To avoid contamination of the sample with skin ribonuclease it is advisable to wear disposable plastic surgical gloves for all manipulations.

In addition to hydrolytic enzymes, chemical constituents of solutions can also affect the integrity of macromolecules. Extremes of pH must be avoided since both acid and alkaline conditions can lead to degradation, denaturation or modification of macromolecules such as RNA, DNA and proteins. Nucleic acids can also be degraded by heavy metal ions and, moreover, they can bind to nucleic acids changing both their partial specific volume and their sedimentation behaviour. The best approach is to ensure that, when working with nucleic acids, all the solutions are treated with one of the chelating resins that are commercially available.

Finally, some types of very large macromolecules, notably DNA, are very susceptible to physical degradation as a result of shear stress within the solution. Hence, in manipulating DNA solutions, care must be taken not to shear the DNA strands.

3. RATE-ZONAL SEPARATIONS OF MACROMOLECULES

3.1 Introduction

Because single species of macromolecules such as RNA and proteins are homogeneous with respect to molecular weight, rate-zonal centrifugation is ideal for not only purifying these types of macromolecule but also the rate at which they sediment is a property of these particular particles and can be used to characterise them in terms of molecular weight and partial specific volume.

A description of the theoretical basis of rate-zonal sedimentation and the definition of sedimentation coefficients has been described in detail in Chapter 1. In addition, Chapter 2 describes the preparation of the various types of gradient used for rate-zonal separations and the way in which resolution of the gradient can be optimised. In this part of the chapter no attempt will be made to describe in detail the various types of rate-zonal fractionation, but instead more general advice will be given, together with indications of the pitfalls that the reader may encounter.

3.2 Choice of Gradient Medium and Gradient Shape

A detailed description of the various physico-chemical properties of the various gradient media have been described in Section 5 of Chapter 1. The usual guideline followed is that unless there is a specific reason for choosing an alternative medium it is simplest to use sucrose gradients. Sucrose solutions are ideal in that they are not very dense, they exhibit intermediate viscosity and, most importantly of all, their physico-chemical characteristics are extremely well documented. The interaction of sucrose with most macromolecules is minimal. Some enzymes interact with sucrose whilst others are unstable in sucrose solutions. In both cases the usual alternative is to use glycerol gradients, a 5 – 20% (w/w) glycerol gradient has a similar viscosity profile to a

5 − 20% (w/w) sucrose gradient at 5°C. Occasionally other gradient media, such as metrizamide and CsCl, have been used for rate-zonal centrifugation but their use is not very frequent.

The choice of gradient shape has been discussed in detail in Section 3.3 of Chapter 2. Usually for separating macromolecules one uses simple linear gradients since there is little or no advantage in using other shapes. As described in Section 3.3 of Chapter 2 one can usually improve the resolution of gradients by making them steeper, although steeper gradients have a smaller sample capacity and one only obtains optimal results using the high performance swing-out rotors (e.g., Beckman SW60 or equivalent rotor).

3.3 Choice of Rotor and Centrifugation Conditions

A detailed discussion of the choice of rotors is given in Section 2.2 of Chapter 2. Briefly, it is found that, of the various rotors available, zonal rotors have the greatest potential for optimising the resolution because of the absence of any wall effects which disrupt the sample zones as they sediment through the gradient. Of the non-zonal rotors, the rotor of choice for most separations is the swing-out rotor because disruption of the zones as a result of wall effects is not serious. Vertical rotors also give excellent resolution for rate-zonal separations especially of large macromolecules such as RNA or DNA. However, for macromolecules with molecular weights of less than 2×10^5 daltons there is significant broadening of the sample bands because of the enhanced diffusion that occurs in vertical rotors. The extensive wall effects that are generated in fixed-angle rotors make them generally unsuitable for many rate-zonal separations. It appears that while rate-zonal separations of small proteins can be carried out using fixed-angle rotors, for larger macromolecules and macromolecular complexes significant losses of material occur as a result of wall effects.

Usually the centrifugation conditions for rate-zonal separations of macromolecules are chosen to minimise the diffusion of the sample zones, that is to make the run as short as possible. Hence one usually uses the maximum possible centrifugal force. However, an alternative approach to shorten the run is to use shorter sedimentation distances as in short column gradients (1) although there may be some loss of resolution when the gradient is fractionated.

In some cases there are limitations to the centrifugal force that one can use. A well-documented case is the anomalous sedimentation of high molecular weight DNA in sucrose gradients in which the observed sedimentation coefficient is dependent on the centrifugal force applied (2,3); this appears to be a reflection of conformational changes that occur with increasing centrifugal force (4). Another limitation is that some macromolecular complexes are labile to the hydrostatic pressure generated within the gradient. The hydrostatic pressure generated in the gradient depends on the length of the liquid column and the centrifugal force applied. The hydrostatic pressure generated can be sufficient to disrupt ribosomes (5), microtubules (6) and histone complexes (7). Such dissociations obviously lead to anomalous sedimentation patterns.

4. EXPERIMENTAL PROCEDURES USED FOR RATE-ZONAL SEPARATIONS

Since most rate-zonal separations are carried out using sucrose gradients, only these will be described in the following sections. However, when fractionating proteins, carrying out separations on $5-20\%$ glycerol gradients may help to enhance the retention of the native activity of enzymes.

4.1 Separations on Non-denaturing Gradients

4.1.1 *Introduction*

Non-denaturing sucrose gradients are widely used for separating all types of macromolecules and macromolecular complexes on both analytical and preparative scales. Since, to all intents and purposes, the protocols for the separations are identical, irrespective of the type of macromolecule, only generalised descriptions of the use of these gradients will be given. Sucrose gradients can be used quantitatively for estimating the *s*-values and molecular weights of macromolecules; these quantitative aspects are described in detail in Chapter 4. In addition, it is possible to use sedimentation in sucrose gradients to calculate the partial specific volume of macromolecules.

4.1.2 *Choice and Preparation of Gradients*

The choice of gradient depends on the performance of the rotor. Gradients of $5-20\%$ (w/w) sucrose will give adequate resolution of macromolecules in most types of swing-out and vertical rotors. Another advantage of $5-20\%$ (w/w) sucrose gradients is that they are essentially isokinetic allowing one to use macromolecules of known sedimentation coefficients as markers. When using long-bucket swing-out rotors one often obtains better resolution using $10-40\%$ (w/w) sucrose gradients. If one uses a very high performance rotor (e.g., Beckman SW60) then one can obtain good resolution using a steep $(10-60\%$ w/w) gradient provided that the density does not exceed that of the sample. Usually one uses a linear gradient since little if any enhancement of resolution is obtained using other more complex gradient profiles.

Having decided on the type of gradient, it should be prepared by diffusion or using a gradient maker as described in Section 5.1 of Chapter 2. For swing-out rotors, unless using short-column gradients (1) in thick-walled tubes, the top of the gradient should be about $3-5$ mm from the top of the tube. Care must be taken to pre-equilibrate the gradients to the same temperature as is used for the centrifugal separation.

4.1.3. *Preparation and Loading of the Sample*

The sample is usually prepared in the same buffer as the gradient. However, especially when working with large macromolecules or macromolecular complexes (e.g., ribosomes), it is often advantageous to omit sucrose from the sample as this allows concentration (zone sharpening) of the sample molecules at the interface at the top of the gradient.

Care must be taken to preserve the native state of the sample material. For most samples this involves keeping the sample cool to minimise degradation of the macromolecules. In contrast, in the case of RNA, aggregation of the molecules can occur. Hence RNA samples may need to be dissociated by heating the sample at 65°C for 10 min prior to loading onto gradients or separated in denaturing gradients as described in Section 4.2.

If one wishes to obtain optimum resolution it is particularly important that the volume of the sample is not excessive. The volume of the sample should be kept as small as possible, typically $0.2 - 0.5$ ml for a 1.0 cm diameter tube and $1.0 - 2.0$ ml for a 2.5 cm diameter tube. If larger samples are used there is usually a serious loss of resolution.

In terms of the amount of material loaded there is, unfortunately, no simple way to predict the sample capacity of gradients since this depends on the heterogeneity of the sample, volume of the gradient and the steepness of the gradient. As a starting point one should try loading $10 - 20$ μg of sample material for each millilitre of gradient. If a high degree of resolution is not required it is possible to load a substantially greater amount of sample.

As described in Section 6.1.2 of Chapter 2, great care is required in layering the sample onto the gradient in order to form a narrow band of sample at the top of the gradient. The best method to use will vary from one person to another, usually one uses a Pasteur pipette or a syringe. In the case of vertical rotors the degree of resolution may be enhanced by placing a buffer overlay over the sample as this should help to protect the sample layer against perturbation during reorientation of the gradient.

4.1.4 *Centrifugation and Analysis of Gradients*

To avoid degradation of samples and to minimise variations in the temperature of the gradients during centrifugation it is important to pre-equilibrate the sample, gradients, rotor and centrifuge bowl prior to centrifugation at the temperature at which centrifugation is to be carried out.

The rotor should be accelerated slowly to about 1000 r.p.m. to avoid perturbation of the sample layer and gradient. Slow acceleration is particularly important when using vertical rotors or low-viscosity gradients in tubes greater than 1 cm in diameter. The length of run can usually be deduced from already published procedures or, more accurately, by calculation (see Section 6.1.4 of Chapter 2) or by using the simulation procedures described in Section 4 of Chapter 4. At the end of the run the rotor should be allowed to coast to a halt without braking. The rotor must be removed gently from the centrifuge and, prior to unloading, the tubes should be kept in a vibration-free environment at the same temperature as that used for centrifugation; this latter precaution helps to minimise convectional disturbances. Gradients should be unloaded by upward displacement avoiding density inversions during unloading, as described in Section 7.3 of Chapter 2. The distribution of sample throughout the gradient can be analysed using appropriate methods.

4.1.5 *Determination of Partial Specific Volumes by Sedimentation in Sucrose Gradients*

Edelstein and Schachman (8) first showed that the partial specific volume ($\bar{v}$) of macromolecules could be determined by sedimentation in H_2O and D_2O. The original treatment was devised for analytical centrifuges but subsequently a number of methods have been devised which allow the calculation of partial specific volumes using sucrose gradients (9 – 11).

The method involves preparing 5 – 20% (w/w) sucrose gradients, which are essentially isokinetic, in H_2O and D_2O. The samples are loaded onto the gradients, centrifuged and the distance sedimented measured in both cases. It can be shown (9) that the ratio of the distances sedimented in sucrose/H_2O (r_H) and sucrose/D_2O (r_D) gradients is:

$$\frac{r_H}{r_D} = \frac{k_H (1 - \bar{v}\varrho_H)}{k_D (1 - \bar{v}\varrho_D)}$$

where ϱ_H and ϱ_D are the average densities of the H_2O and D_2O medium, respectively. As an approximation, the average density through which the macromolecules have passed can be used, that is, the density at a position $r/2$ in the respective gradients. The factors k_H and k_D are constants for each medium. Hence, using this relationship it is relatively simple to calculate the partial specific volume, $\bar{v}$. Similar expressions have been used by other workers (10,11).

The advantage of these methods for determining partial specific volumes is that it is not necessary to isolate large amounts of pure macromolecule, indeed, provided that one can assay the macromolecule (e.g., by enzymic activity), crude mixtures can be applied to the gradients. These techniques have also been widely used for the study of membrane proteins which have dissociated with Triton X-100.

4.2 Separations on Denaturing Gradients

4.2.1 *Introduction*

Some types of macromolecules are prone to aggregate and this gives rise to artifactual sedimentation patterns. In addition, the variable secondary structure of macromolecules means that if one wishes to compare the molecular weights of macromolecules then it is necessary to convert the molecules into random coil conformation. The aggregation of RNA can be reversed by heating to 65°C for 10 min at neutral pH. However, on cooling, the secondary structure of the RNA tends to reform. Although the amount of secondary structure of the RNA can be minimised by sedimentation in the absence of salt, in order to convert RNA to random coil conformation it is necessary to sediment the RNA in a denaturing medium. Denaturing gradients are used mainly for fractionating RNA and, to a lesser extent, DNA and hence here discussion will be limited to these two types of macromolecule.

4.2.2 *Choice and Preparation of Gradients*

Two approaches have been used in devising denaturing gradients, these are either using a denaturing solvent [e.g., dimethyl sulphoxide (DMSO)] or adding an additional solute to the sucrose gradient to denature the macromolecules (e.g., alkaline sucrose gradients for separating denatured DNA). Ideally, the denaturing agents should not irreversibly modify the macromolecules. Although in the past formaldehyde, which reacts with amino groups, has been used to minimise the aggregation of RNA, a much more satisfactory method of preparing denaturing gradients is to use either DMSO or formamide to form an essentially non-aqueous sucrose gradient. Only small volumes of aqueous concentrated buffer and EDTA are added in order to maximise the concentration of denaturant; the final concentrations of buffer and EDTA used are usually 10 mM and 1 mM, respectively. Gradients of both $0 - 10\%$ and $5 - 20\%$ sucrose in 99% DMSO or 85% formamide have been used (12,13). For DNA, gradients containing 0.1 M NaOH have been used for separating denatured DNA (15); these cannot be used for RNA since it is degraded in alkaline conditions. Gradients can be prepared using any of the usual techniques (see Section 5.1 of Chapter 2). However, polycarbonate tubes should not be used since they are attacked by DMSO and alkalis, instead one must use polyallomer tubes which are more inert.

Having decided on the type of gradient, it should be prepared by diffusion or using a gradient maker as described in Section 5.1 of Chapter 2. For swing-out rotors, unless using short-column gradients (1) in thick-walled tubes, the top of the gradient should be about 3 mm from the top of the tube. Care must be taken to pre-equilibrate the gradients to the same temperature as is used for the centrifugal separation.

4.2.3 *Preparation and Loading of Samples*

The samples are dissolved either in the denaturing solvent or in a similar denaturing medium; as an example, mixtures of dimethyl formamide and DMSO (3:5 v/v) have been used. As in the case of non-denaturing gradients, only a small volume of sample can be layered onto the gradient (Section 4.1.3); $100 - 200$ μg of RNA can be loaded onto a 12 ml gradient. Great care must be taken, especially with $0 - 10\%$ sucrose gradients in order to obtain a thin sample layer.

4.2.4 *Centrifugation and Analysis of Gradients*

Gradients are centrifuged at 25°C in order to maximise the denaturation of the macromolecules. If one is using alkaline sucrose gradients it is advisable to avoid using aluminium rotors since they are rapidly attacked by alkali; if such rotors are used they should be rinsed out immediately after use. Similar precautions in terms of slow acceleration and deceleration of the rotor should be taken in order to ensure that the gradient is not disrupted during centrifugation (Section 4.1.4).

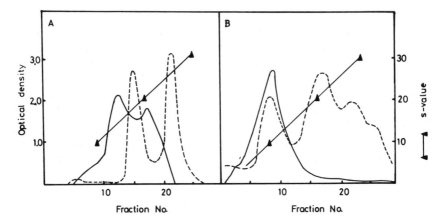

Figure 1. Rate-zonal centrifugation of RNA in non-denaturing and denaturing sucrose gradients. Samples of **(A)** rRNA and **(B)** globin 9S mRNA were centrifuged either in sucrose (– – –) or formamide-sucrose (——) gradients at 25°C in a swing-out rotor and the sedimentation pattern of the RNA was monitored. Data derived from ref. 13.

Gradients are unloaded by upward displacement. Formamide, dimethyl formamide and DMSO all absorb at 260 nm and hence it is necessary to measure the distribution of RNA by the absorption at 280 nm. RNA can be precipitated from fractions at $-20°C$ by the addition of a tenth volume of 30% (w/v) potassium acetate and three volumes of ethanol. *Figure 1* compares the sedimentation patterns of RNA in non-denaturing and denaturing gradients.

4.3 Use of Rate-zonal Centrifugation to Study the Interactions of Macromolecules

4.3.1 *Introduction*

When macromolecules interact their sedimentation rates are altered. The degree of change in sedimentation rates depends on the nature of the macromolecules, that is, the size, density and conformation of each and also the nature of the complex in terms of its composition and conformation. It is necessary to differentiate between non-cooperative, independent, binding and cooperative binding where the binding of one macromolecule to the complex enhances the binding of another. Additionally, unless the reaction is essentially a non-equilibrium reaction (e.g., histones binding to DNA in the absence of salt), it is often desirable to investigate the association constant of the reaction.

Historically, most work has been done studying the interactions between proteins and nucleic acids, hence in this section only these types of interaction will be considered.

4.3.2 *Sizing of Nucleic Acids*

When one isolates total DNA from most bacterial or eukaryotic cells one usually obtains a polydisperse population of DNA molecules sedimenting between 10S and 30S, that is, with molecular weights up to 20×10^6. In the case of purifying specific species of RNA, it is usually possible to isolate populations essentially homogeneous in size. However, in some instances the size of the RNA as isolated is too large for studying interactions with proteins. In the case of DNA, sonication for 60 sec will reduce the DNA to a fairly homogeneous size of about 6S ($\cong 2 \times 10^5$). Alternatively, more specific degradation can be carried out using appropriate restriction nucleases. In the cases of RNA, random breakage of RNA can be carried out by partial alkaline digestion (e.g., by incubating the RNA between pH 9 and 10 for a few hours). Alternatively, one can use partial enzymic digestion using either pancreatic RNase A or RNase T1.

4.3.3 *Experimental Protocols for Studying Non-equilibrium Interactions of Macromolecules*

In this case, the complex, once formed, is stable irrespective of the concentrations of the individual macromolecules. Typically, the macromolecules are mixed together in the appropriate ionic environment and separated by centrifugation using a simple linear sucrose gradient of the same ionic composition. The nature of the binding can be determined by varying the ratio of macromolecules. In the case of histones binding to DNA, at low ratios of histones to DNA it is possible to find DNA fully complexed with histones (i.e., forming 1:1 complexes) and also free DNA not associated with histones, indicating that histones bind to DNA in a highly cooperative manner.

4.3.4. *Experimental Protocols for Studying Equilibrium Interactions between Macromolecules*

Several methods have been devised for studying the reversible interactions of macromolecules. A boundary method has been devised for preparative centrifuges (16) but it is limited in that it cannot be used if the binding constant is greater than $10^7 \ M^{-1}$ and it requires fairly large amounts of material. However, two other methods have been devised which involve zone sedimentation. The first method, developed by Yamamoto and Alberts (17), was devised to investigate the binding of hormone receptors to DNA and has been described as 'sedimentation partition chromatography'. A schematic representation of the method is shown in *Figure 2*. It involves sedimenting a mobile component, in this case DNA, through a narrow zone of the other macromolecules (the hormone receptors) which, because of their smaller size (5S compared with the 18 – 20S DNA) sediment more slowly. For practical purposes it can be envisaged as the receptor sedimenting in the gradient in the presence of a constant concentration of DNA. While the receptor is bound to the DNA its sedimentation rate is similar to that of the DNA, hence the change in sedimentation rate will reflect the degree of interaction and this can be expressed in terms of the percentage of time that the receptors are associated with the DNA. *Figure 3* shows a typical set of data and the method of calculation is given in the legend.

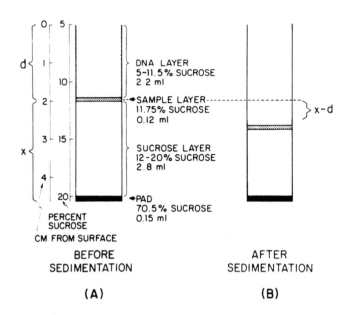

Figure 2. Use of sedimentation partition chromatography to study the interaction of macromolecules. (**A**) Schematic representation of a sucrose gradient for sedimentation partition chromatography before beginning sedimentation. Volumes, concentrations and dimensions given are optimal for sedimentation of 20S DNA with 5S estradiol receptor in cellulose nitrate tubes (1.2 x 5 cm). (**B**) Same gradient after centrifugation at 114 000 g for 14 h. Note that the 'front' of the DNA layer has just reached the pad, while the slowly moving receptor zone is still within the trailing edge of the DNA layer. From ref. 17 with permission of the authors and publishers.

Another approach has been to study the sedimentation of the complex from the sample zone containing the original mixture of DNA and proteins through a sucrose gradient (18). Once the complex sediments out of the sample zone then the protein present is a reflection of the affinity of the protein for the DNA; the greater affinity of the protein for the DNA the greater is the amount of protein that migrates through the gradient. It can be shown that the relationship between the concentration of the protein in the sample (P_0) and the nth fraction (P_n) is given by:

$$\frac{P_n}{P_0} = \left(\frac{KL}{1 + KL}\right)^n$$

where K is the association constant and L is the concentration of *free* nucleic acid binding sites. Rearranging the previous equation gives:

$$KL = \left[\left(\frac{P_n}{P_0}\right)^{-\frac{1}{n}} - 1\right]^{-1}$$

The validity of this approach requires that the complexes do not sediment faster than the time required for the binding equilibrium to be established and also the concentration of the protein must be much lower than the number of binding sites present.

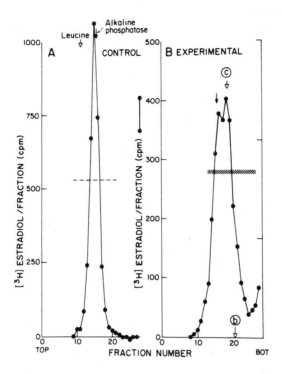

Figure 3. Sedimentation partition chromatography in sucrose gradients. Gradients are prepared and centrifuged as described in *Figure 2*. Closed arrows denote the peak position of the alkaline phosphatase marker activity (10 µg/gradient), co-layered with receptor. **(A)** Control gradient; contains 5S receptor, but no DNA; open arrow represents the peak of [14C]leucine radioactivity used to mark the starting location of the zone of receptor and alkaline phosphatase. **(B)** Experimental gradient; contains 5S receptor (●——●) and mechanically sheared 18 – 20S rat uterine DNA (82 µg/ml). Gradient fractions containing DNA at the end of sedimentation are marked by cross-hatching, as determined by A_{260} readings. The DNA concentration across this region (ideally constant) varies by as much as 30 – 40%, perhaps due to concentration effects on the sedimentation rate. Therefore, the DNA concentration used for the calculation of K_{RD} was the average of the concentrations which had sedimented through the receptor peak. For the calculation of K_{RD}, most of the heterogeneity in the receptor peak resulting from released hormone and/or inactive receptor is corrected for by adopting the following convention. First, the peak width for a homogeneous band is measured at one-half peak height in the control gradient (see broken line in **Panel A**). This value is usually 2.5 – 3.0 fractions. That number multiplied by 1.5 (an arbitrary allowance for heterogeneity) is denoted a. Next, the fraction number of the receptor 'front' at one-half the peak height in the experimental gradient is determined. This fraction number is denoted b (see **Panel B**). The centre of the peak of active receptor in this gradient is then taken as Fraction b – (a/2). In this example, a = 2.5 x 1.5 = 3.75, and b = 20.2; thus, the active receptor peak is taken to be at fraction 20.2 – (3.75/2) = 18.3 (denoted c in **Panel B**). Thus, in this experiment, the interaction of DNA with receptor caused the receptor to sediment 18.3 – 15.0 = 3.3 fractions further than the control (note that the alkaline phosphatase marker peak is used to align the gradients; in general, it is found in the same fraction of each gradient). Since the DNA sedimented 14.0 fractions in this time, the receptor must have been bound 3.3/14.0 = 0.235 = 23.5% of the time, and free 76.5% of the time. But K_{RD} = (free/bound) x (DNA sites). For DNA, 1 µg per ml = 1.5 µM DNA sites, assuming one site per base pair. Thus for these data,

$$K_{RD} = \left(\frac{76.5}{23.5}\right) 82 \ \mu g \ ml^{-1} \ x \ 1.5 \ \mu M \ \mu g^{-1} \ ml = 4 \ x \ 10^{-4} \ M$$

From ref. 17 with permission of the authors and publishers.

The advantage of these methods is that it is possible to examine the inter-action of macromolecules using a wide range of ionic environments simply by changing the ionic composition of the sample and gradient.

4.3.5 *Sources of Error in the Determination of Association Constants*

One common problem with all rate-zonal separations is that of aggregation, in that if, for example, the protein ligand tends to aggregate it sediments faster leading to an erroneously high value of the association constant. Care must also be taken in estimating the actual concentration of macromolecules during sedimentation, since the actual concentrations used are very low. Both nucleic acids and proteins are adsorbed to plastics to a variable degree depending on the centrifuge tube material, in this respect polyallomer is better than cellulose nitrate. Adsorption of macromolecules can also be decreased by rinsing the tubes with a solution of an inert protein such as bovine serum albumin before use since this blocks the adsorption sites. In the zone sedimentation method, the range of the ratio of P_n/P_o should be within the range $0.2-0.6$ otherwise errors appear and will be greatly amplified in the calculation. Sedimentation artifacts may also be present in that wall effects may disrupt the sedimentation of zones, especially in long-bucket swing-out rotors. In addition, differences in the partial specific volumes of the constituent macromolecules and the com-plex may lead to hydrostatic pressure effects modifying the association cons-tant.

5. ISOPYCNIC SEPARATIONS OF MACROMOLECULES

5.1 **Introduction**

Whilst rate-zonal separations are ideal for separating particles with defined but varying sizes or conformations, isopycnic centrifugation offers the possibility of other types of separation. Isopycnic centrifugation is able not only to separate different types of macromolecules, for example, DNA, RNA and protein, but also one can separate macromolecules on the basis of their compositions. Examples of separations based on composition are the separa-tion of DNA on the basis of its content of $G+C$ and the separation of density-labelled nucleic acids and proteins; none of these types of separations can be carried out using rate-zonal methods. An additional feature of isopycnic cen-trifugation is that it is possible to separate different conformations of DNA. This second part of the chapter describes some of the more important pro-tocols used for the isopycnic fractionations of macromolecules and macro-molecular complexes.

5.2 **Factors Affecting the Density of Macromolecules in Solution**

The key factor affecting the buoyant density of macromolecules in solution is the degree of hydration. In solution, clusters of water molecules are associated with all solute molecules including the sample. The amount of water associated with the sample depends on the amount of water available, that is the water activity of the solution. This is primarily dependent on the concentration of the gradient solute. In CsCl gradients which contain $3-6$ M CsCl, most of the water molecules are associated with the hydration shells of the caesium and chloride ions and hence macromolecules such as nucleic acids are poorly hydrated and hence one observes densities close to 1.7 g/cm^3 as compared with the non-hydrated density of DNA (caesium salt) of about 2.3 g/cm^3. In contrast, the concentrations of metrizamide and Nycodenz in gradients are usually less than 0.5 M and hence much more water is available to hydrate nucleic acids and, for example, in these media DNA bands at about 1.1 g/cm^3.

The amount of water that associates with proteins, even in solutions of high water activity, is relatively low and so proteins band at similar densities in all types of gradients.

5.3 Choice of Gradient Medium and Gradient Shape

The physico-chemical properties and the applications of the various types of gradient media have been described in detail in Section 5 of Chapter 1. In choosing a gradient medium a number of factors should be considered. Firstly, the maximum density of the gradient medium must be sufficient to band the sample, for example, RNA can only be banded in CsCl gradients if one uses impractically high temperatures ($\cong 50°$C) to increase the solubility of CsCl. Usually RNA pellets because it bands at densities greater than 1.9 g/cm^3 which is greater than the maximum achievable with CsCl gradients at $25°$C. In Cs$_2$SO$_4$ gradients RNA bands at about 1.6 g/cm^3, much lighter than in CsCl gradients. However, the slope of equilibrium Cs$_2$SO$_4$ gradients is much steeper than for CsCl gradients. Hence, although Cs$_2$SO$_4$ gradients are ideal for separating different types of macromolecules such as DNA, RNA and protein, they are less suitable for subfractionating a single type of macromolecule. Chaotropic salts, especially the trichloroacetate and trifluoroacetate salts of caesium and rubidium, can be used for separating native and denatured DNA as well as being able to separate DNA, RNA and protein (24,30). These gradient media also have the advantage that they are not precipitated by ethanol as are CsCl and Cs$_2$SO$_4$. Hence it is very important to choose the gradient medium which will be most suitable in order to obtain the type of separation required.

The other important point to be considered is whether the gradient medium will adversely affect the sample. For example, high ionic strength solutions can partially denature and precipitate proteins. Similarly, gradient media of high ionic strength dissociate nucleoproteins. Although Cs$_2$SO$_4$ gradients have been used to isolate stable nucleoprotein complexes such as RNA polymerase molecules bound to DNA, usually nucleoproteins must be fixed with formaldehyde prior to centrifugation. There are a number of problems associated with fixation in terms not only of the artifacts arising from fixation but also of the subsequent analysis of the nucleoprotein complexes. Alternatively, one can use one of the nonionic iodinated density gradient media such as metrizamide or Nycodenz (see Section 5.2.4 of Chapter 1) for analysing native nucleoproteins; nonionic gradient media have also proved very useful for studying the interactions between proteins and nucleic acids (20).

Preforming gradients using one of the standard methods (see Section 5 of Chapter 2) is possible and will significantly shorten the time required for isopycnic banding (see Section 5.5). However, because of the prolonged centrifugation time required to band macromolecules, most low molecular weight gradient media (e.g. CsCl and Cs_2SO_4) adopt their equilibrium shapes during centrifugation. Hence the density range of the gradient will reflect the properties of the gradient medium, the centrifugal force applied and the type of rotor used. However some other types of gradient media used for macromolecules, for example metrizamide and Nycodenz, diffuse much more slowly and it is possible to manipulate the shape of the gradient to optimise the resolution over a particular range of densities.

5.4 **Preparation of Gradient Media and Reagents**

Prior to preparation of the gradient, the gradient medium should be prepared properly to ensure that it does not produce artifactual results. The major problem with most gradient media, particularly when dealing with nucleic acids, is the presence of heavy metal ions. Heavy metal ions can form complexes with nucleic acids and so alter their behaviour considerably. So, unless one uses very high grade commercial preparations, gradient media need to be purified. One method is to prepare a concentrated solution and filter it through a Whatman No. 1 or GF/C (or equivalent) filter and then pass the filtrate through a bed of neutralised chelating resin (e.g., Chelex, Bio-Rad Laboratories) at about 1 ml of packed resin per 100 ml of solution for most gradient media; for non-ionic gradient media such as Nycodenz or metrizamide, the level of heavy metal contamination is usually considerably less and so a smaller column can be used. The eluate can be used directly or converted to a solid by rotary evaporation and freeze drying.

In addition to the gradient medium, all buffers should be passed over a column of chelating resin and all glassware and centrifuge tubes washed thoroughly with water that has also been passed through a column of chelating resin. This purified water should also be used for the gradient medium. In addition, $1 - 10$ mM EDTA is sometimes added to the gradient. If urea is being used in denaturing gradients, or to band RNA in Cs_2SO_4, it should also be passed through a mixed-bed deionising column such as Dowex AG501-X8(D); this should be general practice for all urea solutions as it removes cyanate ions which can modify and degrade macromolecules. The addition of $1 - 10$ mM Tris-HCl to gradients helps to neutralise any cyanate ions formed from the urea during centrifugation.

5.5 **Choice of Rotors and Centrifugation Conditions**

As described in Section 2.3 of Chapter 2, for isopycnic separations optimal separations will be achieved using either vertical rotors or shallow angle fixed-angle rotors. Fixed-angle rotors are usually better if material is likely to pellet during centrifugation. In both of these types of rotors gradients form faster, samples band more quickly and gradients have a greater capacity than equivalent swing-out rotors. However, it is important to remember that rotors usually need to be derated when using dense solutions (>1.2 g/cm^3) according to the equation given in Section 6.2.3 of Chapter 2. Special care should be taken, especially when using gradients of caesium salts in swing-out rotors, to ensure that the centrifugation conditions do not result in precipitation of the gradient solute at the bottom of the gradient since the dense crystals that form can overstress the rotor.

The centrifugation speed used is usually designed to be sufficient to band the macromolecules and to maintain the correct density range of the gradient. The equations which allow one to predict the gradient are given in Section 4.2.2. of Chapter 2. An alternative approach is to band the macromolecules rapidly using a high centrifugal force and then to reduce the speed to allow the gradient to become shallower; this is known as gradient relaxation and was first described by Anet and Strayer (41). The equation used to predict the exact length of time to form equilibrium gradients is given in Section 6.2.4 of Chapter 2 and the same section also gives the equations that can be used to calculate the length of time required for particles to reach their isopycnic positions in gradients. The actual time required for particles to reach equilibrium depends on whether one uses self-forming or preforming gradients; particles reach equilibrium faster in preformed gradients. The time required for a particle of density, ϱ_p, and a sedimentation coefficient of $s_{20,w}$ banding at a distance of r cm from the axis of rotation is calculated using the following equations:

Self-forming gradients

$$t\,(h) = \frac{9.83 \times 10^{13} \times \beta° \, (\varrho_p - 1)}{N^4 \, r^2 \, s_{20,w}}$$

Preformed gradients

$$t\,(h) = \frac{2.53 \times 10^{11} \times (\varrho_p - 1)}{N^2 \, r \, s_{20,w} \, d\varrho/dr} \left[\ln\left(\frac{r - r_t}{r_t}\right) + 4.61 \right]$$

where N is the rotor speed, $d\varrho/dr$ the gradient slope, and r_t is the distance from the axis of rotation to the meniscus. Typically particles reach equilibrium about twice as fast in preformed gradients as compared with self-forming gradients.

6. EXPERIMENTAL PROCEDURES USED FOR ISOPYCNIC SEPARATIONS

Table 1 summarises the buoyant densities of most macromolecules in the principle gradient media. It is obvious that, as described in Section 5.5, some gradient media are more suitable than others for some particular types of separation.

Table 1. Buoyant Densities of Macromolecules in Various Density Gradient Media

Gradient media	Gradient[a] slope	Native DNA	Denatured DNA	RNA	Poly-saccharides	Proteins
CsCl	1.00	1.71	1.73	1.9	1.62	1.33
Cs$_2$SO$_4$	1.75	1.43	1.45	1.64	–	–
CsTCA[b]	0.97	1.58	–	1.75	–	1.52
CsTFA[c]	0.80	1.62	1.70	1.7 – 1.9	–	1.2 – 1.5
RbCl	0.51	–	–	–	–	1.29
RbTCA[b]	0.86	1.50	1.68	–	–	–
KI	0.66	1.49	1.52	1.6	–	–
NaI	0.39	1.52	1.55	1.6 – 1.7	–	–
Metrizamide	–	1.12	1.14	1.17	1.28	1.27
Nycodenz	–	1.13	1.17	1.18	1.28	1.27

[a]Gradient slope expressed relative to that of equilibrium gradients of CsCl as calculated from the $\beta°$ values corresponding, as closely as possible, to the value relevant to the banding of nucleic acids.
[b]TCA: trichloroacetate salt.
[c]CsTFA: caesium trifluoroacetate. CsTFA is a registered trademark of Pharmacia Fine Chemicals A.B.

6.1 Fractionation of DNA on the Basis of Base Composition

The classical method for separating types of DNA with different base compositions is using CsCl gradients. These gradients can also be used to analyse the G + C content of DNA on the basis of buoyant density, because G.C base pairs are hydrated less than A.T base pairs.

In these analytical gradients it is usually advantageous to add at least one internal marker DNA to the gradient; usually a bacterial DNA, either from *Escherichia coli* (1.710 g/cm³) or from *Micrococcus lysodeikticus* (1.731 g/cm³). The protocol for these gradients is as follows for a gradient volume of 5.0 ml and an initial density of 1.707 g/cm³ at 20°C. This is achieved by adding together:

4.85 g CsCl
50 μl 1.0 M Tris-HCl (pH 7.4)
25 μl 0.2 M EDTA (pH 7.4)
Sample of DNA (5 – 50 μg)
Water [the volume of solvent in this gradient is 3.8 ml and hence the volume of water to be added is (3.8 – 0.075 – sample) ml].

The density of the gradient is estimated from the refractive index of the solution. The density is calculated from the relationship:

$$\varrho_{25°C} = 10.8601\eta_{25°C} - 13.4974 \quad \text{(when } \eta_{H_2O} = 1.3325)$$

The refractive index corresponding to the correct starting density in this case is 1.4000.

If the density of the solution is too high, then buffer must be added. The volume to be added (in ml) can be calculated using the relationship:

$$\text{Volume} = \Delta\varrho \times V \times 1.52$$

where $\Delta\varrho$ is the difference between the actual and the required densities and V is the volume of the gradient.

If the density is too low, then solid CsCl must be added. The weight to be added (in grams) can be calculated using the relationship:

$$\text{Weight of CsCl} = \Delta\varrho \times V \times 1.32$$

Caesium chloride solutions are extremely well characterised and there is a simple relationship between density (ϱ_{25}) and CsCl concentration (x) in terms of % (w/w) namely:

$$\varrho_{25} = \frac{138.11}{137.48 - x}$$

The buoyant density of *E. coli* DNA is only slightly greater than that of the isoconcentration point, therefore the DNA will band in the top half of the gradient which is the shallowest part and so the resolution of the gradient is maximised.

The solutions are placed in $10-13$ ml thick-walled polycarbonate centrifuge tubes, balanced and centrifuged in a fixed-angle rotor. If thin-walled tubes are used the balanced gradients must be overlayered with liquid paraffin and capped prior to centrifugation. Tubes centrifuged in vertical rotors must always be securely sealed. The actual centrifugation conditions depend on the size of gradient and the nature of the sample as well as the type of rotor. Typically, fixed-angle rotors should be centrifuged at 100 000 g for 48 h at 20°C. To obtain a similar density profile in vertical rotors it is necessary to use a greater centrifugal force than equivalent fixed-angle rotors but the centrifugation time required is usually less than half as long as fixed-angle rotors. Using a discontinuous gradient in a vertical rotor, separations can be achieved in $2-4$ h (21).

Gradients should be unloaded by upward displacement (see Section 7.3.2 of Chapter 2) into at least 25 fractions and for optimal resolution into 50 fractions. First measure the refractive index of each fraction and calculate the density using the equation:

$$\text{Density (25°C)} = 10.8601\eta_{25°C} - 13.497$$

Locate the position of the DNA. It is usually easiest to do this by measuring the optical density of the diluted fractions at 260 nm. Alternatively, if the DNA is radioactive, one can use scintillation counting to analyse the gradients. The dioxan-based scintillation fluids are not suitable for counting CsCl solutions as this gradient medium produces heavy quenching of all widely used radioisotopes (^{3}H, ^{14}C, ^{35}S and ^{32}P). On the other hand, if fractions are acid precipitated, then the problem of sample self absorption can arise when using ^{3}H-labelled samples, that is a thick precipitate can be sufficient shielding to absorb a significant fraction of the emitted β-particles. A Triton-toluene based scintillant, which is water miscible, is acceptable and the drop in efficiency of counting is small. The Triton-toluene scintillant is prepared by dissolving 4.0 g of PPO and 0.05 g of POPOP in a litre of toluene and adding 500 ml of Triton X-100. This can be used for most applications.

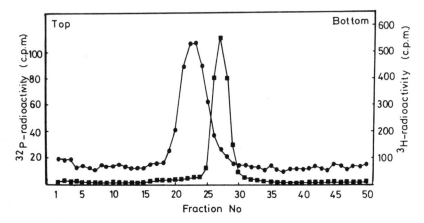

Figure 4. Separation of DNA on the basis of base composition in CsCl gradients. Yeast [^{32}P]nuclear DNA (1.699 g/cm³) and *E. coli* [^{3}H]-DNA (1.710 g/cm³) were banded in a self-forming CsCl gradient (initial density of 1.707 g/cm³) as described in the text in an MSE 10 x 10 ml titanium rotor at 35 000 r.p.m. for 66 h. The distribution of ^{32}P (●——●) and ^{3}H (■——■) were determined in each case.

The base composition of the DNA can be calculated from the density (ϱ) of the DNA using the equation:

$$\% \, (G+C) = \frac{\varrho - 1.660}{0.098} \times 100$$

If, on analysis of the gradient, the marker DNA appears to band slightly denser than expected and it is known not to result from heavy metal contamination then this is probably a reflection that some diffusion of the gradient has taken place after centrifugation and prior to its fractionation. In this case the observed buoyant density of the unknown DNA must be normalised prior to calculating its base composition. *Figure 4* illustrates a typical separation of DNA on the basis of base composition using CsCl gradients.

Enhanced separations of DNA on the basis of base composition can be obtained by introducing sequence-specific ligands into CsCl gradients. Ligands such as actinomycin D, which is specific for G.C base pairs, and netropsin, which is specific for A.T base pairs, bind to the DNA and reduce its density. Using these techniques it is possible to isolate minor satellite species from eukaryotic DNA (22). Alternatively, one can use Cs_2SO_4 gradients and add mercuric or silver ions to the gradient, the former bind preferentially to A.T base pairs and the latter to G.C base pairs. Usually these ions are added such that they are about the same molar concentrations as the nucleotides. This method can also be used to subfractionate a number of minor satellite species from eukaryotic DNA (23).

6.2 Fractionations of Different Conformations of DNA

Although native DNAs of the same base composition and molecular weight but of differing conformation sediment at different rates in a rate-zonal separation, in a CsCl isopycnic separation all of the DNA molecules band at the same density. However, an alternative approach is to band DNA in CsCl gradients in the presence of ethidium bromide or propidium iodide which intercalate between the bases of the DNA. The intercalation of ethidium bromide reduces the density of the DNA. The amount of ethidium bromide that intercalates depends on the conformation of the DNA, in that superhelical plasmid DNA binds less ethidium bromide than linear DNA; this method can also be adapted to determine the superhelicity of DNA (36,37). In practical terms, in the presence of saturating amounts of ethidium bromide, the buoyant density of linear DNA is decreased by 0.19 g/cm^3 while that of plasmid DNA is only decreased by 0.13 g/cm^3.

Typically the plasmid preparation[1] is dissolved in 0.1 M Tris-HCl (pH 8.0), 10 mM EDTA and 1.0 g of CsCl (analytical grade) is added for each millilitre of solution. When the CsCl has dissolved, add sufficient ethidium bromide to give a final concentration of 150 μg/ml. Once the ethidium bromide has been added protect the solution from light to prevent photo-nicking of the DNA.

The gradients (5 − 8 ml) are centrifuged in a fixed-angle rotor at 100 000 g for 24 − 36 h at 15°C. After centrifugation the DNA bands can be visualised by illumination with u.v. light, preferably at 365 nm since this does less damage to the DNA than short-wave u.v. light. Illumination shows the linear DNA floating at the top of the gradient while the supercoiled plasmid bands close to the middle of the gradient. A second band of DNA of relaxed closed circular DNA may be found banding above the band of supercoiled DNA. Any RNA in the sample is pelleted during centrifugation.

When unloading the plasmid from the gradient, care must be taken not to damage the plasmid and not to contaminate it with linear DNA. If thin-walled tubes have been used it is possible to remove the plasmid by inserting a hypodermic syringe needle through the side of the tube and extracting it into a hypodermic syringe. Alternatively, gradients may be dripped out from the bottom of the tube though when using this method care must be taken because in some cases there may be a pellet of RNA at the bottom of the tube. For thick-walled tubes, if care is taken, it is possible to remove the plasmid from the top using a Pasteur pipette. Alternatively, it is possible to use an upward displacement technique using the apparatus with two concentric tubes as described in Section 7.4 of Chapter 2. The ethidium bromide can be removed from the DNA simply by extracting the plasmid fraction with an equal volume of iso-amyl alcohol or isopropyl alcohol pre-saturated with CsCl solution.

An alternative approach for separating plasmid and linear DNA is to use caesium trifluoroacetate gradients. In these gradients, in the absence of inter-calating agents such as ethidium bromide, supercoiled plasmid DNA bands denser than linear DNA (*Figure 5*). The centrifugation conditions used for these gradients is given in the legend to *Figure 5*. However, the degree of separation between the linear and supercoiled DNA molecules is less than that

---

[1]In order to obtain sufficient plasmid from bacteria it is often necessary to select bacteria with the plasmid (e.g., by antibiotic resistance) and to amplify it by chloramphenicol treatment.

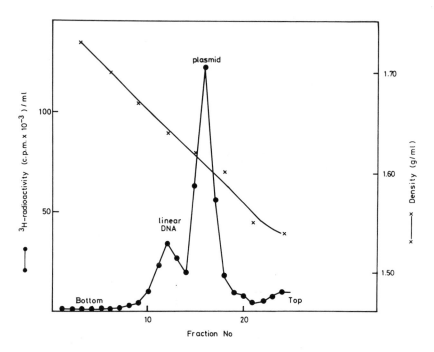

Figure 5. Separation of plasmids on caesium trifluoroacetate gradients. A self-forming gradient of caesium trifluoroacetate of initial density 1.62 g/cm³ was prepared by mixing 3.1 ml of a stock solution (ϱ_i = 2.00 g/cm³) with 1.9 ml of gradient diluent containing the plasmid DNA. The gradients were centrifuged at 275 000 g for 15 h at 14°C in a Beckman VTi65 vertical rotor and the distribution of the DNA was determined by liquid scintillation counting. Personal communication, K. Andersson and R. Hjorth.

achieved using ethidium bromide/CsCl gradients. An additional problem is that, in the absence of ethidium bromide, the position of the plasmid must be detected by u.v. absorption; this involves scanning the whole gradient which may be inconvenient.

6.3 Separation of Single and Double-stranded DNA and RNA-DNA Hybrids

Native and denatured DNA as well as RNA-DNA hybrids can be separated on CsCl gradients as shown in *Figure 6*. In contrast, Cs_2SO_4 gradients give very poor resolution of these macromolecules. An alternative gradient solute which has been used for such separations is NaI. The $\beta°$ values of NaI are larger than those of CsCl and hence the degree of resolution obtained is much greater (*Figure 6*) enabling one to obtain a complete separation of native and denatured DNA. One interesting feature of NaI gradients is that RNA-DNA hybrids band between native and denatured DNA. However, when using NaI gradients, their u.v. absorption make it impossible to monitor the distribution of nucleic acids at 260 nm, although it may be possible to monitor the distribution by absorption measurements at 280 nm or by fluorescence using ethidium bromide. In addition, it is advisable to add a small amount of reducing agent

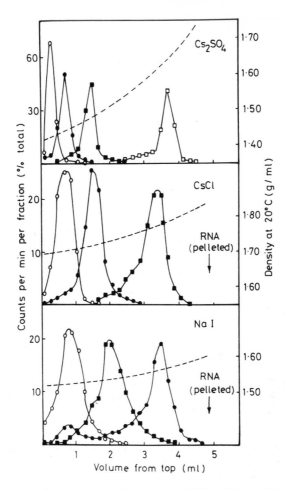

Figure 6. Isopycnic banding of native DNA, denatured DNA, DNA−RNA hybrid and RNA from mouse in density gradients formed from Cs_2SO_4, CsCl and NaI. The ^3H-labelled nucleic acids were mixed with Cs_2SO_4 (initial density 1.54 g/cm^3), CsCl (initial density 1.75 g/cm^3) or NaI (initial density 1.55 g/cm^3), and 4.6 ml portions (5−10 μg of nucleic acid) were centrifuged at 25°C for 68 h at 45 000 r.p.m. (136 000 g) in the MSE 10 x 10 ml titanium fixed-angle rotor; ○——○, native DNA; ●——●, denatured DNA; ■——■, DNA−RNA hybrid, □——□, RNA. Data derived from ref. 19.

(e.g., 2-mercaptoethanol) to both the gradients and samples for liquid scintillation counting to prevent interference arising from the formation of free iodine.

Although Cs_2SO_4 gradients give a very poor separation of native and denatured DNA, the degree of resolution can be greatly enhanced by adding ligands to the gradient that are specific for one type of DNA or another. It is found that silver and mercuric ions are bound preferentially to denatured DNA and hence Cs_2SO_4 gradients containing silver or mercuric ions can be used to obtain a good resolution between native and denatured DNA.

6.4 **Fractionation of RNA**

While RNA is too dense to band in CsCl gradients except at elevated temperatures, it will band in Cs_2SO_4 gradients. However, many species of RNA aggregate and even precipitate in Cs_2SO_4 gradients. A number of different ways of avoiding the aggregation of RNA have been tried. The addition of formaldehyde to the gradient stops the aggregation but formaldehyde also covalently modifies the bases of the RNA. The addition of DMSO to Cs_2SO_4 gradients helps to reduce aggregation and in Cs_2SO_4 gradients containing 10% DMSO one can separate double-stranded RNA, single-stranded RNA and double-stranded DNA (25). Instead of DMSO it is possible to add 4 M urea to the gradient. In the author's experience the recipe given in Section 6.7 is suitable for banding RNA. The aggregation of RNA can also be avoided by using gradients that contain a mixture of CsCl and Cs_2SO_4 (see Section 6.6). The other caesium salts that have been used for banding RNA are caesium formate and caesium trifluoroacetate.

Gradients of KI can also be used for banding RNA (28). No aggregation occurs in these media and, because their $\beta°$ values are much larger than those of Cs_2SO_4, the gradients are much shallower and a correspondingly greater degree of resolution is obtainable. A suitable gradient for separating RNA is obtained by dissolving the sample in 15 mM sodium citrate, 10 mM sodium bisulphite, 5 mM sodium phosphate (pH 7.0) and adding 1.0 g of KI for each millilitre of solution to give a refractive index of 1.4290. The solution is placed into centrifuge tubes and the RNA is banded by centrifugation at 135 000 g for 72 h at 23°C. Up to 1 mg of RNA can be loaded onto a 10 ml gradient. If the gradients are to be analysed by scintillation counting, 2-mercaptoethanol must be added to the fractions to prevent the generation of free iodine. Using gradients containing a mixture of KI and NaI at a ratio of 1:4, it appears possible to fractionate RNA on the basis of base composition and secondary structure (26).

6.5 **Fractionations of Proteins, Lipoproteins and Proteoglycans**

Variations in amino acid composition can endow proteins with slightly different densities up to 0.01 g/cm^3. Even greater differences are found when proteins are associated with non-protein components, such as lipids or polysaccharides, and hence isopycnic centrifugation is frequently used for separating different types of protein. Proteins band at much lower densities than nucleic acids in most gradient media and hence a wide range of gradient media can be used for fractionating proteins.

CsCl gradients can be used to separate proteins but, usually, it is advantageous to use some of the other gradient media such as RbCl, NaBr or KBr which form shallower gradients and hence give better resolution.

In the case of serum lipoproteins, these can be separated by flotation on a NaBr gradient of density range $1.003 - 1.200$ g/cm^3. In this type of separation a continuous gradient is formed using a simple gradient maker and 0.32 g of NaBr is added to each millilitre of the sample to adjust its density to 1.22

g/cm^3. The sample is layered under the gradient and the gradients are centrifuged at 95 000 g for 2 h at 4°C.

6.6 Fractionations of Density-labelled Macromolecules

Molecules can be density labelled by incorporating isotopically labelled compounds, for example, deuterated amino acids or nucleotides. In the case of DNA the incorporation of deuterated nucleotides can increase the density by 0.035 g/cm^3; incorporation of [15N]nucleotides also increases the density by a similar amount. An alternative approach is to use compounds which have heavy atoms present. An example of such an analogue is bromodeoxyuridine which is incorporated into DNA in place of thymidine and can increase the density of DNA up to a maximum of 0.08 g/cm^3. Similarly, mercurated nucleotides (27) can be used for the density labelling of DNA and RNA.

Because of the relatively small density differences involved, gradient media should be used that give the greatest resolution. For proteins, RbCl gradients can be used and these are prepared by mixing:

1.75 g RbCl
0.5 ml of 1.0 M Tris-acetate (pH 7.1)
Sample (50 – 200 μg)
Water (3.25 – sample volume) ml

Centrifuge the gradients at 250 000 g for 65 h at 5°C. Gradients should be fractionated into about 50 fractions.

For nucleic acids one can use CsCl or NaI gradients for analysis of density-labelled DNA (see Section 6.1) while mixed $CsCl$-Cs_2SO_4 gradients containing 5% DMSO and caesium formate gradients have been used for fractionating density-labelled RNA (19,29). Typically, a 5.0 ml mixed $CsCl$-Cs_2SO_4 gradient for separating RNA can be prepared by mixing:

1.9 ml saturated CsCl
1.7 ml saturated Cs_2SO_4
1.15 ml RNA sample (5 – 10 μg) in 0.1 M Tris-HCl (pH 7.5), 10 mM EDTA
0.25 ml DMSO

The gradients are centrifuged at 25°C firstly at 170 000 g for 18 h and then at 100 000 g for 48 h in a fixed-angle rotor (19). After the gradients have been fractionated the density profile must be determined by pycnometry since, when using these mixed solute gradients, it is not possible to calculate density from the refractive index.

In addition to using ionic gradient media for separating density-labelled macromolecules it is also possible to use the nonionic iodinated density gradient media such as metrizamide and Nycodenz. Taking advantage of the stable nature of preformed metrizamide gradients when centrifuged in swing-out rotors, it is possible using shallow D_2O/metrizamide gradients to obtain a high degree of resolution. In addition, mercury-labelled RNA has been separated in metrizamide gradients and bromodeoxyuridine-labelled DNA in Nycodenz gradients using self-forming gradients of 28% (w/v) Nycodenz or metrizamide gradients centrifuged at 63 000 g for 44 h at 5°C.

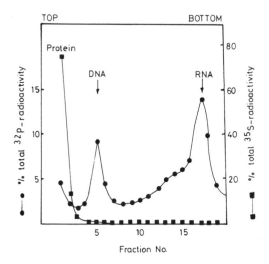

Figure 7. Isopycnic separation of macromolecules in Cs_2SO_4-urea gradients. Yeast mitochondrial nucleoids (39) in which the proteins were labelled with [^{35}S]sulphate (■——■) and the nucleic acids with [^{32}P]phosphate (●——— ●) were fractionated on a Cs_2SO_4-urea gradient as described in the text.

6.7 Separation of Proteins, DNA and RNA

Frequently it is desirable to separate and analyse the macromolecular components of native complexes such as chromatin or nuclei. Hence one requires a highly dissociating environment. Proteins from chromatin have been separated from nucleic acids using CsCl gradients containing 4 M urea although better yields of proteins can be obtained if CsCl gradients contain both 4 M urea and 0.5 M guanidine hydrochloride. While these types of gradients are useful for obtaining proteins free of nucleic acids, if one wishes to band all three types of macromolecule it is necessary to use gradient media, such as Cs_2SO_4 or caesium trifluoroacetate, which form steep gradients. It has been found that the inclusion of 4 M urea in Cs_2SO_4 gradients not only increases the degree of dissociation of nucleoproteins but it also prevents the aggregation and precipitation of RNA (see Section 6.4). For a gradient volume of 5.0 ml and containing 60% (w/v) Cs_2SO_4 in 4 M urea, add together:

    3.00 g Cs_2SO_4

    2.5 ml 8 M urea (deionised by passage over a mixed-bed resin)

    50 μl 1.0 M Tris-HCl (pH 7.4)

    25 μl 0.2 M EDTA (pH 7.4)

    Sample; the volume of sample and water required to bring the gradient to volume is 1.25 ml.

The gradients should be centrifuged at 130 000 g for 64 h at 5°C; a typical example of the type of separation obtainable is shown in *Figure 7*.

In the case of caesium trifluoroacetate, the chaotropic nature of this medium ensures complete disruption of nucleoproteins into their constituent macromolecules. Caesium trifluoroacetate forms shallower gradients than

Cs_2SO_4 solutions but even so it is possible to band proteins, DNA and RNA on a single gradient. It is not necessary to add urea to these gradients.

6.8 Isolation of Nucleoprotein Complexes

6.8.1 *Introduction*

Ionic gradient media such as CsCl disrupt virtually all nucleoprotein structures and hence if one wishes to preserve the native composition of nucleoproteins it is necessary to fix the complexes using cross-linking agents such as formaldehyde or glutaraldehyde which covalently bind the components together. However, fixation does preclude further analysis of nucleoproteins. The dissociative nature of Cs_2SO_4 gradients has, however, been used to isolate salt-resistant nucleoprotein complexes of, for example, RNA polymerase transcribing DNA and 'core' ribosomal particles.

In order to isolate native nucleoproteins it is necessary to use nonionic gradient media and for this purpose the iodinated gradient media, metrizamide and Nycodenz, have proved to be ideal. Nonionic iodinated gradient media do not appear to dissociate nucleoprotein complexes. Not only do metrizamide and Nycodenz not affect the high order structures of metaphase chromosomes but also they do not disrupt the nucleosome structure of chromatin (31). In addition, this type of medium has been used to fractionate different types of ribonucleoprotein (35).

6.8.2 *Isopycnic Separations of Chromatin and Nucleoids*

In the absence of ions, chromatin is readily soluble in solutions of metrizamide and Nycodenz and high molecular weight chromatin bands as a single component. When chromatin is banded in metrizamide gradients containing submillimolar amounts of ions the buoyant density of the chromatin depends on the amount of chromatin loaded onto the gradient. Increasing the amount of chromatin loaded onto a metrizamide gradient increases the buoyant density; a similar but smaller effect is also seen in the case of Nycodenz gradients (*Figure 8*). This effect can be explained in terms of aggregation of the chromatin as it forms a band in the gradient. The higher the concentration of the chromatin in the band the higher is the degree of aggregation, leading to a decrease in the hydration of the chromatin. It is not clear why this effect of loading is smaller in the case of Nycodenz gradients. This effect is seen not only in the case of chromatin but nucleoids from organelles also show the same effect. However, no such effect is seen with ribonucleoproteins in either medium.

Increasing the concentration of NaCl in the gradient increases the buoyant density of chromatin reaching a maximum of 0.14 M NaCl; at higher concentrations of salt the buoyant density decreases as a result of the proteins becoming dissociated from the chromatin. The presence of millimolar amounts of $MgCl_2$ also increases the buoyant density of chromatin in metrizamide gradients. In both cases the increases in buoyant density of chromatin in the presence of NaCl and $MgCl_2$ would appear to be a reflection of the lower solubility and higher degree of aggregation in these ionic environments. Hence

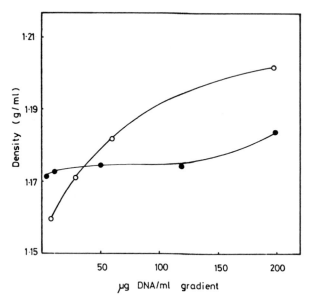

Figure 8. Effect of loading on the buoyant density of chromatin in metrizamide and Nycodenz gradients. Chromatin isolated from mouse-liver nuclei was solubilised and loaded separately onto metrizamide and Nycodenz gradients and centrifuged. The variation of buoyant density with loading in metrizamide (○——○) and Nycodenz (●——●) gradients is shown. Figure reproduced from ref. 40.

chromatin is less hydrated in these ionic environments. As might be expected, there is a smaller effect of loading on the buoyant density of chromatin when the chromatin is aggregated. The observed changes in the degree of aggregation of chromatin as a result of concentration or the presence of ions has been found to be fully reversible.

During replication, changes occur in the structure and composition of chromatin close to the replication fork. When chromatin from dividing cells is fractionated on metrizamide gradients a small amount of the chromatin bands at a marginally higher density than the bulk of the chromatin. Pulse labelling with thymidine reveals that the denser chromatin is associated with the newly replicated DNA (*Figure 9*). The similarity of the fractionations of replicating chromatin obtained with such varied cell types as sea urchin embryos and a line of tissue culture cells derived from mouse leukaemia cells suggests that isopycnic fractionation in metrizamide or Nycodenz gradients may prove to be a general method for the isolation of fractions enriched in replicating chromatin.

Other investigations have used alkaline metrizamide gradients to investigate the deposition of histones on newly replicated chromatin (33). The method used is to denature the chromatin at pH 12.5, the strands of DNA separate but the histones remain associated with the DNA. The denatured chromatin is then loaded onto a self-forming gradient of 27% (w/v) metrizamide dissolved in 20% (v/v) triethanolamine giving a final pH of pH 10.5. The use of triethanolamine instead of NaOH to adjust the pH of the gradient avoids the possibility

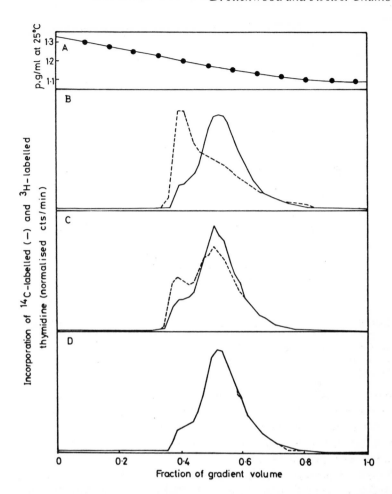

Figure 9. Separation of newly replicated chromatin from bulk chromatin in metrizamide gradients (**A**). Cells were labelled for 24 h with [14C]thymidine (——) and then for 1 min (**B**), 10 min (**C**), or 100 min (**D**) with [3H]thymidine (– – –). After digestion with micrococcal nuclease at 5.33×10^7 units/nucleus and removal of acid-soluble material (2.4% 14C; 8.8, 2.4, 2.4% 3H), the total nucleosomal fraction was centrifuged in metrizamide gradients. Figure reproduced from ref. 40.

of the histones dissociating and becoming redistributed along the DNA.

The conditions for the purification of a number of mitochondrial nucleoids have been thoroughly investigated and the experimental procedures that are used generally involved lysis of the mitochondria with Triton X-100 followed by centrifugation on metrizamide gradients. The buoyant density of mitochondrial nucleoids in metrizamide gradients is dependent on the amount of nucleoid loaded onto the gradient. As in the case of chromatin, increasing the amount of nucleoid loaded onto the gradient increases the buoyant density. Hence it is possible to ensure that the nucleoids band at a density completely separate from possible contaminants, particularly membrane material. Since the buoyant density of nucleoids can be manipulated to a lesser extent in

121

Table 2. Buoyant Density of Ribonucleoprotein Particles in Metrizamide Gradients

Ribonucleoprotein particles	Gradient system			
	Metrizamide H₂O (g/cm³)	*Metrizamide H₂O + Mg²⁺ (g/cm³)*	*Metrizamide D₂O + Mg²⁺ (g/cm³)*	*CsCl[a] (g/cm³)*
80S ribosomes	1.26	1.31	1.34	1.56
60S ribosomes	1.22	1.32	1.33	1.57
40S ribosomes	1.22	1.24	1.26	1.49
polysomes	1.24	1.35	1.38	1.53
cytoplasmic mRNP	1.21	1.21	1.23	1.39

[a]Nucleoproteins fixed prior to centrifugation

Nycodenz gradients, it would appear that, in this case, it may be more appropriate to use metrizamide gradients rather than Nycodenz gradients.

6.8.3 *Isopycnic Separations of Ribonucleoproteins*

When ribonucleoproteins are banded in solutions of nonionic iodinated density gradient media their buoyant density depends on the ionic environment (32,34,35). In the absence of magnesium ions, the buoyant density of ribonucleoproteins reflects the relative amounts of protein and RNA present in the complex. However, in the presence of millimolar amounts of magnesium ions some types of ribonucleoprotein bind more magnesium ions than others. As shown in *Table 2*, the binding of magnesium ions has a dramatic effect in increasing the buoyant density of ribonucleoproteins. It is notable that ribonucleoprotein particles containing mRNA bind very small amounts of magnesium ions and hence band at a lower density than the ribosomes and their subunits. This method can be used to isolate these messenger ribonucleoprotein particles (mRNP). In some cases, the separations of mRNP can be enhanced by using deuterium oxide instead of water as the gradient solvent (32).

When banding ribonucleoproteins, some of the less tightly bound proteins that are in a dynamic equilibrium with the particles dissociate and hence ribonucleoprotein particles banded without fixation will, after centrifugation, be associated with those proteins which are permanently bound to the particles. An additional factor to be remembered is that the hydrostatic pressure generated in the gradient may be sufficient to disrupt the ribonucleoprotein particles (5,35).

6.9 Use of Isopycnic Centrifugation to Study Interactions of Macromolecules

Interactions between nucleic acids and proteins can be divided into two main types, namely ionic interactions and sequence-specific interactions. Ionic interactions involve the formation of ionic links between basic proteins and the negatively-charged phosphate groups of the nucleic acid; these complexes once formed tend to be fairly stable. The sequence-specific interactions between nucleic acids and proteins tend to be less stable, for example, the half-lives of complexes between RNA polymerases and DNA can be of the order of an hour or less, but in other cases completely stable complexes are formed. Of the isopycnic gradient media available, only the nonionic iodinated gradient media have proved suitable for studying the interaction of macromolecules.

6.9.1 *Loading of Gradients*

While basic proteins such as histones bind avidly to DNA, other proteins, particularly those that exhibit selective binding, do not form stable complexes with DNA. Hence, in these circumstances, it is important to be able to separate the non-bound proteins from the DNA/protein complexes. The problem is that many of the proteins have low molecular weights and hence, take a long while to reach their isopycnic positions if the proteins are originally dispersed throughout the gradient solution. The way to overcome this problem is to load the sample into the bottom of the gradient, then during centrifugation free DNA and DNA/protein complexes float upwards while the proteins remain in the loading zone at the bottom of the gradient. The key to the successful application of this technique is to use DNA large enough to float up to the top of the gradient but small enough such that the density of the DNA is modified significantly when proteins are attached to it (20,31).

6.9.2 *Factors that Affect the Interaction of Proteins with DNA*

(i) *Proteins.* Different types of nuclear proteins bind with different affinities to DNA. Hence, for example, proteins easily extracted from nuclei generally, as might be expected, bind more weakly to DNA. In addition, the methods used to prepare the proteins can also affect their DNA-binding activity. For some studies described elsewhere (20,31) the proteins were prepared by the hydroxy-apatite procedure which is one of the more gentle isolation procedures. Proteins extracted with salt or acid also appear to retain many of their functional activities. In contrast, procedures which involve a high degree of denaturation of the proteins, for example, extraction with SDS or phenol solutions, are likely to produce much less active protein fractions.

(ii) *Nucleic acid contamination.* Most methods of preparing nuclear proteins yield protein fractions which are contaminated to a greater or lesser extent with nucleic acids, particularly RNA. The presence of RNA inhibits the interaction of proteins with the DNA apparently as a result of its preferential binding to the proteins. The proteins also seem to bind preferentially to single-stranded DNA in the same way as they do to RNA. Hence, in preparing protein fractions, care should be taken to ensure that the contamination with nucleic acids is minimal; this can be done by purifying the proteins by isopycnic centrifugation on urea-Cs_2SO_4 gradients (Section 6.7). The DNA used for binding studies should be treated with nuclease S1 to remove any single-stranded fragments of DNA.

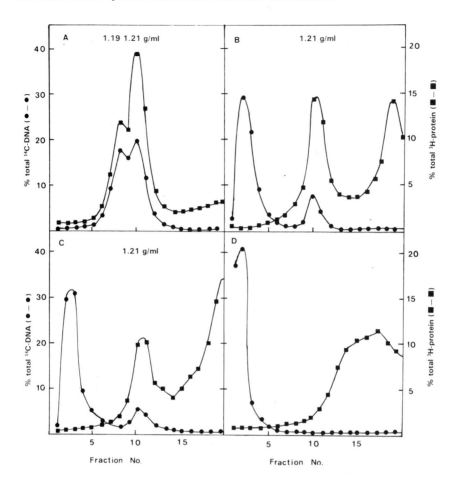

Figure 10. The effect of ionic strength on the binding of non-histone proteins to mouse DNA. The proteins were reconstituted to DNA from 2 M NaCl, 5 M urea into a solution containing either **(A)** no salt; **(B)** 0.14 M NaCl; **(C)** 0.2 M NaCl and **(D)** 0.3 M NaCl and separated on metrizamide gradients containing the same concentration of salt as described in the text. The gradients were fractionated and the density and distribution of DNA (●——●) and protein (■——■) in each fraction were determined. The density of each complex in g/cm³ is shown in each case. Data from ref. 20.

(iii) *Ionic environment.* When chromatin non-histone proteins are either mixed with DNA or reconstituted to DNA by dialysis procedures into dilute buffer nearly all of the proteins are bound to all of the DNA (*Figure 10A*). In contrast, when the mixture of DNA and proteins is dialysed into 0.14 M NaCl the proteins bind cooperatively to the DNA; the protein-associated DNA is clearly separated from the non-complexed DNA (*Figure 10B*). A similar pattern of binding is observed when the proteins are dialysed into 0.2 M NaCl (*Figure 10C*), while in 0.3 M NaCl no stable complexes are formed (*Figure 10D*). Depending on the type of proteins under study the stability of the DNA/protein complexes may vary (20).

124

7. REFERENCES

1. Griffith,O.M. (1978) *Anal. Biochem.,* **90**, 435.
2. Chia,D. and Schumaker,V.N. (1974) *Biochem. Biophys. Res. Commun.,* **56**, 241.
3. Kovacic,R.T. and Van Holde,K.E. (1977) *Biochemistry (Wash.),* **16**, 1490.
4. Clark,R.W. and Lange,C.S. (1980) *Biopolymers,* **19**, 945.
5. Hauge,J.G. (1971) *FEBS Lett.,* **17**, 168.
6. Marcum,J.M. and Borisy,G.G. (1978) *J. Biol. Chem.,* **253**, 2852.
7. Jamaluddin,M. and Philip,M. (1982) *FEBS Lett.,* **150**, 429.
8. Edelstein,S.J. and Schachman,H.K. (1967) *J. Biol. Chem.,* **242**, 306.
9. Meunier,J.C., Olsen,R.W. and Changeux,J.P. (1972) *FEBS Lett.,* **24**, 63.
10. Clarke,S. (1975) *J. Biol. Chem.,* **250**, 5459.
11. Sadler,J.E. (1979) *J. Biol. Chem.,* **254**, 4443.
12. Ruiz-Carillo,A., Beato,M., Schutz,G., Feigelson,P. and Allfrey,V.G. (1973) *Proc. Natl. Acad. Sci. USA,* **70**, 3641.
13. MacNaughton,M., Freeman,K.B. and Bishop,J.O. (1974) *Cell,* **1**, 117.
14. Rickwood,D. (1983) in *Iodinated Density Gradient Media: A Practical Approach,* Rickwood,D. (ed.), IRL Press, Oxford and Washington, p. 1.
15. Studier,F.W. (1965) *J. Mol. Biol.,* **11**, 373.
16. Jensen,D.E. and Von Hippel,P.H. (1977) *Anal. Biochem.,* **80**, 267.
17. Yamamoto,K.R. and Alberts,B. (1974) *J. Biol. Chem.,* **249**, 7076.
18. Draper,D.E. and Von Hippel,P.H. (1979) *Biochemistry (Wash.),* **18**, 753.
19. Birnie,G.D. (1978) in *Centrifugal Separations in Molecular and Cell Biology,* Birnie,G.D. and Rickwood,D. (eds.), Butterworths, London and Boston, p. 169.
20. Rickwood,D. and MacGillivray,A.J. (1977) *Exp. Cell Res.,* **104**, 287.
21. Wells,J.R. and Brunk,C.F. (1979) *Anal. Biochem.,* **97**, 196.
22. Tatti,K.M., Hudspeth,M.E.S., Johnson,P.H. and Grossman,L.I. (1978) *Anal. Biochem.,* **89**, 561.
23. Wilson,V.L., Rinehart,F.P. and Schmid,C.W. (1976) *Anal. Biochem.,* **73**, 350.
24. Burke,R.L. and Bauer,W.R. (1980) *Nucleic Acids Res.,* **8**, 1145.
25. Williams,A.E. and Vinograd,J. (1971) *Biochim. Biophys. Acta,* **228**, 423.
26. Andrean,B.A.G. and De Kloet,S.R. (1973) *Arch. Biochem. Biophys.,* **156**, 373.
27. Hanausek-Walaszek,M., Walaszek,Z. and Chorazy,M. (1981) *Mol. Biol. Rep.,* **7**, 57.
28. Wolf,H. (1975) *Anal. Biochem.,* **68**, 505.
29. Grainger,R.M. and Wessells,N.K. (1974) *Proc. Natl. Acad. Sci. USA,* **71**, 4747.
30. Burke,R.L., Anderson,P.J. and Bauer,W.R. (1978) *Anal. Biochem.,* **86**, 264.
31. Ford,T. and Rickwood,D. (1983) in *Iodinated Density Gradient Media: A Practical Approach,* Rickwood,D. (ed.), IRL Press, Oxford and Washington, p. 23.
32. Buckingham,M.E. and Gros,F. (1976) in *Biological Separations in Iodinated Density Gradient Media,* Rickwood,D. (ed.), IRL Press Oxford and Washington, p. 71.
33. Russev,G. and Tsanev,R. (1976) *Nucleic Acids Res.,* **3**, 697.
34. Rickwood,D. and Jones,C. (1981) *Biochim. Biophys. Acta,* **654**, 26.
35. Houssais,J.F. (1983) in *Iodinated Density Gradient Media: A Practical Approach,* Rickwood,D. (ed.), IRL Press, Oxford and Washington, p. 43.
36. Upholt,W.B. (1977) *Science (Wash.),* **195**, 891.
37. Burke,R.L. and Bauer,W. (1977) *J. Biol. Chem.,* **252**, 291.
38. Flamm,W.G., Bond,H.E. and Burr,H.E. (1966) *Biochim. Biophys. Acta,* **129**, 310.
39. Chambers,J.A.A., Rickwood,D. and Barat,M. (1981) *Exp. Cell Res.,* **133**, 1.
40. Rickwood,D. and Ford,T. (1983) in *Iodinated Density Gradient Media: A Practical Approach,* Rickwood,D. (ed.), IRL Press, Oxford and Washington, p. 69.
41. Anet,R. and Stryer,D.R. (1969) *Biochem. Biophys. Res. Commun.,* **34**, 328.

CHAPTER 4

Measurement of Sedimentation Coefficients and Computer Simulation of Rate-zonal Separations

B.D. YOUNG

1. INTRODUCTION

The separation of macromolecules of different sizes can be readily accomplished by rate-zonal centrifugation. In this technique, macromolecules are layered on top of a density gradient and are sedimented through the gradient by centrifugation. This is called rate-zonal centrifugation because molecules of a similar size move through the gradient medium as a discrete zone. One of the most useful applications of this technique is the physical separation of molecules of different sedimentation rates. The preparation, centrifugation and unloading of such gradients is considered elsewhere in this volume. In general, larger molecules sediment faster than smaller molecules, although there is not always a simple relationship between molecular weight and rate of sedimentation. For this reason it is often convenient to describe macromolecules in terms of their sedimentation characteristics which can be determined by rate-zonal centrifugation.

Often, rate-zonal separations in preparative centrifuges are used to separate particles using conditions similar to those used by other workers. In such cases, the type of gradient, length of run and centrifugal force used are based on those used previously. This type of approach is suitable if it is sufficient to obtain a separation rather than obtain quantitative data. If one wishes to obtain very accurate sedimentation data it is necessary to use an analytical centrifuge as described in Chapter 8. However, analytical centrifuges are not always available and often sufficiently accurate data can be achieved using preparative ultracentrifuges. Quantitative estimations of sedimentation coefficients can be obtained using isokinetic gradients together with sedimentation markers, as described in Section 3.2. However, a number of other, and usually more versatile, techniques have been developed and these are also described in detail in this chapter.

One recurring problem in rate-zonal centrifugation when attempting new separations is the estimation of the exact centrifugation conditions required for the separation of particles. An extremely valuable approach to this problem is the use of computer simulation. As described in Section 4, using the computer programs given in this chapter it is possible to simulate the sedimentation patterns of particles using a wide range of gradient shapes, length of run and centrifugal force. Hence it is possible to optimise both the gradient shape and centrifugal conditions to obtain the best degree of separation without the necessity of carrying out several empirical determinations.

2. SEDIMENTATION COEFFICIENTS

The rate at which a macromolecule sediments is characterised by its sedimentation coefficient. This has been defined by Svedberg and Pederson (1) as the sedimentation velocity in unit field strength:

$$s = \frac{1}{\omega^2 r} \cdot \frac{dr}{dt} \qquad \text{Equation 1}$$

where r is the distance of the molecule from the axis of rotation at time t and ω is the angular rotor speed in radians/sec. Sedimentation coefficients are expressed in Svedberg units; one Svedberg unit is equal to 10^{-13} sec. The sedimentation coefficient defined by Equation 1 is dependent not only on the characteristics of the macromolecule but also on the experimental conditions and on the properties of the centrifugation medium. However, it is frequently desirable to compare sedimentation coefficients obtained under different experimental conditions and in different sedimentation media. In order to permit standardisation, Svedberg and Pederson (1) defined a standard sedimentation coefficient which one would obtain for a given macromolecule in water at 20°C ($s_{20,w}$). It must be emphasised that, in practice, water alone at 20°C is not used as the medium for centrifugation because a density gradient such as that provided by sucrose is necessary to ensure the stability of the zones of particles during centrifugation. Hence, sedimentation coefficients are determined in density gradient media and then standardised to their $s_{20,w}$ values. The sedimentation coefficient ($s_{T,m}$) of a particle sedimenting at temperature T through a homogeneous medium (m) of density $\varrho_{T,m}$ and viscosity $\eta_{T,m}$ can be converted to the corresponding $s_{20,w}$ by the following equation:

$$s_{20,w} = s_{T,m} \cdot \frac{\eta_{T,m}(\varrho_p + \varrho_{20,w})}{\eta_{20,w}(\varrho_p - \varrho_{T,m})} \qquad \text{Equation 2}$$

where $\varrho_{20,w}$ and $\eta_{20,w}$ are the density and viscosity of water at 20°C and ϱ_p is the density of the macromolecule in that medium. Combining Equation 1 and Equation 2 one obtains the relationship:

$$s_{20,w} = \frac{1}{\omega^2 r} \cdot \frac{dr}{dt} \cdot \frac{\eta_{T,m}(\varrho_p - \varrho_{20,w})}{\eta_{20,w}(\varrho_p - \varrho_{T,m})} \qquad \text{Equation 3}$$

Thus, in principle, the standardised sedimentation coefficient ($s_{20,w}$) may be determined directly from the experimental data.

3. METHODS OF MEASURING SEDIMENTATION COEFFICIENTS

3.1 **Introduction**

There are two main methods by which sedimentation coefficients may be estimated. Firstly, the sedimentation of an unknown macromolecule may be compared with that of a known standard. If one uses isokinetic gradients then the distance migrated by a macromolecule is proportional to its $s_{20,w}$ value and thus the sedimentation coefficient can be determined by comparision with the migration of macromolecules of known size. If a suitable standard macromolecule is not available, it is possible to calculate the $s_{20,w}$ value of a macromolecule from a detailed knowledge of the conditions of centrifugation. If computer facilities are not available, sedimentation coefficients can be calculated by the use of pre-computed tables devised by McEwen (2) and given in Appendix A. Alternatively, the computer programs listed in Appendices B and C of this chapter can be used to calculate sedimentation coefficients.

3.2 **Isokinetic Gradients**

An isokinetic gradient is a gradient in which at constant rotor speed a given macromolecule moves at a constant velocity independently of the distance sedimented. Such a condition will be met if, as the particle sediments, the increasing centrifugal force is balanced by the increasing density and viscosity of the gradient medium. Fortunately, isokinetic conditions can be obtained using relatively simple gradients. It has been shown that an isokinetic gradient can be obtained if one uses a linear gradient between 5% and 20% (w/w) sucrose (3,4). Similarly, glycerol gradients are isokinetic if the concentration varies linearly between 10% and 30% (5). Noll (6) showed that isokinetic gradients can be obtained over a wider range of sucrose concentrations if convex exponential gradients are used; a typical mixing apparatus is described in Section 5.1.4 of Chapter 2 which can produce such gradients.

The real advantage of isokinetic gradients is that after centrifugation the distance through which a particle has sedimented is directly proportional to its sedimentation coefficient. All of the particles are sedimented under the same conditions and thus any variations in the centrifugal conditions will affect all particles equally. Hence this result is independent of any variations in acceleration, deceleration, rotor speed, and, even more importantly, rotor temperature which may have occurred during centrifugation. If a swing-out rotor is used and the gradient is fractionated from the top into equal volume fractions then the distance sedimented and therefore the sedimentation coefficient of a particle will be proportional to its fraction number. Hence, the sedimentation coefficient (s_2) of a macromolecule may be estimated by comparision of its migration (l_2) in the gradient with that of a macromolecule of known sedimentation coefficient (s_1, l_1) by the equation:

$$s_2 = s_1 \cdot \frac{l_2}{l_1} \qquad \text{Equation 4}$$

In general, reference macromolecules should be of the same type as the unknown macromolecule. Thus it is inadvisable to use a protein marker to measure the sedimentation coefficient of RNA molecules. Furthermore, the reference and unknown macromolecules should have reasonably similar molecular weights and should be centrifuged in the same tube or at least simultaneously in the same rotor.

In an isokinetic gradient the rate of sedimentation of a particle is proportional only to its sedimentation coefficient and to the square of the rotor speed. Hence the distance sedimented by a particle in time t is given by:

$$r - r_m = \alpha_{20,w} \omega^2 t \qquad \text{Equation 5}$$

where r and r_m are the distances from the axis of rotation to the particle and to the top of the gradient, respectively. The proportionality constant α has been calculated for the centrifugation of proteins in $5-20\%$ (w/w) linear sucrose gradients in a variety of rotors (5). Thus the sedimentation coefficient of an unknown protein can be calculated directly from Equation 5 without the use of reference proteins. However, where possible, a suitable reference protein should be included as a further check on the calculations.

In sector-shaped cells or zonal rotors the radial increase in volume is proportional to the square of the radius and thus, if equal volume fractions are collected, the radial distance through which a macromolecule has sedimented is not proportional to its fraction number. This difficulty for zonal rotors has been resolved by Pollack and Price (7) who proposed the use of equivolumetric gradients. In such a gradient the volume through which a particle migrates is proportional to the sedimentation coefficient of the particle. Hence the proportionality between $s_{20,w}$ and equal volume fraction number which holds for isokinetic gradients in swing-out rotors is also true for equivolumetric gradients in zonal rotors.

3.3 Pre-computed Tables

A useful set of tables has been computed by McEwen (2) which allow direct calculation of $s_{20,w}$ values without the use of reference macromolecules. The only restrictions are that one must use a linear sucrose gradient at 0°C, 5°C or 20°C. The user is referred to the original publication for a detailed description of the technique and only a brief summary is presented here.

Equation 3 may, by integration, be expressed thus:

$$s_{20,w} \int \omega_2 dt = \int \frac{\eta_{T,m}(\varrho_p - \varrho_{20,w})}{\eta_{20,w}(\varrho_p - \varrho_{T,m})} \cdot \frac{1}{r} \cdot dr \qquad \text{Equation 6}$$

To simplify the calculation, the acceleration and deceleration times are assumed to be negligible. Therefore, the time integral becomes $s_{20,w} \omega_2 t$ where t is the total centrifugation time. Empirical equations have been derived which express the viscosity ($\eta_{T,m}$) and density ($\varrho_{T,m}$) of sucrose solution as functions of sucrose concentration. If a linear gradient is assumed, then the radius can also be expressed as a function of the sucrose concentration (z) and thus:

$$\frac{dz}{dr} = \frac{z - z_0}{T}$$

Equation 7

where z_0 is the extrapolated value of the sucrose concentration at the centre of rotation. Thus the right-hand integral of Equation 6 may be expressed in terms of sucrose concentration and may be evaluated if the temperature, particle density in sucrose and z_0 are known. McEwen (2) has computed values of this integral for particle densities from 1.1 to 1.9 g/cm³, temperatures 0°C, 5°C and 20°C and for values of z_0 from $+5$ to -100. If during centrifugation a particle moves from sucrose concentration z_1 at time t_1 to sucrose concentration z_2 at time t_2 we have from Equation 6:

$$s_{20,w} \omega^2 (t_2 - t_1) = I(z_2) - I(z_1)$$

Equation 8

assuming constant angular velocity. The values of $I(z_2)$ are obtained from the tables from a knowledge of the particle density, the temperature and z_0.

The appropriate value of z_0 can be estimated from a knowledge of the dimensions of the rotor used. The simplest method is to use graphical extrapolation from two known sucrose concentrations and radii. Usually z_0 has a negative value, but if shallow gradients are used it can have a positive value. Frequently z_0 will not coincide exactly with one of the values used for the computation of the tables. In this case the integral values may be estimated by interpolation between the listed values. However such interpolation is only valid if the integration procedure used to calculate the tables has been started from the same point. The tables involved different starting points according to whether $z_0 < 0$, $z_0 = 0$, or $z_0 > 0$ and therefore it is incorrect to interpolate z_0 between values zero and 5. In order to solve this problem the tables have been recalculated in the region of z_0 values between zero and -10 and are listed in Appendix A of this chapter. Hence if it is found that $0 > z_0 > -10$ the tables in Appendix A should be used, otherwise the original tables (2) should be used.

An alternative technique which uses pre-computed tables for the calculation of sedimentation coefficients in B14 and B15 zonal rotors has been developed by Funding and Steensgaard. The user is referred to the original publication for a detailed description and a worked example (8). In this approach Equation 6 is simplified by the introduction of the function 'sedim(Y)' which is defined by the relationship:

$$\text{sedim}(Y_k) = \frac{\eta_{T,mk} (\varrho_p - \varrho_{20,w})}{\eta_{20,w} (\varrho_p - \varrho_{T,m,k})}$$

Equation 9

where Y is the sucrose concentration in fraction k.

In fact, provided the sucrose concentration is known in each fraction, the right-hand integral in Equation 6 can be approximated by treating it as a step function, thus converting it to a summation over the fractions. Tabulated values of the sedim function are available (8) for particle densities from 1.1 to 2.0 g/cm³, temperatures from 4°C to 16°C and sucrose concentrations from 0% (w/w) to 44.5% (w/w). Also given are the data relating the radius to volume for the B14 and B15 rotors. Furthermore, it can be shown that if linear acceleration and deceleration is assumed then the force-time integral can be approximated as:

$$\int \omega^2 dt = (\pi N)^2 \cdot (P + (A + D)/3)15 \qquad \text{Equation 10}$$

where N is the rotor speed in revolutions per minute, A is the acceleration time, D is the deceleration time and P is the run time in minutes.

This technique has the advantage that, provided the change in sucrose concentration between fractions is kept reasonably small, any shape of gradient can be accommodated. Thus it is possible to calculate sedimentation coefficients even if the gradient is not isokinetic or equivolumetric.

3.4 Computer Programs

The semi-manual technique of Funding and Steensgaard described elsewhere (8) was devised only for the calculation of sedimentation coefficients in zonal rotors. In principle, however, this technique can be applied to any type of rotor if the relationship between radius and volume is known. Computer programs based on this technique have been developed for the calculation of sedimentation coefficients in swing-out rotors (9) and in vertical rotors (10). Both programs use Equation 6 to calculate $s_{20,w}$ for each fraction of the sucrose gradient.

An implicit assumption in Equation 6 is that there is no interaction between the particle and the gradient medium. This is justified in the case of nucleic acids and sucrose, but may not be so with other macromolecules and gradient media. Although there is no simple analytical solution to Equation 6 the right-hand integral can be treated as the summation of a step function where each step corresponds to one of the fractions into which the gradient is unloaded. The sucrose concentration can be determined by measurement of the refractive index of each fraction. The viscosity and density of the medium as a function of the sucrose concentration and temperature is computed from the formulae of Barber (11). The volume-radius relationship can be computed from a knowledge of the tube and rotor for dimensions and hence the right-hand integral can be accurately approximated by a step function. The time integral in Equation 6 may be obtained directly from certain types of centrifuges. Alternatively, it may be approximated using Equation 10. The molecular weight (M) of macromolecules can also be computed from $s_{20,w}$ values using the empirical relationships that have been derived (*Table 1*).

The relationship between volume or fraction number and radius is relatively simple for swing-out rotors. However, in vertical rotors this relationship is more complicated and requires the following analysis. Consider a flat-topped tube of radius r_1, of cylindrical length l and whose centre is at a distance r_2 from the centre of rotation. For simplicity one assumes that the tube has a flat top and a hemispherical bottom. It is necessary to know the position of equal volume fractions when the gradient is orientated in the centrifugal field. To do this, one considers a vertical plane through a point at a distance x from the

Table 1. Empirical Relationships Between Sedimentation Coefficients and Molecular Weights of Macromolecules.

Macromolecule	Conditions	Relationship	Mol. wt. range[a]
Proteins	Neutral	$s = 0.00242\ M^{0.67}$	$10^4 - 10^7$
Native RNA	Neutral	$s = 2.61 + 0.022\ M^{0.43}$	–
Linear native DNA	Neutral	$s = 2.8 + 0.00834\ M^{0.479}$	$10^6 - 10^8$
Circular native DNA	Neutral	$s = 2.7 + 0.01759\ M^{0.445}$	$10^6 - 10^7$
Supercoiled native DNA	Neutral	$s = 7.44 + 0.00243\ M^{0.58}$	$10^6 - 10^7$
Single-stranded DNA	Neutral	$s = 0.0105\ M^{0.549}$	$10^6 - 10^8$
Single-stranded DNA	Alkaline	$s = 0.0528\ M^{0.4}$	$10^6 - 10^8$
Chromatin	Neutral	$s = 0.011\ M^{0.554}$	–

[a]Indicates the range of molecular weights for which the relationship has been tested; it may also be valid outside these ranges.

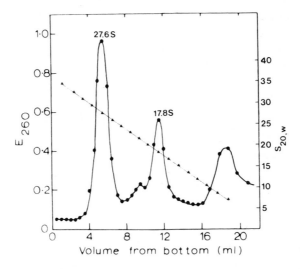

Figure 1. Fractionation of human rRNA by rate-zonal sedimentation through a $15 - 30\%$ (w/w) sucrose gradient. Human rRNA was dissolved in Buffer 1 (0.1 M NaCl, 10 mM Tris-HCl, pH 7.4, 1 mM EDTA, 0.5% SDS). Portions of this solution (1.0 ml) containing not more than 500 μg of RNA, were loaded onto 20 ml $15 - 30\%$ (w/w) sucrose gradients in Buffer 1 and centrifuged at 23 000 r.p.m. (55 000 g) for 17 h at 20°C in an MSE 3 x 25 ml swing-out rotor. The gradients were unloaded from the bottom by displacement with liquid paraffin and the percentage sucrose in each fraction was estimated from the refractive index. Sedimentation coefficients were computed using the computer program listed in Appendix B, assuming a particle density of 1.7 g/cm³. Data reproduced from ref. 9 with the permission of the authors and publishers.

wall of the tube on the radius from the centre of rotation through the centre of the tube. For the purpose of this calculation, one can define a term p which is equal to x/r_1 and therefore the value of p varies between 0 and 2. To calculate the total volume included by such a plane one uses the expression:

volume = volume (cylinder) + volume (hemisphere)

$$= [\cos^{-1}(1 - p) - (1 - p)(2p - p^2)^{1/2}] \cdot r_1^2 \cdot l + (3 - p) \cdot p^2 \cdot \pi \cdot r_1^3/6$$

Equation 11

where p is expressed in radians. For tubes with hemispherical tops, for example, the Beckman 'Quickseal' tubes, the second term is doubled. For practical purposes, it can be assumed that the MSE tube cap approximates to a flat surface. Hence it is possible to determine the volume as a function of radius. However, in order to solve Equation 6 it is necesary to know the radius as a function of volume. Equation 11 cannot easily be rearranged and therefore it is necessary to use a successive approximation technique to calculate p from a given volume. The complete progam is listed in Appendix C.

3.4.1 *Examples of the use of Computer Programs for the Calculation of Sedimentation Coefficients*

The listings of the computer programs for the calculation of sedimentation coefficients in swing-out and vertical rotors are given in Appendices B and C, respectively. The application of both programs is illustrated in the following sections and more complete information on the experiments can be obtained from Young and Krumlauf (9) and Young and Rickwood (10).

(i) *Swing-out rotors.* The accuracy of the calculation of sedimentation coefficients and molecular weight has been tested in swing-out rotors using both RNA and DNA. First, human rRNA was centrifuged on a linear $15 - 30\%$ (w/w) sucrose gradient and the computer program was used to calculate the sedimentation coefficients of each fraction (*Figure 1*). The calculated sedimentation coefficient of each peak is shown and these agree closely with the accepted sedimentation coefficients of 28S and 18S for human rRNA. This experiment has been repeated using different rotors with varying run times and speeds; in each case identical results were obtained.

As a further test of the program, the calculated size of DNA fragments after centrifugation was compared with that estimated by gel electrophoresis. Randomly sheared DNA was sedimented on a $5 - 25\%$ (w/w) sucrose gradient and a range of different size fractions taken. The program was used to estimate the molecular weight of each fraction. Five of the fractions were re-run on $5 - 25\%$ (w/w) sucrose gradients and the molecular weights of the peaks computed (*Figure 2*). A sample from each of 10 fractions was run on an agarose slab gel and the molecular weights were estimated by comparison of the positions of the marker bands (*Figure 3*). These results are summarised in *Table 2* and it can be seen that there is close agreement between the molecular weights estimated by the program and those found by gel electrophoresis.

(ii) *Vertical rotors.* The program described in Appendix C can be tested by centrifuging a macromolecule of known sedimentation coefficient in both vertical

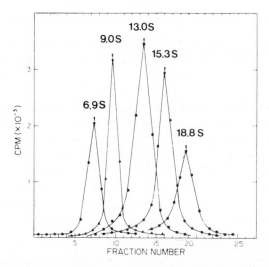

Figure 2. Size profile of isolated DNA fragments. DNA with a broad size distribution was prepared (9). Portions of this DNA solution (0.5 ml) were loaded onto 11.0 ml gradients of 5 – 25% (w/w) sucrose in Buffer 2 (1.0 M NaCl, 0.01 M Tris-HCl, pH 8.0, 0.005 M EDTA) and centrifuged at 27 500 r.p.m. (92 400 *g*)for 14.2 h at 20°C in a Beckman SW41 swing-out rotor. The DNA in each fraction was collected by precipitation with ethanol, and redissolved in 0.5 ml of 10 mM Tris-HCl (pH 7.8), 5 mM NaCl, 2 mM EDTA. Aliquots of five of these fractions were readjusted in Buffer 2 and 0.1 ml portions of these samples were rerun on 11.0 ml 5 – 25% (w/w) sucrose gradient in Buffer 2 at 28 750 r.p.m. (101 000 *g*)for 13.4 h at 20°C in a Beckman SW41 rotor. The sedimentation coefficients were determined using the computer program in Appendix B, assuming a particle density of 1.3 g/cm³. Data reproduced from ref. 9 with the permission of the authors and publishers.

Table 2. Comparison of Size Determination of DNA Fragment Classes.

Preparative gradient		Second gradient		Agarose gel
$s_{20,w}$	kbp[a]	$s_{20,w}$	kbp[a]	(kbp)
27.2	23.6	—	—	19.0
23.5	15.5	—	—	13.9
20.9	11.0	—	—	10.4
18.4	7.6	18.8	8.1	—
17.2	6.2	—	—	5.9
16.0	5.1	—	—	5.0
14.9	4.1	15.3	4.5	—
13.8	3.3	—	—	3.5
12.7	2.7	13.0	2.8	3.0
11.6	2.0	—	—	2.3
9.5	1.1	—	—	1.6
8.5	0.8	9.0	1.0	—
7.5	0.6	—	—	0.7
6.5	0.4	6.9	0.4	—

[a]Determinations using Studier's equations (12) and a conversion factor of 660 for one base pair.

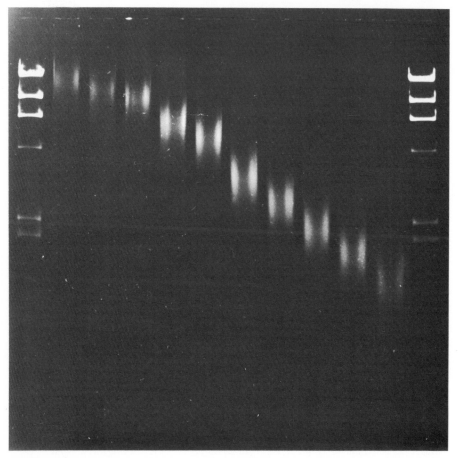

Figure 3. Electrophoresis of size-fractionated DNA fragments. DNA samples isolated from preparative sucrose gradients (*Figure 2*) were sized by electrophoresis on a 1% agarose gel and visualised by ethidium bromide staining (9). **Lanes 1** and **12** are *Hind*III digests of λ1857. **Lanes 2 – 11** represent DNA size clases of 19.0, 13.9, 10.4, 5.9, 5,0, 3.5, 3.0, 2.3, 1.6 and 0.7 kbp, respectively. Data reproduced from ref. 9 with permission of the authors and publishers.

and swing-out rotors. Catalase which has an $s_{20,w}$ value of 11.45 (13) was centrifuged on 5 – 20% (w/w) sucrose gradients in the MSE 6 x 38 ml swing-out rotor and in the MSE 8 x 35 ml vertical rotor. The results are shown in *Figure 4*. The sedimentation coefficients in the swing-out rotor were calcuated using the program described in Appendix B. The calculated $s_{20,w}$ values of each peak are 11.0S in the swing-out rotor and 11.7S in the vertical rotor. These values are close to the published value and confirm that the program in Appendix C is giving values comparable with those obtained from the swing-out rotor program. It is also clear from *Figure 4* that the band-width obtained in the vertical

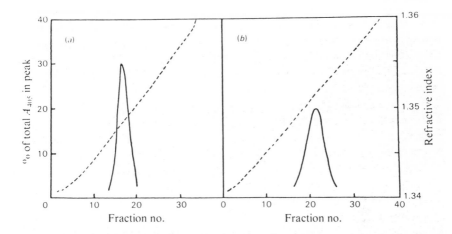

Figure 4. Sedimentation of catalase in a 5 − 20% (w/w) sucrose gradient. Gradients of 5 − 20% (w/w) sucrose were loaded with 1.0 ml of catalase and centrifuged in either **(A)** the MSE 6 x 38 ml rotor centrifuged at 24 000 r.p.m. (79 200 *g*) for 16.5 h at 22°C, or **(B)** the MSE 8 x 35 ml vertical rotor centrifuged at 45 000 r.p.m. (167 500 *g*) for 3.4 h at 22°C. Data reproduced from ref. 10 with the permission of the authors and publishers.

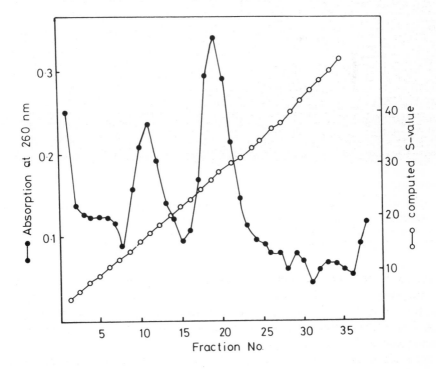

Figure 5. Sedimentation of rRNA in a 5 − 20% (w/w) sucrose gradient in the MSE 8 x 35 ml vertical rotor. Mouse cell rRNA was centrifuged in a 5 − 20% (w/w) sucrose gradient at 45 000 r.p.m. (167 500 *g*) for 60 min at 24°C. Data reproduced from ref. 10 with the permission of the authors and publishers.

rotor is greater than that of a comparable swing-out rotor. This is to be expected since the much greater surface area of the sample band during centrifugation enhances diffusion of the sample and, in addition, the reorientation of the gradient may cause some disturbance which is reflected in band broadening.

As a further test of the program, rRNA was centrifuged in a vertical rotor and the results are shown in *Figure 5*. The main peaks (28S and 18S) lie at positions calculated by the program to correspond to 27.1S and 17S, respectively. These small differences may be due to slight degradation of the RNA or to inaccurate temperature control. It is clear, however, that the program gives a reasonably accurate estimate considering the errors involved in unloading and fractionating such gradients.

During centrifugation in a vertical rotor, each zone is effectively orientated on the surface of a cylinder with a radius equal to the distance to the centre of rotation. However, the above analysis makes the approximation that the zones can be considered as flat surfaces. In fact, provided that the distance to the centre of rotation is sufficiently greater than the internal tube radius, this approximation does not introduce significant errors. It can be shown that with the vertical rotors currently in use the error is less than 3% on the final calculated sedimentation coefficients. Hence, in the program the formula in Equation 11 has been chosen in view of its greater simplicity.

4. SIMULATION TECHNIQUES

Frequently the centrifugation conditions used are based on those used in earlier work for similar samples. However, sometimes it may be necessary to use very different rotors or gradients different from those used originally. Alternatively, one may wish to try and improve the quality of the separation over that achieved in the original work by using a different type of gradient.

Using the computer programs given in Appendices B and C it is possible to simulate rate-zonal experiments, hence avoiding the necessity of empirical approaches. To do the simulation one simply chooses the type of centrifugation conditions that are to be tested in terms of gradient profile, rotor, speed, time and temperature; include in the input realistic values for times of acceleration and deceleration. Depending on whether the rotor to be used is a swing-out or vertical rotor use the program given in Appendix B or C, respectively. The computer will give a sedimentation profile of the gradient and, knowing the sedimentation coefficient of the actual particles, it is possible to determine exactly how far the particle would have sedimented down the gradient using these particular conditions. If the particle does not sediment sufficiently using the chosen conditions, then it is possible to do other simulations modifying either the speed or time of centrifugation; in this respect it should be remembered that changing the speed has a relatively greater effect than changing the length of the run (see Equation 1). Alternatively, it is possible to alter the steepness of the gradient or to use concave or convex gradients, prepared as described in Section 5 of Chapter 2, in order to optimise the resolution of particles of particular sizes.

5. TROUBLESHOOTING

This section will not discuss the general problems of rate-zonal centrifugation but rather it will deal specifically with those factors which can affect the accuracy of the calculation of sedimentation coefficients. However, it is important to realise that the estimates of sedimentation coefficients calculated using the method described in this chapter will not usually be as accurate as that obtained by analytical centrifugation as described in Chapter 8.

If, after carrying out a rate-zonal experiment, calculation reveals an unexpected result then it is possible that errors have arisen either in doing the experimental work or in the subsequent calculation. Experimental errors can arise from a number of sources but perhaps the most important of these is that of temperature. As the temperature decreases, the viscosity of sucrose solutions increases rapidly, for example, the viscosity of a 10% (w/w) sucrose solution changes by more than 15% between 5°C and 10°C. Most preparative ultracentrifuges use infar-red sensors to measure and regulate the temperature of rotors. However, in practice most preparative centrifuges regulate the temperature only to within 1°C. An additional problem is that the temperature calibration of the instrument may be inaccurate and hence the rotor will be maintained at an incorrect temperature. This latter problem can be checked by measuring the temperature of a blank gradient using a pre-equilibrated thermometer at the end of the run. Both problems can be overcome by running a standard marker of known sedimentation coefficient and, by measuring the observed sedimentation coefficient, it is possible to calculate the average temperature during the centrifugation run. The other problem is that of inaccuracy of speed measurement. The speed of most d.c. motor-driven machines is only accurate to within 1% of the set value and hence it is more accurate to use the odometer readings at the beginning and end of the run to obtain a completely accurate estimate of the total number of revolutions. In the case of induction-drive machines the speed control is very much more accurate and hence this source of error should be negligible.

Most of the other errors involved in the estimation of sedimentation coefficients are human errors. In the case of manual methods, arithmetical errors may arise, while estimations using the computer program errors are restricted to the incorrect input of data. Errors of input may be as a result of incorrect centrifugation data being used or incorrect estimation of the sucrose concentrations of fractions. The other possible source of error is the assignment of the density of the particles. The density value that should be used for calculations is the density of the particle in the sucrose gradient solution. The precise densities of particles in sucrose gradients are not known, usually values of 1.3 g/cm^3 are used for protein and DNA while the value used for RNA is usually 1.7 g/cm^3. However, putting various density values into the computer program reveals that it is not critical to use a very exact value for the density in order to obtain a reasonably accurate value for the sedimentation coefficient.

6. REFERENCES

1. Svedberg,T. and Pederson,K.O. (1940) *The Ultracentrifuge*, published by Clarendon Press.
2. McEwen,C.R. (1967) *Anal. Biochem.,* **20**, 114.
3. Martin,R.G. and Ames,B.N. (1961) *J. Biol. Chem.,* **236**, 1372.
4. Burgi,E. and Hershey,A.D. (1963) *Biophys. J.,* **3**, 309.
5. Fritsch,A. (1973) *Anal. Biochem.,* **55**, 57.
6. Noll,H. (1967) *Nature,* **215**, 360.
7. Pollack,M.S. and Price,C.A. (1971) *Anal. Biochem.,* **42**, 38.
8. Steensgaard,J., Møller,N.P.H. and Funding,L. (1978) in *Centrifugal Separations in Molecular and Cell Biology*, Birnie,G.D. and Rickwood,D. (eds.), Butterworths, London, p. 115.
9. Young,B.D. and Krumlauf,R. (1981) *Anal. Biochem.,* **115**, 97.
10. Young,B.D. and Rickwood,D. (1981) *J. Biochem. Biophys. Methods,* **5**, 95.
11. Barber,E.J. (1966) *Natl. Cancer Inst. Monogr.,* **21**, 219.
12. Studier,F.W. (1965) *J. Mol. Biol.,* **11**, 373.
13. Barlow,G.H. and Margoliash,E. (1969) *Biochem. Biophys. Acta,* **188**, 159.

APPENDIX A

Pre-computed Tables for Calculating Sedimentation Coefficients

SUCROSE GRADIENT CENTRIFUGATION

TIME INTEGRAL VALUES FOR PARTICLE DENSITY 1.1 AT TEMPERATURE 5.0 DEG C

PERCENTAGE SUCROSE	ZO= 0.0	ZO= -1.0	ZO= -2.0	ZO= -3.0	ZO= -4.0	ZO= -5.0	ZO= -6.0	ZO= -7.0	ZO= -8.0	ZO= -9.0	ZO= -10.0
0	0.0000	0.0000	0.0000	0.0000	0.0000	0.0000	0.0000	0.0000	0.0000	0.0000	0.0000
2	0.0000	1.8010	1.1317	0.8348	0.6632	0.5508	0.4711	0.4117	0.3657	0.3290	0.2989
4	1.3045	2.7634	1.8964	1.4699	1.2065	1.0256	0.8929	0.7911	0.7105	0.6649	0.5905
6	2.1979	3.5054	2.5312	2.0247	1.6993	1.4689	1.2957	1.1603	1.0512	0.9613	0.8858
8	2.9510	4.1637	3.1159	2.5507	2.1774	1.9070	1.7001	1.5358	1.4016	1.2897	1.1949
10	3.6538	4.8004	3.6946	3.0810	2.6668	2.3615	2.1242	1.9334	1.7758	1.6432	1.5297
12	4.3747	5.4565	4.3002	3.6434	3.1917	2.8535	2.5873	2.3707	2.1901	2.0368	1.9046
14	5.1488	6.1753	4.9711	4.2723	3.7836	3.4126	3.1170	2.8739	2.6694	2.4943	2.3422
16	6.0420	7.0128	5.7593	5.0168	4.4889	4.0826	3.7552	3.4831	3.2521	3.0528	2.8784
18	7.1627	8.0712	6.7621	5.9695	5.3964	4.9489	4.5839	4.2773	4.0146	3.7860	3.5845
20	8.7619	9.5907	8.2095	7.3513	6.7182	6.2158	5.8002	5.4470	5.1411	4.8723	4.6334

SUCROSE GRADIENT CENTRIFUGATION

TIME INTEGRAL VALUES FOR PARTICLE DENSITY 1.2 AT TEMPERATURE 5.0 DEG C

PERCENTAGE SUCROSE	ZO= 0.0	ZO= -1.0	ZO= -2.0	ZO= -3.0	ZO= -4.0	ZO= -5.0	ZO= -6.0	ZO= -7.0	ZO= -8.0	ZO= -9.0	ZO= -10.0
0	0.0000	0.0000	0.0000	0.0000	0.0000	0.0000	0.0000	0.0000	0.0000	0.0000	0.0000
2	0.0000	1.7571	1.1023	0.8126	0.6453	0.5358	0.4582	0.4004	0.3556	0.3198	0.2906
4	1.2156	2.6532	1.8141	1.4035	1.1508	0.9774	0.8505	0.7532	0.6762	0.6136	0.5618
6	2.0052	3.3086	2.3748	1.8935	1.5859	1.3688	1.2062	1.0792	0.9770	0.8929	0.8224
8	2.6312	3.8559	2.8607	2.3305	1.9931	1.7328	1.5420	1.3910	1.2680	1.1657	1.0791
10	3.1777	4.3474	3.3073	2.7799	2.3608	2.0835	1.8693	1.6978	1.5567	1.4384	1.3374
12	3.6834	4.8108	3.7350	3.1369	2.7314	2.4309	2.1962	2.0065	1.8492	1.7162	1.6020
14	4.1718	5.2643	4.1582	3.5536	3.1047	2.7634	2.5302	2.3238	2.1514	2.0046	1.8779
16	4.6590	5.7210	4.5880	3.9395	3.4892	3.1187	2.8781	2.6659	2.4690	2.3090	2.1701
18	5.1588	6.1930	5.0352	4.3643	3.8938	3.5349	3.2475	3.0099	2.8089	2.6358	2.4848
20	5.6845	6.6924	5.5108	4.8183	4.3280	3.9510	3.6470	3.3940	3.1787	2.9925	2.8292

SUCROSE GRADIENT CENTRIFUGATION

TIME INTEGRAL VALUES FOR PARTICLE DENSITY 1.3 AT TEMPERATURE 5.0 DEG C

PERCENTAGE SUCROSE	ZO= 0.0	ZO= -1.0	ZO= -2.0	ZO= -3.0	ZO= -4.0	ZO= -5.0	ZO= -6.0	ZO= -7.0	ZO= -8.0	ZO= -9.0	ZO= -10.0
0	0.0000	0.0000	0.0000	0.0000	0.0000	0.0000	0.0000	0.0000	0.0000	0.0000	0.0000
2	0.0000	1.7429	1.0929	0.8054	0.6396	0.5309	0.4541	0.3968	0.3524	0.3169	0.2880
4	1.1805	2.6199	1.7835	1.3829	1.1335	0.9625	0.8373	0.7415	0.6656	0.6040	0.5529
6	1.9485	3.2498	2.3281	1.8545	1.5522	1.3392	1.1776	1.0551	0.9550	0.8727	0.8036
8	2.5410	3.7676	2.7879	2.2680	1.9280	1.6835	1.4974	1.3502	1.2304	1.1308	1.0465
10	3.0487	4.2242	3.2029	2.6423	2.2789	2.0093	1.8014	1.6352	1.4986	1.3841	1.2865
12	3.5091	4.6461	3.5922	3.0047	2.6163	2.3255	2.0990	1.9162	1.7648	1.6370	1.5273
14	3.9437	5.0496	3.9668	3.3627	2.9485	2.6382	2.3962	2.1985	2.0337	1.8936	1.7728
16	4.3666	5.4460	4.3418	3.7150	3.2822	2.9563	2.6981	2.4867	2.3093	2.1578	2.0264
18	4.7882	5.8442	4.7191	4.0734	3.6235	3.2821	3.0098	2.7853	2.5960	2.4334	2.2918
20	5.2176	6.2521	5.1075	4.4442	3.9782	3.6220	3.3361	3.0991	2.8982	2.7248	2.5731

SUCROSE GRADIENT CENTRIFUGATION

TIME INTEGRAL VALUES FOR PARTICLE DENSITY 1.4 AT TEMPERATURE 5.0 DEG C

PERCENTAGE SUCROSE	ZO= 0.0	ZO= -1.0	ZO= -2.0	ZO= -3.0	ZO= -4.0	ZO= -5.0	ZO= -6.0	ZO= -7.0	ZO= -8.0	ZO= -9.0	ZO= -10.0
0	0.0000	0.0000	0.0000	0.0000	0.0000	0.0000	0.0000	0.0000	0.0000	0.0000	0.0000
2	0.0000	1.7359	1.0882	0.8019	0.6367	0.5286	0.4520	0.3950	0.3508	0.3155	0.2867
4	1.1754	2.6021	1.7760	1.3729	1.1251	0.9552	0.8310	0.7358	0.6604	0.5993	0.5485
6	1.9214	3.2214	2.3057	1.8357	1.5361	1.3249	1.1669	1.0436	0.9445	0.8630	0.7946
8	2.4904	3.7256	2.7534	2.2304	1.9020	1.6602	1.4763	1.3309	1.2126	1.1143	1.0312
10	2.9886	4.1665	3.1541	2.6056	2.2408	1.9749	1.7699	1.6061	1.4716	1.3589	1.2628
12	3.4293	4.5703	3.5267	2.9515	2.5637	2.2774	2.0547	1.8750	1.7264	1.6009	1.4933
14	3.8412	4.9527	3.8836	3.2660	2.8785	2.5746	2.3363	2.1426	1.9812	1.8441	1.7260
16	4.2377	5.3245	4.2293	3.6164	3.1915	2.8721	2.6195	2.4129	2.2397	2.0910	1.9638
18	4.6288	5.6937	4.5833	3.9488	3.5080	3.1742	2.9084	2.6898	2.5055	2.3475	2.2099
20	5.0221	6.0674	4.9391	4.2884	3.8329	3.4855	3.2073	2.9772	2.7823	2.6143	2.4676

SUCROSE GRADIENT CENTRIFUGATION

TIME INTEGRAL VALUES FOR PARTICLE DENSITY 1.5 AT TEMPERATURE 5.0 DEG C

PERCENTAGE SUCROSE	ZO= 0.0	ZO= -1.0	ZO= -2.0	ZO= -3.0	ZO= -4.0	ZO= -5.0	ZO= -6.0	ZO= -7.0	ZO= -8.0	ZO= -9.0	ZO=-10.0
0	0.0000	0.0000	0.0000	0.0000	0.0000	0.0000	0.0000	0.0000	0.0000	0.0000	0.0000
2	0.0000	1.7317	1.0854	0.7998	0.6350	0.5271	0.4508	0.3939	0.3498	0.3146	0.2859
4	1.1677	2.5922	1.7686	1.3669	1.1201	0.9509	0.8272	0.7324	0.6574	0.5965	0.5460
6	1.9055	3.2047	2.2924	1.8247	1.5265	1.3165	1.1594	1.0368	0.9383	0.8573	0.7894
8	2.4736	3.7011	2.7333	2.2211	1.8868	1.6467	1.4640	1.3197	1.2023	1.1047	1.0222
10	2.9539	4.1331	3.1258	2.5809	2.2188	1.9549	1.7517	1.5893	1.4560	1.3443	1.2492
12	3.3834	4.5267	3.4891	2.9181	2.5335	2.2499	2.0293	1.8514	1.7043	1.5802	1.4739
14	3.7828	4.8974	3.8351	3.2424	2.8387	2.5381	2.3023	2.1108	1.9514	1.8160	1.6994
16	4.1651	5.2558	4.1723	3.5609	3.1404	2.8247	2.5753	2.3714	2.2006	2.0549	1.9287
18	4.5398	5.6097	4.5075	3.8794	3.4437	3.1142	2.8522	2.6367	2.4553	2.2998	2.1645
20	4.9142	5.9653	4.8463	4.2027	3.7530	3.4106	3.1367	2.9103	2.7187	2.5538	2.4098

SUCROSE GRADIENT CENTRIFUGATION

TIME INTEGRAL VALUES FOR PARTICLE DENSITY 1.6 AT TEMPERATURE 5.0 DEG C

PERCENTAGE SUCROSE	ZO= 0.0	ZO= -1.0	ZO= -2.0	ZO= -3.0	ZO= -4.0	ZO= -5.0	ZO= -6.0	ZO= -7.0	ZO= -8.0	ZO= -9.0	ZO=-10.0
0	0.0000	0.0000	0.0000	0.0000	0.0000	0.0000	0.0000	0.0000	0.0000	0.0000	0.0000
2	0.0000	1.7290	1.0835	0.7984	0.6339	0.5262	0.4500	0.3932	0.3492	0.3140	0.2854
4	1.1625	2.5856	1.7637	1.3630	1.1168	0.9481	0.8247	0.7302	0.6554	0.5946	0.5443
6	1.8950	3.1937	2.2837	1.8174	1.5203	1.3110	1.1544	1.0324	0.9342	0.8535	0.7859
8	2.4573	3.6850	2.7201	2.2098	1.8769	1.6378	1.4560	1.3123	1.1955	1.0984	1.0163
10	2.9312	4.1102	3.1074	2.5648	2.2044	1.9419	1.7398	1.5783	1.4458	1.3348	1.2403
12	3.3537	4.4983	3.4646	2.8964	2.5139	2.2320	2.0128	1.8362	1.6901	1.5669	1.4613
14	3.7451	4.8617	3.8037	3.2143	2.8131	2.5145	2.2804	2.0904	1.9322	1.7979	1.6823
16	4.1184	5.2117	4.1331	3.5253	3.1077	2.7944	2.5470	2.3448	2.1756	2.0312	1.9062
18	4.4829	5.5559	4.4592	3.8351	3.4027	3.0760	2.8163	2.6029	2.4233	2.2694	2.1356
20	4.8457	5.9005	4.7874	4.1484	3.7024	3.3632	3.0920	2.8680	2.6786	2.5156	2.3733

143

SUCROSE GRADIENT CENTRIFUGATION

TIME INTEGRAL VALUES FOR PARTICLE DENSITY 1.7 AT TEMPERATURE 5.0 DEG C

PERCENTAGE SUCROSE	ZO= 0.0	ZO= -1.0	ZO= -2.0	ZO= -3.0	ZO= -4.0	ZO= -5.0	ZO= -6.0	ZO= -7.0	ZO= -8.0	ZO= -9.0	ZO= -10.0
0	0.0000	0.0000	0.0000	0.0000	0.0000	0.0000	0.0000	0.0000	0.0000	0.0000	0.0000
2	0.0000	1.7270	1.0822	0.7974	0.6331	0.5255	0.4494	0.3927	0.3487	0.3136	0.2850
4	1.1562	2.5809	1.7603	1.3602	1.1144	0.9461	0.8229	0.7286	0.6539	0.5933	0.5431
6	1.8876	3.1858	2.2776	1.8122	1.5159	1.3071	1.1510	1.0292	0.9314	0.8509	0.7834
8	2.4453	3.6736	2.7107	2.2018	1.8698	1.6315	1.4503	1.3071	1.1907	1.0940	1.0122
10	2.9153	4.0958	3.0944	2.5534	2.1943	1.9327	1.7314	1.5706	1.4387	1.3282	1.2341
12	3.3328	4.4784	3.4475	2.8812	2.5002	2.2195	2.0013	1.8254	1.6801	1.5575	1.4524
14	3.7187	4.8367	3.7818	3.1946	2.7951	2.4980	2.2651	2.0761	1.9188	1.7853	1.6704
16	4.0859	5.1809	4.1057	3.5005	3.0849	2.7733	2.5273	2.3263	2.1581	2.0147	1.8906
18	4.4435	5.5186	4.4256	3.8044	3.3743	3.0495	2.7915	2.5795	2.4012	2.2484	2.1156
20	4.7984	5.8557	4.7467	4.1109	3.6675	3.3304	3.0612	2.8388	2.6509	2.4892	2.3481

SUCROSE GRADIENT CENTRIFUGATION

TIME INTEGRAL VALUES FOR PARTICLE DENSITY 1.8 AT TEMPERATURE 5.0 DEG C

PERCENTAGE SUCROSE	ZO= 0.0	ZO= -1.0	ZO= -2.0	ZO= -3.0	ZO= -4.0	ZO= -5.0	ZO= -6.0	ZO= -7.0	ZO= -8.0	ZO= -9.0	ZO= -10.0
0	0.0000	0.0000	0.0000	0.0000	0.0000	0.0000	0.0000	0.0000	0.0000	0.0000	0.0000
2	0.0300	1.7255	1.0812	0.7967	0.6325	0.5250	0.4490	0.3923	0.3484	0.3133	0.2847
4	1.1562	2.5775	1.7577	1.3581	1.1127	0.9446	0.8216	0.7274	0.6529	0.5923	0.5422
6	1.8821	3.1800	2.2730	1.8084	1.5125	1.3042	1.1484	1.0269	0.9292	0.8489	0.7816
8	2.4373	3.6651	2.7038	2.1959	1.8646	1.6268	1.4461	1.3033	1.1871	1.0907	1.0091
10	2.9035	4.0848	3.0848	2.5450	2.1868	1.9259	1.7252	1.5649	1.4334	1.3232	1.2294
12	3.3174	4.4637	3.4348	2.8699	2.4900	2.2102	1.9927	1.8175	1.6727	1.5505	1.4459
14	3.6992	4.8182	3.7656	3.1801	2.7819	2.4858	2.2538	2.0656	1.9089	1.7760	1.6616
16	4.0612	5.1582	4.0856	3.4822	3.0681	2.7577	2.5128	2.3127	2.1453	2.0026	1.8791
18	4.4145	5.4911	4.4010	3.7818	3.3535	3.0301	2.7733	2.5624	2.3850	2.2330	2.1010
20	4.7637	5.8229	4.7168	4.0834	3.6419	3.3065	3.0386	2.8175	2.6307	2.4699	2.3297

SUCROSE GRADIENT CENTRIFUGATION

TIME INTEGRAL VALUES FOR PARTICLE DENSITY 1.9 AT TEMPERATURE 5.0 DEG C

PERCENTAGE SUCROSE	Z0= 0.0	Z0= -1.0	Z0= -2.0	Z0= -3.0	Z0= -4.0	Z0= -5.0	Z0= -6.0	Z0= -7.0	Z0= -8.0	Z0= -9.0	Z0= -10.0
0	0.0000	0.0000	0.0000	0.0000	0.0000	0.0000	0.0000	0.0000	0.0000	0.0000	0.0000
2	0.0000	1.7244	1.0805	0.7961	0.6320	0.5246	0.4487	0.3920	0.3481	0.3131	0.2845
4	1.1541	2.5747	1.7557	1.3565	1.1113	0.9434	0.8205	0.7265	0.6520	0.5916	0.5415
6	1.8779	3.1755	2.2694	1.8054	1.5100	1.3020	1.1463	1.0250	0.9275	0.8474	0.7802
8	2.4307	3.6586	2.6984	2.1913	1.8606	1.6232	1.4428	1.3003	1.1844	1.0861	1.0068
10	2.8943	4.0756	3.0773	2.5385	2.1810	1.9207	1.7204	1.5605	1.4293	1.3194	1.2258
12	3.3055	4.4523	3.4250	2.8613	2.4822	2.2030	1.9861	1.8114	1.6670	1.5452	1.4409
14	3.6843	4.8040	3.7532	3.1689	2.7717	2.4765	2.2451	2.0575	1.9013	1.7689	1.6548
16	4.0435	5.1408	4.0701	3.4682	3.0553	2.7458	2.5017	2.3023	2.1355	1.9933	1.8703
18	4.3923	5.4701	4.3821	3.7646	3.3375	3.0152	2.7593	2.5492	2.3726	2.2212	2.0897
20	4.7372	5.7977	4.6941	4.0624	3.6224	3.2882	3.0214	2.8012	2.6152	2.4552	2.3157

SUCROSE GRADIENT CENTRIFUGATION

TIME INTEGRAL VALUES FOR PARTICLE DENSITY 2.0 AT TEMPERATURE 5.0 DEG C

PERCENTAGE SUCROSE	Z0= 0.0	Z0= -1.0	Z0= -2.0	Z0= -3.0	Z0= -4.0	Z0= -5.0	Z0= -6.0	Z0= -7.0	Z0= -8.0	Z0= -9.0	Z0= -10.0
0	0.0000	0.0000	0.0000	0.0000	0.0000	0.0000	0.0000	0.0000	0.0000	0.0000	0.0000
2	0.0000	1.7235	1.0799	0.7956	0.6317	0.5243	0.4484	0.3918	0.3479	0.3129	0.2843
4	1.1526	2.5726	1.7541	1.3552	1.1103	0.9425	0.8197	0.7258	0.6514	0.5910	0.5409
6	1.8745	3.1719	2.2666	1.8031	1.5080	1.3002	1.1447	1.0236	0.9262	0.8461	0.7790
8	2.4255	3.6534	2.6941	2.1876	1.8574	1.6204	1.4402	1.2979	1.1822	1.0861	1.0049
10	2.8371	4.0686	3.0714	2.5333	2.1764	1.9166	1.7166	1.5570	1.4260	1.3194	1.2230
12	3.2960	4.4433	3.4172	2.8544	2.4760	2.1974	1.9809	1.8066	1.6625	1.5410	1.4369
14	3.6724	4.7927	3.7433	3.1600	2.7637	2.4690	2.2383	2.0510	1.8953	1.7632	1.6494
16	4.0290	5.1216	4.0579	3.4571	3.0451	2.7364	2.4929	2.2941	2.1277	1.9933	1.8633
18	4.3747	5.4535	4.3672	3.7509	3.3249	3.0035	2.7483	2.5389	2.3627	2.2119	2.0809
20	4.7162	5.7779	4.6762	4.0459	3.6070	3.2738	3.0078	2.7884	2.6030	2.4436	2.3046

SUCROSE GRADIENT CENTRIFUGATION

TIME INTEGRAL VALUES FOR PARTICLE DENSITY 1.1 AT TEMPERATURE 20.0 DEG C

PERCENTAGE SUCROSE	ZO= 0.0	ZO= -1.0	ZO= -2.0	ZO= -3.0	ZO= -4.0	ZO= -5.0	ZO= -6.0	ZO= -7.0	ZO= -8.0	ZO= -9.0	ZO= -10.0
0	0.0000	0.0000	0.0000	0.0000	0.0000	0.0000	0.0000	0.0000	0.0000	0.0000	0.0000
2	0.0000	1.1704	0.7353	0.5423	0.4308	0.3577	0.3060	0.2674	0.2375	0.2137	0.1942
4	0.8405	1.7904	1.2279	0.9514	0.7807	0.6636	0.5776	0.5118	0.4596	0.4171	0.3819
6	1.4100	2.2634	1.6325	1.3050	1.0949	0.9461	0.8344	0.7471	0.6767	0.6188	0.5701
8	1.8847	2.6783	2.0010	1.6663	1.3961	1.2222	1.0892	0.9837	0.8975	0.8257	0.7649
10	2.3248	3.0742	2.3605	1.9663	1.7005	1.5040	1.3529	1.2309	1.1302	1.0455	0.9731
12	2.7633	3.4761	2.7317	2.3107	2.0219	1.8061	1.6365	1.4987	1.3839	1.2865	1.2027
14	3.2273	3.9069	3.1300	2.6877	2.3767	2.1412	1.9540	1.8003	1.6712	1.5607	1.4649
16	3.7488	4.3958	3.5940	3.1223	2.7884	2.5324	2.3266	2.1560	2.0114	1.8868	1.7779
18	4.3785	4.9906	4.1575	3.6576	3.2983	3.0191	2.7922	2.6022	2.4398	2.2987	2.1746
20	5.2198	5.7899	4.9183	4.3844	3.9936	3.6855	3.4319	3.2174	3.0322	2.8700	2.7263

SUCROSE GRADIENT CENTRIFUGATION

TIME INTEGRAL VALUES FOR PARTICLE DENSITY 1.2 AT TEMPERATURE 20.0 DEG C

PERCENTAGE SUCROSE	ZO= 0.0	ZO= -1.0	ZO= -2.0	ZO= -3.0	ZO= -4.0	ZO= -5.0	ZO= -6.0	ZO= -7.0	ZO= -8.0	ZO= -9.0	ZO= -10.0
0	0.0000	0.0000	0.0000	0.0000	0.0000	0.0000	0.0000	0.0000	0.0000	0.0000	0.0000
2	0.0000	1.1520	0.7226	0.5326	0.4230	0.3511	0.3003	0.2624	0.2330	0.2096	0.1905
4	0.7922	1.7359	1.1863	0.9176	0.7522	0.6389	0.5559	0.4923	0.4419	0.4010	0.3671
6	1.3031	2.1601	1.5491	1.2346	1.0338	0.8921	0.7860	0.7031	0.6365	0.5817	0.5357
8	1.7052	2.5115	1.8612	1.5153	1.2888	1.1258	1.0017	0.9034	0.8234	0.7569	0.7006
10	2.0533	2.8246	2.1457	1.7761	1.5295	1.3493	1.2102	1.0989	1.0073	0.9306	0.8651
12	2.3728	3.1174	2.4159	2.0270	1.7636	1.5687	1.4167	1.2939	1.1921	1.1061	1.0323
14	2.6783	3.4010	2.6806	2.2751	1.9971	1.7892	1.6256	1.4923	1.3811	1.2865	1.2048
16	2.9799	3.6838	2.9467	2.5264	2.2352	2.0154	1.8410	1.6979	1.5759	1.4750	1.3858
18	3.2861	3.9729	3.2206	2.7866	2.4830	2.2519	2.0672	1.9148	1.7859	1.6751	1.5785
20	3.6047	4.2751	3.5084	3.0613	2.7458	2.5038	2.3090	2.1472	2.0097	1.8909	1.7869

SUCROSE GRADIENT CENTRIFUGATION

TIME INTEGRAL VALUES FOR PARTICLE DENSITY 1.3 AT TEMPERATURE 20.0 DEG C

PERCENTAGE SUCROSE	ZO= 0.0	ZO= -1.0	ZO= -2.0	ZO= -3.0	ZO= -4.0	ZO= -5.0	ZO= -6.0	ZO= -7.0	ZO= -8.0	ZO= -9.0	ZO= -10.0
0	0.0000	0.0000	0.0000	0.0000	0.0000	0.0000	0.0000	0.0000	0.0000	0.0000	0.0000
2	0.0000	1.1460	0.7184	0.5294	0.4204	0.3490	0.2985	0.2608	0.2316	0.2083	0.1893
4	0.7772	1.7188	1.1733	0.9070	0.7433	0.6311	0.5490	0.4862	0.4364	0.3960	0.3625
6	1.2711	2.1288	1.5239	1.2134	1.0154	0.8759	0.7714	0.6899	0.6244	0.5706	0.5254
8	1.6536	2.4630	1.8207	1.4804	1.2580	1.0982	0.9766	0.8804	0.8022	0.7371	0.6822
10	1.9782	2.7557	2.0867	1.7241	1.4829	1.3070	1.1714	1.0630	0.9741	0.8995	0.8359
12	2.2719	3.0240	2.3343	1.9540	1.6975	1.5081	1.3607	1.2418	1.1434	1.0603	0.9891
14	2.5459	3.2735	2.5715	2.1766	1.9069	1.7059	1.5481	1.4198	1.3129	1.2222	1.1439
16	2.8102	3.5262	2.8050	2.3968	2.1156	1.9041	1.7368	1.6000	1.4853	1.3873	1.3024
18	3.0714	3.7729	3.0387	2.6188	2.3270	2.1059	1.9298	1.7849	1.6628	1.5580	1.4669
20	3.3343	4.0231	3.2769	2.8462	2.5445	2.3144	2.1300	1.9774	1.8481	1.7367	1.6394

SUCROSE GRADIENT CENTRIFUGATION

TIME INTEGRAL VALUES FOR PARTICLE DENSITY 1.4 AT TEMPERATURE 20.0 DEG C

PERCENTAGE SUCROSE	ZO= 0.0	ZO= -1.0	ZO= -2.0	ZO= -3.0	ZO= -4.0	ZO= -5.0	ZO= -6.0	ZO= -7.0	ZO= -8.0	ZO= -9.0	ZO= -10.0
0	0.0000	0.0000	0.0000	0.0000	0.0000	0.0000	0.0000	0.0000	0.0000	0.0000	0.0000
2	0.0000	1.1430	0.7164	0.5279	0.4191	0.3479	0.2975	0.2600	0.2309	0.2076	0.1887
4	0.7699	1.7104	1.1669	0.9018	0.7390	0.6274	0.5457	0.4832	0.4337	0.3935	0.3602
6	1.2557	2.1137	1.5118	1.2032	1.0066	0.8681	0.7644	0.6836	0.6186	0.5652	0.5204
8	1.6290	2.4399	1.8015	1.4637	1.2433	1.0850	0.9646	0.8695	0.7921	0.7278	0.6734
10	1.9441	2.7232	2.0590	1.6997	1.4611	1.2872	1.1533	1.0463	0.9585	0.8850	0.8223
12	2.2252	2.9808	2.2967	1.9204	1.6670	1.4802	1.3350	1.2179	1.1210	1.0394	0.9694
14	2.4954	3.2228	2.5225	2.1321	1.8662	1.6684	1.5132	1.3872	1.2823	1.1933	1.1166
16	2.7347	3.4561	2.7421	2.3354	2.0627	1.8550	1.6909	1.5568	1.4445	1.3467	1.2658
18	2.9780	3.6859	2.9597	2.5462	2.2596	2.0429	1.8707	1.7291	1.6099	1.5078	1.4190
20	3.2205	3.9162	3.1791	2.7556	2.4598	2.2348	2.0549	1.9062	1.7805	1.6723	1.5778

SUCROSE GRADIENT CENTRIFUGATION

TIME INTEGRAL VALUES FOR PARTICLE DENSITY 1.5 AT TEMPERATURE 20.0 DEG C

PERCENTAGE SUCROSE	Z0= 0.0	Z0= -1.0	Z0= -2.0	Z0= -3.0	Z0= -4.0	Z0= -5.0	Z0= -6.0	Z0= -7.0	Z0= -8.0	Z0= -9.0	Z0= -10.0
0	0.0000	0.0000	0.0000	0.0000	0.0000	0.0000	0.0000	0.0000	0.0000	0.0000	0.0000
2	0.0000	1.1413	0.7152	0.5270	0.4184	0.3473	0.2970	0.2595	0.2304	0.2073	0.1883
4	0.7656	1.7054	1.1631	0.8988	0.7364	0.6251	0.5437	0.4814	0.4321	0.3920	0.3588
6	1.2467	2.1047	1.5046	1.1972	1.0013	0.8634	0.7603	0.6798	0.6152	0.5620	0.5175
8	1.6147	2.4263	1.7902	1.4540	1.2347	1.0773	0.9577	0.8631	0.7862	0.7223	0.6683
10	1.9239	2.7044	2.0429	1.6856	1.4481	1.2757	1.1425	1.0366	0.9495	0.8765	0.8144
12	2.1943	2.9559	2.2750	1.9010	1.6495	1.4642	1.3202	1.2041	1.1082	1.0273	0.9580
14	2.4515	3.1910	2.4943	2.1067	1.8430	1.6469	1.4933	1.3686	1.2648	1.1768	1.1010
16	2.6920	3.4164	2.7065	2.3070	2.0328	1.8272	1.6650	1.5325	1.4216	1.3270	1.2452
18	2.9257	3.6370	2.9155	2.5056	2.2219	2.0077	1.8376	1.6979	1.5804	1.4797	1.3923
20	3.1571	3.8569	3.1249	2.7054	2.4131	2.1909	2.0135	1.8670	1.7432	1.6367	1.5438

SUCROSE GRADIENT CENTRIFUGATION

TIME INTEGRAL VALUES FOR PARTICLE DENSITY 1.6 AT TEMPERATURE 20.0 DEG C

PERCENTAGE SUCROSE	Z0= 0.0	Z0= -1.0	Z0= -2.0	Z0= -3.0	Z0= -4.0	Z0= -5.0	Z0= -6.0	Z0= -7.0	Z0= -8.0	Z0= -9.0	Z0= -10.0
0	0.0000	0.0000	0.0000	0.0000	0.0000	0.0000	0.0000	0.0000	0.0000	0.0000	0.0000
2	0.0000	1.1401	0.7144	0.5263	0.4179	0.3369	0.2966	0.2592	0.2302	0.2070	0.1881
4	0.7628	1.7021	1.1606	0.8967	0.7347	0.6236	0.5424	0.4802	0.4310	0.3911	0.3579
6	1.2407	2.0988	1.4999	1.1932	0.9979	0.8604	0.7576	0.6774	0.6130	0.5600	0.5156
8	1.6053	2.4174	1.7823	1.4476	1.2291	1.0723	0.9531	0.8589	0.7823	0.7188	0.6650
10	1.9106	2.6920	2.0323	1.6763	1.4401	1.2682	1.1359	1.0303	0.9436	0.8711	0.8093
12	2.1809	2.9397	2.2609	1.8845	1.6381	1.4538	1.3106	1.1952	1.0999	1.0195	0.9506
14	2.4293	3.1703	2.4761	2.0902	1.8250	1.6331	1.4804	1.3566	1.2536	1.1662	1.0909
16	2.6645	3.3907	2.6835	2.2861	2.0136	1.8094	1.6483	1.5168	1.4068	1.3131	1.2319
18	2.8922	3.6058	2.8872	2.4796	2.1979	1.9853	1.8166	1.6781	1.5616	1.4619	1.3752
20	3.1167	3.8191	3.0904	2.6736	2.3834	2.1631	1.9872	1.8421	1.7196	1.6142	1.5223

148

SUCROSE GRADIENT CENTRIFUGATION

TIME INTEGRAL VALUES FOR PARTICLE DENSITY 1.7 AT TEMPERATURE 20.0 DEG C

PERCENTAGE SUCROSE	ZO= 0.0	ZO=-1.0	ZO=-2.0	ZO=-3.0	ZO=-4.0	ZO=-5.0	ZO=-6.0	ZO=-7.0	ZO=-8.0	ZO=-9.0	ZO=-10.0
0	0.0000	0.0000	0.0000	0.0000	0.0000	0.0000	0.0000	0.0000	0.0000	0.0000	0.0000
2	0.0000	1.1392	0.7138	0.5259	0.4175	0.3466	0.2964	0.2590	0.2300	0.2068	0.1879
4	0.7608	1.6998	1.1588	0.8953	0.7334	0.6226	0.5415	0.4794	0.4303	0.3904	0.3573
6	1.2365	2.0946	1.4965	1.1904	0.9955	0.8583	0.7556	0.6756	0.6114	0.5585	0.5142
8	1.5987	2.4111	1.7776	1.4431	1.2251	1.0687	0.9499	0.8559	0.7796	0.7162	0.6626
10	1.9011	2.6833	2.0249	1.6698	1.4343	1.2629	1.1311	1.0258	0.9395	0.8642	0.8056
12	2.1686	2.9243	2.2510	1.8796	1.6301	1.4465	1.3038	1.1890	1.0940	1.0140	0.9454
14	2.4138	3.1559	2.4634	2.0788	1.8175	1.6234	1.4715	1.3482	1.2457	1.1587	1.0839
16	2.6453	3.3729	2.6676	2.2715	2.0002	1.7970	1.6367	1.5059	1.3966	1.3033	1.2227
18	2.8688	3.5840	2.8676	2.4616	2.1511	1.9697	1.8019	1.6643	1.5485	1.4495	1.3634
20	3.0838	3.7930	3.0666	2.6515	2.3628	2.1433	1.9691	1.8250	1.7033	1.5987	1.5075

SUCROSE GRADIENT CENTRIFUGATION

TIME INTEGRAL VALUES FOR PARTICLE DENSITY 1.8 AT TEMPERATURE 20.0 DEG C

PERCENTAGE SUCROSE	ZO= 0.0	ZO=-1.0	ZO=-2.0	ZO=-3.0	ZO=-4.0	ZO=-5.0	ZO=-6.0	ZO=-7.0	ZO=-8.0	ZO=-9.0	ZO=-10.0
0	0.0000	0.0000	0.0000	0.0000	0.0000	0.0000	0.0000	0.0000	0.0000	0.0000	0.0000
2	0.0000	1.1386	0.7134	0.5256	0.4173	0.3463	0.2962	0.2588	0.2298	0.2067	0.1878
4	0.7592	1.6980	1.1575	0.8942	0.7325	0.6218	0.5408	0.4788	0.4297	0.3899	0.3568
6	1.2333	2.0915	1.4940	1.1883	0.9936	0.8567	0.7542	0.6743	0.6102	0.5574	0.5132
8	1.5937	2.4064	1.7737	1.4398	1.2222	1.0661	0.9475	0.8538	0.7776	0.7143	0.6609
10	1.8944	2.6769	2.0194	1.6650	1.4300	1.2590	1.1275	1.0225	0.9364	0.8643	0.8030
12	2.1595	2.9198	2.2436	1.8731	1.6242	1.4411	1.2988	1.1843	1.0897	1.0099	0.9416
14	2.4023	3.1452	2.4532	2.0703	1.8098	1.6163	1.4648	1.3420	1.2399	1.1533	1.0787
16	2.6315	3.3597	2.6558	2.2609	1.9903	1.7878	1.6282	1.4979	1.3890	1.2962	1.2159
18	2.8517	3.5680	2.8531	2.4483	2.1689	1.9582	1.7912	1.6541	1.5389	1.4404	1.3548
20	3.0663	3.7738	3.0491	2.6354	2.3478	2.1297	1.9557	1.8124	1.6913	1.5873	1.4966

149

SUCROSE GRADIENT CENTRIFUGATION

TIME INTEGRAL VALUES FOR PARTICLE DENSITY 1.9 AT TEMPERATURE 20.0 DEG C

PERCENTAGE SUCROSE	ZO= 0.0	ZO= -1.0	ZO= -2.0	ZO= -3.0	ZO= -4.0	ZO= -5.0	ZO= -6.0	ZO= -7.0	ZO= -8.0	ZO= -9.0	ZO= -10.0
0	0.0000	0.0000	0.0000	0.0000	0.0000	0.0000	0.0000	0.0000	0.0000	0.0000	0.0000
2	0.0000	1.1381	0.7130	0.5253	0.4170	0.3462	0.2960	0.2586	0.2297	0.2066	0.1877
4	0.7581	1.6967	1.1565	0.8934	0.7318	0.6212	0.5403	0.4783	0.4293	0.3895	0.3565
6	1.2309	2.0891	1.4921	1.1866	0.9922	0.8554	0.7531	0.6733	0.6092	0.5565	0.5124
8	1.5899	2.4028	1.7707	1.4372	1.2199	1.0640	0.9456	0.8521	0.7760	0.7129	0.6595
10	1.8891	2.6719	2.0152	1.6612	1.4267	1.2560	1.1247	1.0200	0.9340	0.8621	0.8009
12	2.1525	2.9133	2.2379	1.8681	1.6197	1.4369	1.2950	1.1807	1.0864	1.0068	0.9387
14	2.3935	3.1370	2.4467	2.0637	1.8038	1.6108	1.4597	1.3372	1.2354	1.1490	1.0747
16	2.6202	3.3495	2.6467	2.2526	1.9828	1.7808	1.6216	1.4918	1.3832	1.2907	1.2107
18	2.8385	3.5557	2.8420	2.4382	2.1595	1.9494	1.7829	1.6464	1.5316	1.4334	1.3461
20	3.0526	3.7591	3.0357	2.6230	2.3363	2.1189	1.9456	1.8028	1.6822	1.5786	1.4883

SUCROSE GRADIENT CENTRIFUGATION

TIME INTEGRAL VALUES FOR PARTICLE DENSITY 2.0 AT TEMPERATURE 20.0 DEG C

PERCENTAGE SUCROSE	ZO= 0.0	ZO= -1.0	ZO= -2.0	ZO= -3.0	ZO= -4.0	ZO= -5.0	ZO= -6.0	ZO= -7.0	ZO= -8.0	ZO= -9.0	ZO= -10.0
0	0.0000	0.0000	0.0000	0.0000	0.0000	0.0000	0.0000	0.0000	0.0000	0.0000	0.0000
2	0.0000	1.1377	0.7128	0.5251	0.4169	0.3460	0.2959	0.2585	0.2296	0.2065	0.1876
4	0.7571	1.6956	1.1556	0.8927	0.7313	0.6207	0.5398	0.4779	0.4289	0.3891	0.3562
6	1.2289	2.0872	1.4905	1.1853	0.9911	0.8544	0.7522	0.6725	0.6085	0.5559	0.5118
8	1.5869	2.3999	1.7683	1.4351	1.2181	1.0624	0.9441	0.8507	0.7748	0.7117	0.6584
10	1.8848	2.6679	2.0110	1.6583	1.4240	1.2536	1.1225	1.0179	0.9322	0.8603	0.7992
12	2.1470	2.9081	2.2334	1.8640	1.6160	1.4336	1.2919	1.1779	1.0837	1.0043	0.9363
14	2.3864	3.1415	2.4409	2.0585	1.7991	1.6064	1.4557	1.3334	1.2318	1.1457	1.0716
16	2.6116	3.3415	2.6396	2.2461	1.9766	1.7752	1.6164	1.4869	1.3786	1.2863	1.2066
18	2.8281	3.5460	2.8333	2.4301	2.1520	1.9425	1.7764	1.6402	1.5258	1.4279	1.3429
20	3.0402	3.7475	3.0251	2.6133	2.3272	2.1104	1.9376	1.7952	1.6750	1.5717	1.4818

APPENDIX B

A Program for the Calculation of Sedimentation Coefficients in Swing-out Rotors

PROGRAM NOTES

1. The symbols used in the following listings are:

    NO = Number of rotors
    R1 = D cm (distance from centre of rotation to tube bottom)
    R2 = r cm (internal radius of tube)
    U1 = Total volume
    U2 = Sample volume
    U3 = Fraction volume
    Q   = Average speed
    T1 = Acceleration time
    T2 = Run time
    T3 = Deceleration time
    N1 = Number of rotor used
    N3 = Number of fractions

2. Different rotors may be incorporated into the program by altering the statements number 410 to 520. There is sufficient space for 10 different rotors.

3. The empirical formulae used to calculate density and viscosity from temperature and sucrose concentration are from Barber (1966).

```
10 REM        THIS PROGRAM CALCULATES SEDIMENTATION COEFFICIENTS
20 REM        FOR SUCROSE GRADIENTS IN SWING-OUT ROTORS.
30 REM
40 REM        IF FURTHER DETAILS OF THIS PROGRAM ARE REQUIRED
50 REM        CONTACT -  DR.BRYAN D.YOUNG
60 REM                      BEATSON INSTITUTE,GARSCUBE ESTATE,
70 REM                      BEARSDEN,GLASGOW,SCOTLAND,UK
80 REM
90 REM
100 DIM P(80),M(80)
110 DIM L(80),A$(10)
120 SELECT PRINT 005
130 PRINT HEX(03)
140 PRINT HEX(0A0A0A),TAB(10),"******* S-VALUE PROGRAM *******"
```

```
150 PRINT HEX(0A0A0A)
160 Z1=0
170 DATA 0.0528,0.4,0.0882,0.346
180 INPUT "IS SAMPLE DNA (Y/N)",C$
190 IF C$="N" THEN 250
200 INPUT "ALKALINE GRADIENT (Y/N)",C$
210 RESTORE
220 IF C$="Y" THEN 240
230 RESTORE 3
240 READ Z1,Z2
250 N0=10
260 REM        THE FOLLOWING DATA STATEMENTS CONTAIN THE
270 REM        BASIC INFORMATION FOR EACH ROTOR,i.e. distance
280 REM        FROM CENTRE OF ROTATION TO TUBE BOTTOM AND
290 REM        INTERNAL TUBE RADIUS(CM).
300 REM
310 DATA 12.82,1.08:A$(1)="MSE 3X25 ML"
320 DATA 10.37,0.6:A$(2)="MSE 3X6.5ML"
330 DATA 16.206,1.54:A$(3)="MSE 3X70 ML"
340 DATA 15.81,0.65:A$(4)="MSE 8X14 ML"
350 DATA 15.1,0.65:A$(5)="IEC 6X14 ML"
360 DATA 15.1,1.25:A$(6)="IEC 6X40 ML"
370 DATA 16.5,1.225:A$(7)="MSE 6X38 ML"
380 DATA 10,0.45:A$(8)="IEC 6X4.2 ML"
390 DATA 12.4,0.35:A$(9)="MSE 6X4.2 ML"
400 DATA 8.86,0.625:A$(10)="SW65 3X5ML"
410 FOR N=1 TO N0:PRINT N,A$(N):NEXT N
420 PRINT
430 INPUT "WHICH ROTOR DO YOU REQUIRE",N1
440 IF N1>N0 THEN 430
450 RESTORE  2*N1-1+4
460 READ R1,R2
470 INPUT "TOTAL VOLUME (ML)",U1
480 INPUT "SAMPLE VOLUME",U2
490 INPUT "FRACTION VOLUME",U3
500 N3=U1/U3
510 PRINT
520 INPUT "ACCELERATION TIME(MINUTES)",T1
530 INPUT "RUN TIME (HOURS)",T4
540 T2=T4*60
550 INPUT "DECCELERATION TIME (MINUTES)",T3
560 INPUT "AVERAGE SPEED (RPM)",Q
570 INPUT "TEMPERATURE (DEG.C)",T5
580 INPUT "PARTICLE DENSITY",P5
590 I=15/((T2+(T1+T3)/3)*(#PI*Q)↑2)
600 PRINTUSING 610,N3
```

```
610 %INPUT % SUCROSE OF THE ## FRACTIONS
620 FOR J=1 TO N3
630 PRINT J,
640 INPUT P(J)
650 NEXT J
660 INPUT "DATA OK",C$:IF C$="Y"THEN 690
670 INPUT "FRACTION NO",J:INPUT "% SUCROSE",P(J)
680 GOTO 660
690 J=1
700 L(J)=LOG(R1-R2+(J*U3-U1+2*#PI*(R2↑3)/3)/(#PI*R2↑2))
710 IF EXP(L(J)))R1-R2 THEN 740
720 J=J+1
730 GOTO 700
740 L(J)=0
750 J=J-1
760 M(1)=L(1)-LOG(R1-R2+(U2/2-U1+2*#PI*(R2↑3)/3)/(#PI*R2↑2))
770 FOR J1=2 TO J
780 M(J1)=L(J1)-L(J1-1)
790 NEXT J1
800 SELECT PRINT 215(120)
810 PRINT :PRINT :PRINT :PRINT :PRINT
820 PRINTUSING 830,A$(N1),Q,T4,T5,P5
830 %#############AT ##### RPM FOR ## HOURS AT ## DEG.C /P.D.=#.#
840 PRINT
850 IF Z1=0 THEN 880
860 PRINT "FRACTION NO    % SUCROSE    S VALUE    MOL. WT."
870 GOTO 890
880 PRINT "FRACTION NO    % SUCROSE    S VALUE"
890 S0=0
900 FOR J1=1 TO J
910 GOSUB '2(T5,P(J1),P5)
920 S0=S0+M(J1)*E
930 IF Z1=0 THEN 970
940 PRINTUSING 950,J1,P(J1),S0*I*1E13,EXP((LOG(S0*I*1E13/Z1))/Z2
)
950 %    ##         ##.#      ###.#     #.#↑↑↑↑
960 GOTO 990
970 PRINTUSING 980,J1,P(J1),S0*I*1E13
980 %    ##         ##.#      ###.#
990 NEXT J1
1000 SELECT PRINT 005
1010 END
1020 DEFFN'1(T,P)
1030 REM SUBROUTINE TO CALCULATE DENSITY (D),VISCOSITY (V)
1040 REM FROM TEMPERATURE (T) AND SUCROSE PERCENTAGE (P)
1050 DATA 1.0003698,3.9680504E-5,-5.8513271E-6
```

```
1060 DATA 0.38982371,-1.0578919E-3,1.2392833E-5
1070 DATA 0.17097594,4.7530081E-4,-8.9239737E-6
1080 DATA 18.027525,4.8318329E-4,7.7830857E-5
1090 DATA 342.3,18.032
1100 DATA 212.57059,0.13371672,-2.9276449E-4
1110 DATA 146.06635,25.251728,0.070674842
1120 DATA -1.5018327,9.4112153,-1.1435741E3
1130 DATA 1.0504137E5,-4.6927102E6,1.0323349E8
1140 DATA -1.1028981E9,4.5921911E9,-1.0803314
1150 DATA -2.0003484E1,4.6066898E2,-5.9517023E3
1160 DATA 3.5627216E4,-7.8542145E4,0,0
1170 DATA 2.1169907E2,1.6077073E3,1.6911611E5
1180 DATA -1.4184371E7,6.0654775E8,-1.2985834E10
1190 DATA 1.3532907E11,-5.4970416E11,1.3975568E2
1200 DATA 6.6747329E3,-7.8716105E4,9.0967578E5
1210 DATA -5.5380830E6,1.2451219E7,0,0
1220 RESTORE 25
1230 READ B1,B2,B3,B4,B5,B6,B7,B8,B9
1240 READ A1,A2,A3,M1,M2,C1,C2,C3
1250 READ G1,G2,G3
1260 Y1=P/100
1270 Y=(Y1/M1)/(Y1/M1+(1-Y1)/M2)
1280 IF T)30 THEN 1310
1290 D=B1+B2*T+B3*T↑2+(B4+B5*T+B6*T↑2)*Y1+(B7+B8*T+B9*T↑2)*Y1↑2
1300 GOTO 1330
1310 D0=Y*M1+(1-Y)*M2
1320 D=D0/(Y*(C1+C2*T+C3*T↑2)+(1-Y)*(A1+A2*T+A3*T↑2))
1330 RESTORE 45
1340 IF P<=48 THEN 1360
1350 RESTORE 53
1360 READ D0,D1,D2,D3,D4,D5,D6,D7
1370 A=D0+D1*Y+D2*Y↑2+D3*Y↑3+D4*Y↑4+D5*Y↑5+D6*Y↑6+D7*Y↑7
1380 RESTORE 61
1390 IF P<=48 THEN 1410
1400 RESTORE 69
1410 READ D0,D1,D2,D3,D4,D5,D6,D7
1420 B=D0+D1*Y+D2*Y↑2+D3*Y↑3+D4*Y↑4+D5*Y↑5+D6*Y↑6+D7*Y↑7
1430 C=G1-G2*SQR(1+(Y/G3)↑2)
1440 V=10↑(A+B/(T+C))
1450 RETURN
1460 DEFFN'2(T1,P1,H1)
1470 REM SUBROUTINE TO CALCULATE SEDIM(E) FROM PARTICLE
1480 REM DENSITY(H1), TEMPERATURE (T1) AND % SUCROSE(P1)
1490 GOSUB '1(20,0)
1500 H2=D
```

```
1510 V2=V
1520 GOSUB '1(T1,P1)
1530 H3=D
1540 V3=V
1550 E=((H1-H2)/(H1-H3))*(V3/V2)
1560 RETURN
```

APPENDIX C

A Program for the Calculation of Sedimentation Coefficients in Vertical Rotors

PROGRAM NOTES

1. The symbols used in the following listing are:
    NO = Number of rotors
    R1 = Internal tube radius (cm)
    R2 = Distance from axis of rotation to centre of tube (cm)
    L   = Length of cylindrical part of tube (cm)
    V   = Total volume
    U2 = Sample volume
    U3 = Fraction volume
    Q   = Average speed
    T1 = Acceleration time
    T2 = Run time
    T3 = Deceleration time
    N1 = Number of rotor used
    N3 = Number of fractions

2. Different rotors may be incorporated into the program by altering statements 150 – 240. There is sufficient space for 10 rotors.

```
10 COM P(80),M(80),I,J,T5,P5,Z1,Z2
20 DIM L(80),A$(10)
30 SELECT PRINT 005
40 PRINT HEX(03)
41 PRINT HEX(0A0A0A),TAB(10),"S-VALUE PROGRAM FOR VERTICAL ROTOR
S"
42 PRINT HEX(0A0A0A)
50 Z1=0
60 DATA 0.0528,0.4,0.0882,0.346
70 INPUT "IS SAMPLE DNA (Y/N)",C$
80 IF C$="N" THEN 140
90 INPUT "ALKALINE GRADIENT (Y/N)",C$
100 RESTORE
110 IF C$="Y" THEN 130
120 RESTORE 3
130 READ Z1,Z2
140 N0=4
145 REM LIST OF ROTORS AND THEIR DIMENSIONS
150 DATA 7.217,1.232,7.4:A$(1)="MSE 8X35 ML"
160 DATA 4,0.57,7.85:A$(2)="BMAN 8X5 ML"
170 DATA 7.06,1.17,7.35:A$(3)="BMAN 8X38 ML"
180 DATA 7,1.22,7.2:A$(4)="SVALL 8X38 ML"
190 DATA 0.0,0.0,0:A$(5)="XXXXXXXXXXX"
200 DATA 0.0,0.0,0:A$(6)="XXXXXXXXXXX"
210 DATA 0.0,0.0,0:A$(7)="XXXXXXXXXXX"
220 DATA 0.0,0.0,0:A$(8)="XXXXXXXXXXXXX"
230 DATA 0.0,0.0,0:A$(9)="XXXXXXXXXXXXX"
240 DATA 0.0,0.0,0:A$(10)="XXXXXXXXXXXXX"
250 FOR N=1 TO N0:PRINT N,A$(N):NEXT N
260 PRINT
270 INPUT "WHICH ROTOR DO YOU REQUIRE",N1
280 IF N1>N0 THEN 270
290 RESTORE (N1-1)*3+5
300 READ L,R1,R2
301 B0=1
302 INPUT "HEMISPHERICAL TOP",A$
304 IF A$="N"THEN 305:B0=B0+1
306 P=2:GOSUB 1000:PRINT "TOTAL VOL=",V:V9=V
320 INPUT "SAMPLE VOLUME",U2
330 INPUT "FRACTION VOLUME",U3
340 N3=V9/U3
350 PRINT
360 INPUT "ACCELERATION TIME(MINUTES)",T1
370 INPUT "RUN TIME (HOURS)",T4
380 T2=T4*60
390 INPUT "DECCELERATION TIME (MINUTES)",T3
```

```
400 INPUT "AVERAGE SPEED (RPM)",Q
410 INPUT "TEMPERATURE (DEG.C)",T5
420 INPUT "PARTICLE DENSITY",P5
430 I=15/((T2+(T1+T3)/3)*(#PI*Q)↑2)
440 PRINTUSING 450,N3
450 %INPUT % SUCROSE OF THE ## FRACTIONS
460 FOR J=1 TO N3
461 PRINT J,
470 INPUT P(J)
480 NEXT J
490 J=1
495 V1=J*U3
496 IF V1>V9THEN 540
497 GOSUB 1200
500 L(J)=LOG(R2-R1+P*R1)
510 IF EXP(L(J)))R1+R2 THEN 540
520 J=J+1
530 GOTO 495
540 L(J)=0
550 J=J-1
552 V1=U2/2:GOSUB 1200
555 M(1)=L(1)-LOG(R2-R1+P*R1)
570 FOR J1=2 TO J
580 M(J1)=L(J1)-L(J1-1)
590 NEXT J1
600 SELECT PRINT 01D
610 PRINT :PRINT :PRINT :PRINT :PRINT
620 PRINTUSING 630,A$(N1),Q,T4,T5,P5
630%##############AT ##### RPM FOR ## HOURS AT ## DEG.C /P.D.=#.#
640 PRINT
650 IF Z1=0 THEN 680
660 PRINT "FRACTION NO    % SUCROSE    S VALUE    MOL. WT."
670 GOTO 690
680 PRINT "FRACTION NO    % SUCROSE    S VALUE"
690 GOTO 2000
998 REM SUBROUTINE TO CALCULATE VOLUME (V) FROM
999 REM RADIAL DISTANCE (P) ACROSS THE TUBE
1000V=ARCCOS(1-P)-(1-P)*SQR(2*P-P*P)
1010 V=V+B0*(#PI*P*P*(3-P)*R1)/(6*L)
1020 V=V*R1*R1*L
1030 RETURN
1198 REM SUBROUTINE TO CALCULATE RADIAL DISTANCE (P)
1199 REM ACROSS THE TUBE  FROM VOLUME (V)
1200P=2*V1/V9
1210 GOSUB 1000:V2=V:IF ABS(V1-V2)/V1<0.001THEN 1250
1220 P=P*0.99:GOSUB 1000:D=(V2-V)/(0.01*P)
```

```
1230 P=P/0.99:P=P-(V2-V1)/D
1240 GOTO 1210
1250 RETURN
2000 S0=0
2010 FOR J1=1 TO J
2020 GOSUB '2(T5,P(J1),P5)
2030 S0=S0+M(J1)*E
2040 IF Z1=0 THEN 2080
2050 PRINTUSING 2060,J1,P(J1),S0*I*1E13,EXP((LOG(S0*I*1E13/Z1))/
Z2)
2060 %    ##           ##.#     ###.##    #.#↑↑↑↑
2070 GOTO 2100
2080 PRINTUSING 2090,J1,P(J1),S0*I*1E13
2090 %    ##           ##.#     ###.##
2100 NEXT J1
2110 SELECT PRINT 005
2120 GOTO 10
2140 DEFFN'1(T,P)
2150 REM SUBROUTINE TO CALCULATE DENSITY (D),VISCOSITY (V)
2160 REM FROM TEMPERATURE (T) AND SUCROSE PERCENTAGE (P)
2170 DATA 1.0003698,3.9680504E-5,-5.8513271E-6
2180 DATA 0.38982371,-1.0578919E-3,1.2392833E-5
2190 DATA 0.17097594,4.7530081E-4,-8.9239737E-6
2200 DATA 18.027525,4.8318329E-4,7.7830857E-5
2210 DATA 342.3,18.032
2220 DATA 212.57059,0.13371672,-2.9276449E-4
2230 DATA 146.06635,25.251728,0.070674842
2240 DATA -1.5018327,9.4112153,-1.1435741E3
2250 DATA 1.0504137E5,-4.6927102E6,1.0323349E8
2260 DATA -1.1028981E9,4.5921911E9,-1.0803314
2270 DATA -2.0003484E1,4.6066898E2,-5.9517023E3
2280 DATA 3.5627216E4,-7.8542145E4,0,0
2290 DATA 2.1169907E2,1.6077073E3,1.6911611E5
2300 DATA -1.4184371E7,6.0654775E8,-1.2985834E10
2310 DATA 1.3532907E11,-5.4970416E11,1.3975568E2
2320 DATA 6.6747329E3,-7.8716105E4,9.0967578E5
2330 DATA -5.5380830E6,1.2451219E7,0,0
2340 RESTORE 35
2350 READ B1,B2,B3,B4,B5,B6,B7,B8,B9
2360 READ A1,A2,A3,M1,M2,C1,C2,C3
2370 READ G1,G2,G3
2380 Y1=P/100
2390 Y=(Y1/M1)/(Y1/M1+(1-Y1)/M2)
2400 IF T>30 THEN 2430
2410 D=B1+B2*T+B3*T↑2+(B4+B5*T+B6*T↑2)*Y1+(B7+B8*T+B9*T↑2)*Y1↑2
2420 GOTO 2450
```

```
2430 D0=Y*M1+(1-Y)*M2
2440 D=D0/(Y*(C1+C2*T+C3*T↑2)+(1-Y)*(A1+A2*T+A3*T↑2))
2450 RESTORE 55
2460 IF P<=48 THEN 2480
2470 RESTORE 63
2480 READ D0,D1,D2,D3,D4,D5,D6,D7
2490 A=D0+D1*Y+D2*Y↑2+D3*Y↑3+D4*Y↑4+D5*Y↑5+D6*Y↑6+D7*Y↑7
2500 RESTORE 71
2510 IF P<=48 THEN 2530
2520 RESTORE 79
2530 READ D0,D1,D2,D3,D4,D5,D6,D7
2540 B=D0+D1*Y+D2*Y↑2+D3*Y↑3+D4*Y↑4+D5*Y↑5+D6*Y↑6+D7*Y↑7
2550 C=G1-G2*SQR(1+(Y/G3)↑2)
2560 V=10↑(A+B/(T+C))
2570 RETURN
2580 DEFFN'2(T1,P1,H1)
2590 REM SUBROUTINE TO CALCULATE SEDIM(E) FROM PARTICLE
2600 REM DENSITY(H1), TEMPERATURE (T1) AND % SUCROSE(P1)
2610 GOSUB '1(20,0)
2620 H2=D
2630 V2=V
2640 GOSUB '1(T1,P1)
2650 H3=D
2660 V3=V
2670 E=((H1-H2)/(H1-H3))*(V3/V2)
2680 RETURN
```

CHAPTER 5

Isolation of Subcellular Organelles and Membranes

J. GRAHAM

1. INTRODUCTION

It is neither feasible nor useful to give a comprehensive review of all available methods for isolating all of the various subcellular organelles from a variety of tissues and cultured cells. The methods described in this chapter are only directly applicable to rat liver which is one of the most widely used and most widely available source for studies on most subcellular organelles. Nevertheless, the methods should be applicable to any soft tissue which can be homogenised under conditions identical to those described in Section 2. Brain is an exception to this because the huge amounts of myelin present in this tissue require the use of some variations of the standard methods. Tissue culture cells, as a rule, are not easily disrupted by the liquid shear techniques in isotonic sucrose media as described in Section 2.3. It is frequently necessary to resort to hypotonic media in order to swell the cells and thus make them more susceptible to disruption in Dounce or Potter-Elvehjem homogenisers. This may also lead to the osmotic swelling of the organelles with the consequent reduction in their buoyant density, and increase in their size and, in some cases, damage to their structure.

Although this chapter is not primarily concerned with the isolation of plasma membrane, the preparation of a relatively crude fraction of this membrane will be included for the sake of completeness. Of all the subcellular fractions, plasma membrane is the most difficult to predict in its sedimenting properties from one tissue to another and from one cell to another. While the dimensions and densities of organelles such as nuclei, mitochondria and endoplasmic reticulum vesicles do vary (for example, the diameter of a nucleus may vary from 3 to 12 μm, that of a mitochondrion from 0.5 to 2 μm and that of an endoplasmic reticulum vesicle from 0.05 to 0.3 μm), the dimensions of a plasma membrane fragment may vary from a 0.05 μm vesicle to a 20 μm sheet. Furthermore, the size of plasma membrane fragments in a single homogenate may vary between the same extremes. In the case of rat liver for instance, while the contiguous membrane tends to remain as relatively large sheets, the sinusoidal membrane always vesiculates. Thus, depending on its size, it may either sediment at low centrifugal forces with the nuclei and/or at high centrifugal forces with the microsomes and/or at any intermediate centrifugal force.

For simplicity, the techniques will primarily involve either differential centrifugation alone or in combination with some simple discontinuous gradients. Specialised centrifugation equipment such as zonal or vertical rotors which may not be widely available will not be described in detail here.

2. HOMOGENISATION

2.1 **Homogenisers**

For most soft tissues such as rat liver it is common to use a liquid-shear technique to achieve cell disruption. All liquid shear homogenisers involve the extrusion of the finely chopped tissue suspension (or cell suspension) through the gap between a moving pestle and the wall of a glass outer vessel.

In the Dounce homogeniser, the pestle is in the form of a glass ball attached to a glass shaft which is always manually operated. In the Potter-Elvehjem homogeniser, the pestle is a piston of Teflon attached to a metal shaft (see *Figure 1*). To improve the efficiency and ease of homogenisation, the shaft is normally attached to a variable-speed overhead motor (preferably one with a high torque) so that the pestle can be rotated as it passes through the suspension. The clearance between the pestle and the containing vessel can vary from 0.05 mm (tight-fitting) to 0.6 mm (loose-fitting).

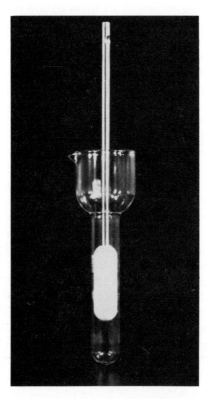

Figure 1. Potter-Elvehjem homogeniser (see text for description).

The efficiency of the homogenisation process and consequently the degree of fragmentation achieved depends on:

(i)   the number of strokes of the pestle;
(ii) the thrust of the pestle;
(iii) the speed of rotation of the pestle;
(iv) the pestle clearance and
(v)   the size and amount of material to be processed.

These parameters, which are normally ill-defined in published methods, must be controlled as carefully as possible if the technique is to be reproducible. The main reason for the inability of a person to reproduce the results of a published method is variation in the homogenisation process. The pestle clearance is particularly ill-defined in published methods, in that nebulous terms such as 'loose-fitting' or 'tight-fitting' are frequently used. Tight-fitting homogenisers (0.05 − 0.08 mm clearance) may lead to excessive fragmentation of organelles and membranes; consequently the centrifugation speeds required to sediment these membranes will become much greater and cross-contamination between subcellular fractions occurs.

For more solid tissues such as skeletal muscle, mechanical shear homogenisers give better results (see *Figure 2*). These devices rely on rotating metal blades or teeth and may be driven from below as in the ordinary domestic liquidiser or from an overhead motor (e.g., Townson & Mercer, Ultra-Turrax, Polytron, Silverson). The former can only be used with relative-

Figure 2. Townson & Mercer homogeniser. A standard overhead motor-driven mechanical shear homogeniser.

ly large volumes of homogenisation medium (100 ml – 1 litre), while the modern forms of these homogenisers also permit the use of sample volumes as small as 2 ml.

For the disruption of a cell suspension rather than an intact tissue, gaseous shear (nitrogen cavitation) may be employed. The stirred sample (5 ml – 1 litre) is contained within a stainless-steel pressure vessel (Artisan Industries or Baskerville & Lindsay Ltd.), pressurised with oxygen-free nitrogen at 42 – 56 kg/cm^2 (600 – 800 p.s.i.) for 10 – 15 min and then expelled through a needle valve. On exposure of the sample to atmospheric pressure, there is an explosive formation of bubbles of nitrogen within the cell cytoplasm leading to cell disruption. The chief advantages of this method over the more widely used liquid-shear techniques are that the disruption process occurs instantaneously; the same forces are experienced by all the sample; no heating can occur and the process is completely reproducible. The disadvantages are that the severity of the process tends to produce fragmentation of internal organelles; the rough endoplasmic reticulum becomes denuded of ribosomes; all membrane vesicles tend to be of the same size and the equipment is expensive and cumbersome to use.

2.2 Homogenisation Media

For the isolation of many organelles (mitochondria, lysosomes, peroxisomes), it is important to maintain an iso-osmotic medium to prevent damage due to osmotic stress. The most commonly used homogenisation medium is 0.25 M sucrose buffered with Tris, Hepes or Tes, at concentrations between 5 mM and 20 mM, to pH 7.4 – 8.0. However, in certain cases, particularly for the isolation of mitochondria, mannitol or sorbitol, either alone or in combination with sucrose, are often used as the osmotic balancers. Generally, ionic salt solutions are avoided since they often remove peripheral proteins from membranes. In the isolation of mitochondria, the inclusion of EGTA to chelate calcium ions is advantageous since this cation can lead to the uncoupling of oxidation from phosphorylation. For the same reason the medium is supplemented with 0.5% bovine serum albumin (BSA) to remove free fatty acids and lysophospholipids.

For the isolation of nuclei it is important to include a variety of cations in order to stabilise the membrane and prevent the swelling and release of DNA. The most common buffer contains 0.25 M sucrose, 50 mM Tris-HCl (pH 7.5), 25 mM KCl and 5 mM MgCl$_2$ (1), others have included 2 mM MgCl$_2$ and 2 mM CaCl$_2$. As long as both magnesium and calcium salts are present, it seems that, at least in the case of rat liver, even a hypotonic buffer such as 1 mM NaHCO$_3$ (2) can be used. In spite of the very low osmolarity of this buffer, rat-liver nuclei remain in a condensed state. This should not be regarded as a universal property of all nuclei, for some cells the presence of KCl is essential if one is to obtain intact nuclei (3).

The presence of magnesium ions aids the isolation of rough endoplasmic reticulum since it stabilises the interactions between the membranes and ribosomes at concentrations as low as 1 mM. In general, divalent cations stabilise membranes against excessive fragmentation but they also tend to cause aggregation between membranes. Indeed, high concentrations of calcium ions (8 mM) cause such massive aggregation of endoplasmic reticulum vesicles that they can be sedimented at low centrifugal forces (4).

Hence it is clear that to obtain, for example, the best preserved preparations of nuclei and mitochondria it is advantageous to use different homogenisation media. When, however, the various subcellular organelles are being analysed for the distribution of a particular enzyme for example, it is neither convenient nor desirable to use different media for the isolation of each organelle. In such instances it should be borne in mind that the products from a single homogenisation may be less than optimally preserved although it may be advantageous in that they are exposed to identical conditions.

2.3 Homogenisation of Rat Liver

The intact liver is a highly vascularised organ and contains a considerable amount of blood. In some cases, particularly for the processing of the nuclear pellet, it is advisable to remove the blood by perfusion prior to homogenisation.

Anaesthetise the rat by intraperitoneal injection of pentobarbitone sodium (6 mg/ml) using 1 ml per 100 g. Meanwhile, using a peristaltic pump, prime an approximately 50 cm length of silicone tubing (i.d. 3 mm) with ice-cold homogenisation medium containing 5 U/ml heparin from a 100 ml reservoir. Open the abdomen with a mid-line incision, displace the intestines to the right and expose the hepatic portal vessel. Insert and secure a 5 cm long 16 gauge cannula and, when it is full of blood, connect it to the primed silicone tubing ensuring that no air bubbles become trapped. Open the thorax and cut the superior vena cava. Perfuse the liver with homogenisation buffer at about 20 ml/min until the liver becomes a uniform light brown colour. Gently pressing the liver aids exsanguination. If this perfusion process is not carried out the rat should be killed by decapitation or cervical dislocation and bled from the neck.

Remove the liver quickly to a chilled, weighed beaker containing 20 ml homogenisation medium. Some experimental animals (e.g., mice and hamsters) possess gall bladders; these should be dissected away from the liver and discarded. After weighing the liver, decant the medium; chop the liver finely using scissors and add homogenisation medium to give a final concentration of 0.25 g liver/ml. This and all subsequent operations must be carried out at $0-4°C$.

For general purposes, the homogenisation medium used is 0.25 M sucrose, 5 mM Tris-HCl (pH 7.4); for nuclear preparations use 0.25 M sucrose, 50 mM Tris-HCl (pH 7.5), 25 mM KCl, 5 mM $MgCl_2$ and for mitochondrial preparations use 200 mM mannitol, 50 mM sucrose, 10 mM Hepes-NaOH (pH 7.4), 1 mM EDTA.

Homogenise the liver in a Potter-Elvehjem homogeniser (clearance 0.3 –

0.5 mm) using 4 – 5 up-and-down strokes of the pestle revolving at 500 r.p.m. In some specific preparations (see later) the number of strokes of the pestle may be modified. In one case — the preparation of large sheets of plasma membrane — the liver should be homogenised in 1 mM $NaHCO_3$ using approximately 15 strokes of the pestle of a loose-fitting Dounce homogeniser.

If the nuclear pellet is to be processed, then filter the homogenate through Nylon bolting cloth (pore size 75 μm) or fine cheese cloth to remove connective tissue, debris and unbroken cells. The filtration may be hastened by gently stirring the homogenate with a glass rod, but on no account should the homogenate be 'squeezed' through the filter.

2.4 Homogenisation of Tissue Culture Cells

It is impossible to give any hard and fast rules for the homogenisation of cells grown in culture. Most cells are not readily disrupted by Dounce or Potter-Elvehjem homogenisation in iso-osmotic sucrose media. If the cells under investigation require more than 15 – 20 strokes of the pestle to disrupt reproducibly more than 90% of the cells in 0.25 M sucrose then either hypotonic media or nitrogen cavitation should be used. The optimal composition of a hypotonic medium depends very much on the cell type. As a guide, the following media may be used: 1 mM $NaHCO_3$ or 5 mM Tris-HCl or 5 mM Hepes-NaOH buffered between pH 7.4 and pH 8.0 supplemented with one or more of the following: 1 – 2 mM $MgCl_2$, 1 – 2 mM $CaCl_2$ and 10 mM KCl. The chosen composition depends largely on the balance between the osmotic fragility of the nuclei and that of the cells. Hypotonic media unsupplemented with cations whilst aiding cell rupture often lead to nuclear rupture; the latter can be prevented by addition of magnesium or calcium ions and sometimes potassium ions to the homogenisation medium. Divalent cations, on the other hand, also tend to stabilise the plasma membrane against rupture. The aims of homogenisation are 3-fold:

(i)   to achieve more than 90% cell rupture with less than 20 up-and-down strokes of the pestle;
(ii) to avoid nuclear breakage;
(iii) to use as high a ratio of cells to homogenisation medium as possible.

The lower the homogenisation medium volume, the higher the concentration of released cytosolic proteins, the greater the protection afforded to the subcellular organelles against lysis by the hypotonic medium. Whatever medium is chosen, after homogenisation, 2 M sucrose should be added to make the final concentration 0.25 M as soon as possible.

If nitrogen cavitation is chosen then first try using 0.25 M sucrose, 5 mM Tris-HCl (pH 7.4), supplementing this medium with cations, as described above, as necessary. The minimum nitrogen pressure, probably not less than 40 kg/cm^2 (600 p.s.i.), to produce at least 90% cell rupture should be used.

3. DIFFERENTIAL CENTRIFUGATION

3.1 **Introduction**

The simplest method for the separation of a homogenate into different fractions is differential centrifugation. This involves sequential centrifugation of the homogenate at increasing speeds to obtain a series of pellets containing material of decreasing sedimentation rate (see Section 3.1 of Chapter 1). The main advantages of differential centrifugation are that it is a rapid and simple technique; is not limited by the volume of the homogenate and the subcellular organelles are not stressed osmotically by exposure to hypertonic gradient media. The main disadvantage is that the behaviour of the particles depends solely upon their sedimentation rate through a medium of a single density and, for a number of reasons, this will lead to the formation of heterogeneous pellets, and the distribution of the same type of organelle between different pellets.

The rate of sedimentation of a particle can be defined by the equation:

$$s = \frac{2r_p^2(\varrho_p - \varrho_m)g}{9\eta}$$

where r_p is the radius of the particle, ϱ_p is its density, ϱ_m is the density of the medium, g is the centrifugal force and η is the viscosity of the liquid. The centrifugal force at the top of the tube in a fixed-angle rotor may be about half as much as that at the bottom, while in a swing-out rotor this factor may be as much as a third: thus the sedimentation of particles at the top of the tube will be $2-3$ times slower than that of identical particles at the bottom of the tube.

The other major factor leading to heterogeneous pellets is the variation within each population of organelles. As shown in *Table 1* there is heterogeneity both in size and density of, for example, plasma membrane sheets, nuclei and mitochondria, and, moreover, except in the case of the density of the nuclei, there is considerable overlap of these parameters for all three types of particle. Thus, merely considering the particle size, the smallest mitochondria will sediment 64 times less rapidly than the largest under the same centrifugal force or, put another way, the smallest mitochondrion will require 64 times greater centrifugal force to sediment at the same rate as will the largest. Centrifugation conditions (1000 g for 10 min) which will pellet all of the nuclei, will cause co-sedimentation of most of the plasma membrane sheets and the largest and densest mitochondria. The least dense and smallest

Table 1. Typical Variations in the Size and Density of Subcellular Components.

Particle	Size (μm)	S-value	Density in sucrose solution (g/cm³)
Plasma membrane sheets	$3-20$	10^2-10^6	$1.15-1.19$
Nuclei	$3-12$	10^6-10^7	>1.30
Mitochondria	$0.5-4$	10^4-10^5	$1.17-1.21$

mitochondria on the other hand will require at least 20 000 g for 20 min to form a pellet.

In addition, slowly-sedimenting material which, on its own would not pellet at a particular low centrifugation speed, may nevertheless pellet at anomalously low speeds as a result of its entrapment by large numbers of rapidly sedimenting particles.

3.2 Fractionation of Rat-liver Homogenates

For this method and all succeeding methods, the volumes of homogenisation medium and gradient media, etc., are designed for a single liver. Assuming that the liver from a young adult male rat weighs approximately 12 g, the homogenate volume should be about 45 ml. A suitable differential centrifugation scheme is shown in *Figure 3*.

For the first centrifugation (C1) use 50 ml plastic conical tubes (the capped type made by Corning are convenient) in a swing-out rotor of a refrigerated low-speed centrifuge.

For C2 and C3 separations one can use 50 ml polycarbonate tubes in an 8 x 50 ml fixed-angle rotor of a high-speed refrigerated centrifuge.

For separation C4 use 50 ml polycarbonate tubes in a 50 ml fixed-angle rotor of an ultracentrifuge. The pellets or part of the pellets, P1, P2 and P3 are not well-packed; hence one should never pour off the supernatants, always use a wide-bore metal cannula (1 mm i.d.) attached to a 20 ml disposable syringe to remove the supernatants.

3.3 Analysis of Fractions

To assess the distribution of the various organelles, one can measure the relative amounts of 5'-nucleotidase (plasma membrane), DNA (nuclei), succinate dehydrogenase (mitochondria), aryl sulphatase (lysosomes), catalase (peroxisomes), galactosyl transferase (Golgi), glucose-6-phosphatase (endoplasmic reticulum), RNA (rough endoplasmic reticulum and ribosomes), in the different fractions. Detailed protocols for assaying these marker enzymes and macromolecules are given in Appendix V. In general, the compositions of the various pellets *(Figure 3)* are as follows:

fraction P1 — nuclei, plasma membrane sheets, heavy mitochondria, unbroken cells plus material from fractions P2 − P4 trapped by these organelles;

fraction P2 — heavy mitochondria, plasma membrane fragments plus material from fractions P3 and P4 trapped by these membrane components;

fraction P3 — mitochondria, lysosomes, peroxisomes, Golgi membranes, some rough endoplasmic reticulum plus material from fraction P4 trapped in this pellet;

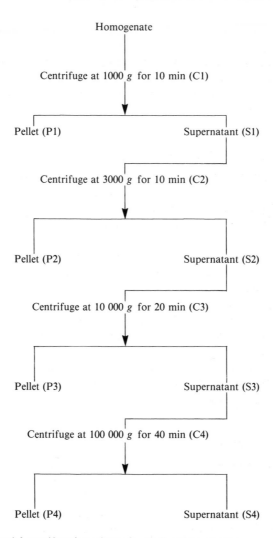

Figure 3. Differential centrifugation scheme for rat liver homogenate.

fraction P4 – membrane vesicles (microsomes) from smooth and rough endoplasmic reticulum, Golgi and plasma membrane (in the case of liver, mainly from the sinusoidal surface of the hepato-cyte);

fraction S4 – all of the soluble components of the cytoplasm.

3.4 Partial Purification of Fractions

The removal of contaminating, more slowly sedimenting, material from a pellet can be achieved by 'washing' the pellet; this involves resuspending the pellet in the homogenisation medium and repelleting using the original centrifugation conditions as before. Fraction P2, for example, can yield relatively pure mitochondria by this method. In other instances the different components of a pellet may possess such sufficiently distinct sedimentation properties that they form visually or operationally discrete layers within the pellet. Similarly, fraction P1 can yield a partially purified plasma membrane fraction while fraction P4 can yield a partially purified smooth microsome fraction.

3.4.1 *Purification of Mitochondria from Fraction P2*

The homogenisation medium used contains 200 mM mannitol, 50 mM sucrose, 1 mM EDTA, 10 mM Hepes-NaOH (pH 7.4). In all operations use ice-cold homogenisation medium. Use the standard homogenisation and centrifugation conditions to produce fraction P2. When removing the supernatant, S2, place the tip of the metal cannula in the meniscus so that the fat layer which forms at the top of the supernatant is efficiently removed and wipe away any remaining fat from the centripetal side of the centrifuge tube using a tissue. It is essential to remove as much fat as possible before resuspending the pellet, P2, since fatty acids cause uncoupling in mitochondria.

Add approximately 20 ml of homogenisation medium to the pellet and resuspend the material by gently drawing into and expelling from a fresh syringe and wide-bore metal cannula. Alternatively, dislodge the pellet using a spatula and resuspend using $3-5$ gentle strokes of the pestle of a loose-fitting Dounce homogeniser. Any aggregated hard-packed material beneath the bulk of the pellet should be discarded. After diluting to 30 ml with homogenisation medium centrifuge the suspension at 3000 g for 10 min and repeat the process twice more. Finally, resuspend the pellet in 10 ml of homogenisation medium. This preparation can be used very successfully in respiration studies with the oxygen electrode and demonstrates a high degree of coupling.

3.4.2 *Partial Purification of Smooth Microsomes from Fraction P4*

The homogenisation medium used contains 0.25 M sucrose, 5 mM Tris-HCl (pH 7.4). The separation is based on the fact that the smooth microsomes tend to form a loosely-packed layer above the hard-packed rough microsomes. After careful aspiration of the S4 fraction, add $2-3$ ml of 0.25 M sucrose, 5 mM Tris-HCl (pH 7.4) to the pellet and swirl it around; some of the more loosely-packed material will resuspend itself in the medium; remove this (fraction P4A) before resuspending the residual pellet (fraction P4B) in $2-3$ ml of fresh homogenising medium. Fraction P4A will be partially enriched in smooth microsomes and fraction P4B will be partially enriched in rough microsomes.

3.4.3 *Partial Purification of Plasma Membrane from Fraction P1*

The homogenisation medium used for this separation is 1 mM $NaHCO_3$ because when the liver is homogenised in 1 mM $NaHCO_3$ using a loose-fitting Dounce homogeniser then it is possible to partially purify large sheets of plasma membrane from fraction P1 using the method of Neville (5).

Homogenise the liver with about 15 strokes of the pestle in about 25 ml of medium. Dilute to 100 ml with 1 mM $NaHCO_3$, stir for 2 min and filter through Nylon bolting cloth (pore size 75 μm) or four layers of surgical gauze prior to obtaining pellet P1. Add 5 ml of medium to pellet P1 and transfer it to the homogeniser. Resuspend using 2−3 gentle strokes; dilute to 15 ml with medium and centrifuge at 1200 g for 10 min in 10 ml glass conical centrifuge tubes in a swing-out rotor. Decelerate the rotor without the brake. The pellet will consist of three clear layers. Resuspend the uppermost mitochondrial layer in the supernatant by gentle stirring and remove as much of this material as possible by aspiration. The middle layer is enriched in plasma membranes and the bottom layer contains nuclei. Carefully add 5 ml of 1 mM $NaHCO_3$ to the remaining bipartite pellet; resuspend the plasma membrane layer by gentle agitation and remove the suspension with as little disturbance of the nuclear layer as possible. Additional purification can be achieved by diluting the suspension to 15 ml and repeating the process.

4. SIMPLE SUCROSE DENSITY BARRIER METHODS

By centrifuging either a whole homogenate or a resuspended differential centrifugation fraction through one or two layers of sucrose solutions of different densities, it is possible to purify a number of subcellular particles satisfactorily. Reference to the sedimentation equation in Section 3.1 shows that increasing the density and viscosity of the medium (ϱ_m and η), the sedimentation of the particle is going to be significantly affected. Moreover, since high concentrations of sucrose exert a significant osmotic pressure, the size of osmotically-sensitive particles will decrease and their density (ϱ_p) increase as they contract due to the outward movement of water. The major problem with simple discontinuous gradients in which all of the particles move downwards is that material builds up transiently at each interface and this can lead to aggregation and entrapment of material. This problem tends to be significantly decreased if the sample itself is adjusted to a higher density by the addition of solid sucrose or a concentrated sucrose solution so that the final density is greater than some particles and less than others: in such a situation the sample is loaded in the middle of the gradient and hence lighter particles will move centripetally (float upwards) while the denser ones will sediment.

4.1 Separation of Rough and Smooth Microsomes from Fraction S3

For the best separations of rough and smooth microsomes it is more convenient and certainly quicker to separate these organelles from a post-mitochondrial supernatant (fraction S3) rather than to have to resuspend pellet P4 prior to separation. The method described below is adapted from that of Bergstrand and Dallner (6). The solutions used besides the homogenisation medium (see Section 3.4.1) are 0.6 M sucrose, 5 mM Tris-HCl (pH 8.0), 15 mM CsCl and 1.3 M sucrose, 5 mM Tris-HCl (pH 8.0), 15 mM CsCl. Prepare the homogenate as usual and obtain fraction S3. For convenience, centrifugation step C2 can be omitted and the S1 fraction centrifuged as in step C3. Theoretically the total homogenate could be centrifuged at 10 000 g for 20 min, omitting steps

C1 and C2; however, the entrapment of small vesicular material by a large amount of rapidly sedimenting nuclei leads to a considerable reduction in the recovery of smooth and rough microsomes.

Using $10-13$ ml thick-walled polycarbonate tubes for an ultracentrifuge fixed-angle rotor, set up the following density barrier: 3 ml of 1.3 M sucrose-Tris-CsCl and 1.5 ml of 0.6 M sucrose-Tris-CsCl. On top, layer the supernatant fraction S3 to fill the tube. Centrifuge this discontinuous gradient at 100 000 g for 90 min.

Two major fractions are obtained: a double-layered band at or just below the 0.6/1.3 M sucrose interface (smooth microsomes) and a pellet which contains all of the rough microsomes. There is a significant amount of material in the 1.3 M sucrose layer itself, containing mainly smooth microsomes although its precise composition tends to vary. Remove the smooth microsomes using a syringe and metal cannula, dilute the suspension with three volumes of 5 mM Tris-HCl (pH 8.0) and pellet by centrifugation at 160 000 g for 30 min. Discard the remaining solution from above the rough microsome pellet and resuspend both pellets in $2-5$ ml of 0.25 M sucrose, 5 mM Tris-HCl (pH 8.0). For additional subfractionations of rough and smooth microsomes, the reader is referred to a methodological review by De Pierre and Dallner (7).

4.2 Purification of Nuclei

4.2.1 *Purification of Nuclei from the Homogenate*

This method, based on that of Blobel and Potter (1) permits the isolation of nuclei directly from the homogenate. The solutions required are 0.25 M sucrose in TKM and 2.3 M sucrose in TKM where TKM is 0.05 M Tris-HCl (pH 7.5), 25 mM KCl, 5 mM $MgCl_2$.

Homogenise the liver using $10-15$ strokes of the pestle in 0.25 M sucrose-TKM. After filtration through gauze add two volumes of 2.3 M sucrose-TKM to bring the sucrose concentration to 1.62 M. It is advisable to use a syringe with wide bore metal cannula or a measuring cylinder when adding dense sucrose. Due to the viscosity of the solutions it is difficult to use a pipette. Whatever method is used, ensure that a sufficient amount of 2.3 M sucrose is added. The final sugar concentration should be checked by refractive index. Transfer 9 ml of this suspension to a 14 ml polycarbonate ultracentrifuge tube and underlay with $3-4$ ml of 2.3 M sucrose-TKM. Centrifuge at 130 000 g for 30 min at 5°C in a swing-out rotor. Remove all material above the pellet of nuclei which can be resuspended in any appropriate buffer.

4.2.2 *Isolation of Nuclei from Pellet P1*

To purify nuclei from the P1 pellet, resuspend the pellet by vortex mixing in the residual buffer and then add an equal volume of 60% (w/w) sucrose-TKM (see Section 4.2.1). Resuspend the material thoroughly using $2-3$ strokes of the pestle of a loose-fitting Dounce homogeniser. Continue adding the 60% sucrose-TKM until the sucrose concentration reaches 56% (w/w). Check the concentration by refractive index. Transfer 9 ml to a 14 ml polycarbonate tube, underlay with $3-4$ ml of the 60% sucrose and centrifuge at 120 000 *g* for 30 min at 5°C in a swing-out rotor. Discard all the material above the pellet and resuspend the latter in any appropriate buffer. Some workers prefer to reduce the membrane contamination of nuclei by washing them with 0.5% Triton X-100 in homogenising medium. This treatment removes the nuclear membrane without disrupting the nuclear structure.

4.3 Purification of Plasma Membrane from Pellet P1

This method is a modification of that of Touster *et al.* (8) and requires the following solutions: 60% (w/w) sucrose, 37.2% (w/w) sucrose, both in 5 mM Tris-HCl (pH 8.0). Prepare the P1 pellet from the filtered homogenate under standard conditions and resuspend in the residual buffer by vortexing. Add 60% sucrose to a final concentration of about 48% (checked by refractive index). Transfer 6 ml to a 14 ml polycarbonate tube; overlay with 6 ml of 37.2% sucrose and 1 ml of homogenisation medium. Centrifuge at 160 000 *g* for $2-4$ h at 5°C in a swing-out rotor.

Plasma membranes collect at the top of the 37.2% layer. Remove the layer using a syringe (plus metal cannula) and harvest by centrifugation at 100 000 *g* for 40 min at 5°C after dilution with three volumes of 5 mM Tris-HCl (pH 8.0).

4.4 Purification of Mitochondria from Pellet P1

This method is adapted from Fleischer *et al.* (9) and requires the following solutions: 0.25 M sucrose, 10 mM Hepes-NaOH (pH 7.5) either alone or supplemented with 1 mM EDTA or 1 mM $MgCl_2$ and 2.4 M sucrose, 10 mM Hepes-NaOH (pH 7.5), 1 mM $MgCl_2$. Prepare pellet P1 from a filtered homogenate using the standard protocol *(Figure 3)*; resuspend the pellet in about 15 ml of 0.25 M sucrose, 10 mM Hepes-NaOH (pH 7.5), 1 mM $MgCl_2$, taking care to avoid resuspending the lowermost red part of the pellet. Make the suspension homogeneous by gentle mixing and add 23 ml of 2.4 M sucrose, 10 mM Hepes-NaOH (pH 7.5), 1 mM $MgCl_2$ so that the final concentration of sucrose is 1.0 M. Check the concentration by refractive index and adjust if necessary. Transfer the suspension to a 50 ml polycarbonate tube, overlay with 8 ml of 0.25 M sucrose, 10 mM Hepes-NaOH (pH 7.5) and centrifuge at 35 000 *g* for 10 min at 5°C. Remove all the supernatant together with any material adhering to the centripetal side of the tube. The pellet is clearly bipartite; run 10 ml of 0.25 M sucrose, 10 mM Hepes-NaOH (pH 7.5), 1 mM EDTA down the side of the tube and by gentle swirling resuspend the upper brown mitochondrial layer of the pellet.

For the isolation of mitochondria from other tissues and a detailed discussion of the problems associated with their isolation see Volume 55 of *Methods in Enzymology*.

5. USE OF DISCONTINUOUS AND CONTINUOUS SUCROSE GRADIENTS

For some purifications a simple barrier system is not adequate and either a more complex discontinuous or continuous gradient must be used.

5.1 Purification of Lysosomes from Pellet P3

This method is adapted from that of Maunsbach (10) and it involves the use of the following solutions: 0.3 M sucrose, 1.1 M sucrose and 2.1 M sucrose, all containing 1 mM EDTA and 5 mM Tris-HCl (pH 7.0).

Prepare two 10 ml linear gradients of $1.1-2.1$ M sucrose in 14 ml polycarbonate tubes. This can be done using a simple gradient maker or by preparing a discontinuous gradient of 2.5 ml each of 1.1 M, 1.4 M, 1.7 M and 2.1 M sucrose and allowing them to stand at 4°C for $12-16$ h, after which time a linear gradient is formed by diffusion as described in Section 5.1.2 of Chapter 2.

The pellet P3 is clearly heterogeneous and to improve the resolution in the subsequent gradient it is advantageous to remove some of the lighter material from the pellet. Gently run about 10 ml of the 0.3 M sucrose solution down the side of the centrifuge tube and gently swirl the contents. Remove the solution and repeat this operation so that the dark brown (lysosomal) button at the bottom of the pellet remains undisturbed. Finally add 4 ml of this medium, scrape off the residual material using a spatula and resuspend it using one or two gentle strokes of the pestle of a loose-fitting Dounce homogeniser. Place 2 ml of the suspension on each gradient and gently stir the gradient/sample interface to reduce the discontinuity in density. Centrifuge the gradients at 95 000 g for $2-5$ h at 5°C in a swing-out rotor.

The lysosomes form a band over the density range $1.20-1.26$ g/cm^3 and the mitochondria band at around $1.17-1.21$ g/cm^3. There is a clear tendency for these bands to overlap but, by judicious sampling, that is, taking the denser lysosomal fractions, relatively pure lysosomes can be harvested. It is virtually impossible to separate lysosomes from peroxisomes by any sucrose gradient technique because of their similar density. This is only possible if the density of lysosomes is artificially reduced by an intraperitoneal injection of Triton WR-1339 (11) about 4 days before sacrifice of the animals.

5.2 Effect of Gradient Centrifugation on Mitochondria

The broadness of the mitochondrial band as described in Section 5.1 may, in part, be due to the partial disruption of mitochondria as a result of hydrostatic pressure within the gradient (12). This phenomenon is not directly associated with osmotic fragility (13). The hydrostatic pressure *(p)* generated at a point x centimetres from the centre of rotation in a gradient is related to the angular velocity of the rotor (ω), the density at the top of the gradient (ϱ_t), the slope of the gradient *(a)* and the radial distance to the meniscus of the gradient (x_t) by the expression:

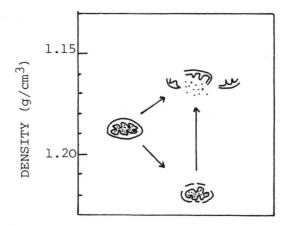

Figure 4. An interpretation of the effect of the disruption of membranes on the banding density of mitochondria.

$$p = \omega^2 \left[a\left(\frac{x^3 - x_t^3}{3}\right) + (\varrho_t - ax_t)\left(\frac{x^2 - x_t^2}{2}\right) \right]$$

The pressure may be considerable at high speeds (12), that is, in excess of 1500 kg/cm² (21 000 p.s.i.); such pressure can disrupt organelle integrity. As shown in *Figure 4*, the mitochondria with a partially disrupted outer membrane are slightly denser than intact mitochondria, while more severely disrupted organelles will be less dense. To minimise hydrostatic pressures it is therefore an advantage to reduce the distance x and the term $(x - x_t)$ as much as possible. Vertical rotors (see Section 4.3.2 of Chapter 1) are therefore very useful in this context since the sedimentation pathlength in such rotors is usually 2.5 cm or less (see Section 6.4.4).

5.3 Isolation of Golgi Membranes from Pellets P2 and P3

This method is adapted from that of Fleischer and Fleischer (14). The method uses the following gradient solutions: 38.7%, 36%, 33%, 29% sucrose all weight/weight and all containing 5 mM Tris-HCl (pH 8.0).

Prepare pellet P2 + P3 by centrifuging the S1 fraction at 10 000 g for 20 min at 5°C (i.e., omit step C2). Resuspend the pellet in about 5 ml of 0.25 M sucrose, 5 mM Tris-HCl (pH 8.0) and add, with constant stirring, solid sucrose to raise the concentration to 43.7% (w/w). Check the concentration with a refractometer. In polycarbonate tubes (for an ultracentrifuge swing-out rotor) layer the following: 3 ml of sample, 4 ml of 38.7%, 2 ml of 36%, 2 ml of 33% and 2 ml of 29% sucrose. Centrifuge the gradients at 160 000 g for 60 min at 5°C in a swing-out rotor. Using a syringe and metal cannula collect the top two bands which contain the Golgi membranes.

It is also possible to use this system to isolate Golgi membranes from the whole homogenate which has been made 43.7% (w/w) with respect to sucrose and then made part of the same discontinuous gradient. The larger volume of the homogenate however makes this approach less convenient.

6. ALTERNATIVE CENTRIFUGATION GRADIENT MEDIA

6.1 Permeability of Membranes to Gradient Solutes

The buoyant density of a particle is affected significantly by its degree of hydration. As a general rule, the higher the water activity of the gradient the lower will be the buoyant density of the particle. This is of particular importance in the case of membrane-bound organelles and membrane vesicles. The water compartment of such particles is significant and, depending on the gradient solute, its buoyant density can be considerably lower than that of the membrane itself.

If the medium is isotonic and the gradient solute molecules can readily pass through the membrane then the particle exhibits a density equal to that of the membrane. In an isotonic solution of a solute which is impermeable to the membrane then the density of the particle will be that of the whole particle, that is, the membrane and water compartments. The membrane permeability of a solute depends on the solute's size, charge and hydrophobicity. Membranes tend to be permeable to small molecules such as glycerol but impermeable to large macromolecules such as Ficoll. In between these two extremes are molecules such as sucrose, metrizamide and Nycodenz which can slowly penetrate some membranes. Percoll, being a colloidal suspension is totally impermeable to all membranes.

6.2 Osmotic Effects of Gradient Media

The most commonly used gradient solution, sucrose, exerts an osmotic pressure of 300 mOsm at a concentration of about 0.25 M, that is, this concentration of sucrose is iso-osmotic with the contents of most mammalian cells. Iso-osmotic conditions would be clearly beneficial in maintaining the integrity of subcellular organelles and in minimising their damage during preparation. Thus the use of a simple differential centrifugation scheme, using a suspension medium of 0.25 M sucrose, while generally not permitting the isolation of subcellular organelles in an uncontaminated state, is the least potentially disruptive.

All the sucrose barrier and gradient systems described in Sections 4 and 5 involve the exposure of organelles and vesicles to hyper-osmotic media. Under these conditions, water will exit from the organelle or vesicle by osmosis, its buoyant density will rise and its size diminish. Moreover, in any gradient, as a particle moves from one sucrose concentration to another it will experience either increasing or decreasing osmotic pressure, thus its sedimentation or flotation properties will change continuously as it sediments.

A number of modern gradient media of low osmolarity will now be considered. Ficoll and dextran will not be considered because their use has become relatively rare, they are extremely viscous, even at relatively low concentrations and, in addition, although they have a very low osmotic pressure below 20% (w/v), above 30% (w/v) it increases almost exponentially.

6.3 Iodinated Gradient Media

6.3.1 *Background*

There are two commercially available iodinated gradient media, metrizamide and Nycodenz, which overcome most osmotic problems. Metrizamide and Nycodenz are both water-soluble derivatives of triiodobenzoic acid; the properties of these media are detailed in Section 5.2.4 of Chapter 1. Nycodenz and metrizamide solutions possess a low osmotic activity which increases virtually linearly with concentration and both give solutions of low viscosity.

Whether an organelle has a different buoyant density in metrizamide or Nycodenz to that in sucrose depends on whether it has an osmotic space or not. Peroxisomes, for example, are devoid of such a space and hence the density of these organelles in sucrose is the same as that in iodinated density-gradient media (15). *Table 2* shows the properties of three subcellular organelles in these media.

6.3.2 *Isolation of Lysosomes and Peroxisomes from Pellet P3*

This method is adapted from Wattiaux *et al.* (16). Prepare fraction P3 in 0.25 M sucrose, 5 mM Tris-HCl (pH 7.4) and, after washing once in this medium, resuspend the pellet by gentle homogenisation in a loose-fitting Dounce homogeniser in the same medium (total volume about 5 ml). Use this suspension in one of the following methods.

(i) *Lysosome preparation.* This method uses the following gradient solutions: 85%, 32.8%, 26.3%, 24.5% and 19.8% metrizamide or Nycodenz. All solutions are weight/volume and all contain 5 mM Tris-HCl (pH 7.4). Mix 1 volume of the P3 suspension with 2 volumes of 85% (w/v) metrizamide or Nycodenz to bring the final concentration to 57% (w/v). Transfer 2.5 ml of this sample to polycarbonate tubes (for an ultracentrifuge swing-out rotor) and overlayer it with a discontinuous gradient of 2 ml each of 32.8%, 26.3% and 24.5% metrizamide or Nycodenz and 3 ml of 19.8% metrizamide or Nycodenz. Centrifuge the gradients for 2 h at 100 000 *g* at 5°C in a swing-out rotor. The band of material at the interface of the top two layers contains

Table 2. Comparison of the Buoyant Density of Organelles in Various Media.

Medium	Buoyant densities of organelles		
	Mitochondria (g/cm³)	Lysosomes (g/cm³)	Peroxisomes (g/cm³)
Sucrose	1.188	1.216	1.231
Metrizamide	1.115	1.140	1.231
Nycodenz	1.132	1.159	1.231
Percoll	1.10	1.04 – 1.10	1.06

purified lysosomes; remove it with a syringe and metal cannula.

By layering the sample at the bottom of the tube underneath the gradient it is possible to subject the mitochondria to high hydrostatic pressure during centrifugation. Under these conditions the mitochondria band at slightly higher densities than lysosomes (15).

(ii) *Preparation of peroxisomes.* This method uses the following gradient solutions: 50%, 45%, 40%, 30%, 20% metrizamide or Nycodenz. All solutions are weight/volume and all contain 5 mM Tris-HCl (pH 7.4). Prepare, in polycarbonate tubes for a 14 ml ultracentrifuge swing-out rotor, a discontinuous gradient of 2 ml each of the above metrizamide or Nycodenz solutions and load 1 − 2 ml of sample (P3 suspension) onto the top. Centrifuge the gradients at 110 000 g for 2.5 h at 5°C. The peroxisomes form a band above the 40/45% interface.

6.4 Percoll Gradients

6.4.1 *Background*

Percoll is a colloidal suspension of silica particles coated with polyvinyl pyrrolidone. Because it is a colloid it exhibits a negligible osmotic activity at all concentrations; osmotic activity is provided by the addition of sucrose or salt to the suspension. Details of the properties of Percoll gradients are given in Section 5.2.5 of Chapter 1. In solutions of Percoll the buoyant density of membrane-bound organelles is considerably less than in either sucrose or metrizamide or Nycodenz (see *Table 2*).

In spite of the overlap of buoyant densities of mitochondria, lysosomes and peroxisomes in Percoll, it is possible to devise conditions which will permit the isolation of one particular organelle.

6.4.2 *Isolation of Mitochondria from a Homogenate*

This method is adapted from that of Reinhardt *et al.* (18) and requires the use of the following solutions: Percoll diluted to 52%, 42%, 31% and 19% (v/v) of the stock solution in a medium containing 210 mM mannitol, 60 mM sucrose, 10 mM KCl and 10 mM Hepes-NaOH (pH 7.4).

Prepare the homogenate in the mannitol-sucrose-KCl-Hepes medium and layer 2 ml over 2 ml each of 19%, 31% and 42% (v/v) Percoll and 3 ml of 52% Percoll in 15 ml tubes and place in a fixed-angle rotor. Centrifuge at 5°C at 37 000 g for 30 sec with an acceleration time of 95 sec and a braking time of 180 sec.

Bands around the positions of the original four interfaces are obtained together with a pellet. The band around the 42%/52% interface contains nearly 60% of the total mitochondria with only a small amount of contamination from lysosomes and peroxisomes. No well-defined peroxisome band is present but the 31%/42% interfacial band shows some enrichment of lysosomes.

6.4.3 *Isolation of Lysosomes from the P3 Pellet*

This method is adapted from that of Pertoft *et al.* (19) and it requires the use of Percoll in 0.25 M sucrose, 5 mM Tris-HCl (pH 7.4) adjusted to a density of 1.077 g/cm^3.

Prepare the homogenate in the standard manner and isolate the pellet P3. Resuspend the pellet in 10 ml of 0.25 M sucrose, 5 mM Tris-HCl (pH 7.4) and layer 5 ml onto 35 ml of the Percoll solution in a 50 ml polycarbonate tube for an 8 x 50 ml fixed-angle rotor and centrifuge at 37 000 *g* for 30 – 40 min at 5°C.

During centrifugation the 35 ml Percoll solution forms a linear gradient from 1.05 to 1.10 g/cm^3 through the top 30 ml and then increases sharply to 1.15 g/cm^3 in the bottom 4 – 5 ml of the gradient. The lysosomes should band at around a density of 1.09 g/cm^3. The precise centrifugation conditions required to generate this particular gradient will vary from one rotor to another and it is essential that the form of the gradient should be checked using the commercially available density marker beads.

6.4.4 *Isolation of Peroxisomes from the S2 Supernatant*

Appelkvist *et al.* (20) noted that peroxisomes are extremely sensitive to hydrostatic pressure. The use of vertical rotors is therefore very beneficial in their preparation since the sedimentation pathlength in these rotors is very short. Even so, these workers recommend the use of 1 mM glutaraldehyde to stabilise peroxisomes. This concentration was found not to inhibit a number of enzymes. Appelkvist *et al.* (20) maintained that Percoll was indeed the best medium for isolating these organelles but cautioned that residual Percoll particles interfered with several reagents used in colourimetric assay procedures.

All the recently published methods for the preparation of peroxisomes have employed, in the first step, vertical rotors. It is very difficult to adapt these methods to the more commonly available angle rotors and the reader is advised that the centrifugation conditions detailed in the following method (20) may require modification. If a vertical rotor is not available, an angle rotor should be chosen in which the angle between the vertical and the inclination of the tube pocket is as acute as possible, that is the tubes are as close to being vertical as possible. The MSE aluminium 8 x 35 ml angle rotor for example has a tube angle of only 21° and is a good substitute for a vertical rotor; while the Titanium 10 x 10 angle rotor, with a tube angle of 35° is much less satisfactory.

Prepare the S2 supernatant under standard conditions from rats which were injected intraperitoneally with 1 ml of Triton WR1339 (170 mg/ml) 4 days before sacrifice. To stabilise the peroxisomes make the S2 supernatant 1 mM with respect to glutaraldehyde and 50 mM with respect to cacodylate buffer (pH 7.4) and incubate at 4°C for 30 min. Layer this suspension over undiluted Percoll and a solution of 40% Percoll, 0.25 M sucrose, 5 mM Tris-HCl (pH 7.4) in the ratio of 2:1:3 in a tube for the Beckman VTi 50 rotor or equivalent vertical rotor or for a suitable high-speed angle rotor and then centrifuge at

50 000 g for about 30 min at 5°C (the precise conditions will depend on the rotor). Using a metal cannula and syringe remove the band above the undiluted Percoll. Mix the suspension with the 40% Percoll and transfer to a 10 ml polycarbonate tube for a fixed-angle rotor and centrifuge at 60 000 g for 30 min at 5°C. Peroxisomes are concentrated in the lowermost band.

7. FRACTIONATION OF TISSUES OTHER THAN RAT LIVER

7.1 Introduction

As indicated in the Introduction, the methods which have been described should be broadly applicable to any soft tissue which can be disrupted by Potter-Elvehjem homogenisation. Even in the case of other tissues such as skeletal muscle which require special homogenisation procedures, the differential centrifugation scheme *(Figure 3)* can be applied and the distribution of the various organelles between pellets P1 and P4 assessed by enzymic analysis. Generally speaking, irrespective of the tissue, the nuclei will always be found in the P1 pellet, the endoplasmic reticulum and other vesicular material in pellet P4 and the lysosomes, mitochondria and peroxisomes in fractions P2 or P3. However, the precise distribution of these organelles amongst the different pellets will vary. If the majority of the aryl sulphatase (lysosomal marker), for example, is found in the P2 pellet rather than the P3 pellet, then the former should be used for isolating lysosomal material. If a large spill-over of lysosomal material into the P4 pellet is observed, then the centrifugation conditions in step C3 should be increased to avoid this. Indeed, once the distribution of organelles between the various differential centrifugation pellets has been established, then the centrifugation conditions of each step can be modified, or additional steps inserted to optimise this initial crude fractionation. In tissues with a well-developed Golgi system, gentle homogenisation helps to avoid vesiculation of the membrane sheets and the bulk of these membranes should sediment in the P3 pellet rather than the P4 fraction. Those parts of the plasma membrane from organised tissues which are not stabilised by junctional complexes or by an extensive cytoskeleton, and all the plasma membrane from tissue culture cells will not usually sediment in the P1 pellet. It is impossible to make any detailed suggestions regarding the recovery of plasma membrane in such instances. Broadly speaking, in these cases the scheme outlined in the following sections should be adopted.

7.2 Isolation of Plasma Membrane from the P2 and P3 Pellets of Tissue Culture Cells

This method uses the following gradient solutions: 20%, 30%, 50% (w/w) sucrose in 5 mM Tris-HCl (pH 8.0) and 10%, 15%, 20%, 25%, 40% (w/v) metrizamide or Nycodenz in 5 mM Tris-HCl (pH 8.0). Resuspend the pellet in 5 ml of 50% sucrose using 2 – 3 gentle strokes of the pestle of a Dounce homogeniser, and adjust the final concentration of sucrose to 40% (w/w). Make up a discontinuous gradient of 3 ml each of 20%, 30%, sample in 40% and 50% sucrose in 14 ml polycarbonate tubes for an ultracentrifuge swing-out rotor and centrifuge the gradients at 100 000 g for 12 – 16 h at 5°C. Collect the bands, dilute each with three volumes of 5 mM Tris-HCl (pH 8.0), then pellet the membranes by centrifugation at 100 000 g for 40 min at 5°C and resuspend in 1 – 2 ml of 0.25 M sucrose, 5 mM Tris-HCl (pH 8.0). Assess the distribution of plasma membrane, lysosomes and mitochondria by assaying the 5'-nucleotidase, aryl sulphatase and succinate dehydrogenase activities, respectively, of the harvested bands as described in Appendix V.

To purify the plasma membrane-rich bands, centrifuge the material at 100 000 g for 30 min at 5°C and resuspend the pellet in 2 ml of 40% metrizamide by gentle Dounce homogenisation. Adjust the final concentration to 30% (w/v) and make a discontinuous gradient of 2.5 ml each of 10%, 15%, 20% and 25% (w/v) metrizamide and 2.5 ml of the sample in a 14 ml tube. Centrifuge in an ultracentrifuge swing-out rotor at 160 000 g for 3 h. Harvest the bands as before and identify the material.

7.3 Isolation of Plasma Membrane from the P4 Pellet of Tissue Culture Cells

This method uses the following gradient solutions: 10%, 15%, 20%, 25%, 30% (w/v) metrizamide or Nycodenz containing 5 mM Tris-HCl (pH 8.0) and 5 mM $MgCl_2$.

If the majority of the plasma membrane is in the P4 pellet then prepare a smooth membrane fraction from the S3 fraction as described in Section 4.1. Dilute the smooth membrane fraction with three volumes of 5 mM Tris-HCl (pH 8.0), centrifuge at 100 000 g for 40 min at 5°C and resuspend the pellet in 2 ml of 0.25 M sucrose, 5 mM Tris-HCl (pH 8.0). Prepare a discontinuous gradient of metrizamide or Nycodenz in a 14 ml polycarbonate tube from 2.2 ml each of the above solutions and layer 1 ml of sample carefully on top. Centrifuge at 160 000 g for 60 min at 5°C in a swing-out rotor. Harvest the bands as above and identify the banded material by enzyme analysis as described in Appendix V (5'-nucleotidase − plasma membrane, and glucose-6-phosphatase or NADPH-cytochrome c reductase − endoplasmic reticulum).

8. ACKNOWLEDGEMENTS

The author would like to thank the Cell Surface Research Fund for financial support.

9. REFERENCES

1. Blobel,G. and Potter,V.R. (1966) *Science (Wash.),* **154**, 1662.
2. Price,M.R., Harris,J.R. and Baldwin,R.W. (1972) *J. Ultrastruct. Res.,* **40**, 178.
3. Graham,J.M. (1982) in *Methodological Surveys (B): Biochemistry, 11; Cancer-Cell Organelles,* Reid,E., Cook,G. and Morre,D.J. (eds.), Ellis Horwood Ltd., Chichester, UK, p. 342.
4. Nigam,V.N., Morais,R. and Karasaki,S. (1971) *Biochim. Biophys. Acta,* **249**, 34.
5. Neville,D.M. (1968) *Biochim. Biophys. Acta,* **154**, 540.
6. Bergstrand,A. and Dallner,G. (1969) *Anal. Biochem.,* **29**, 351.
7. DePierre,J. and Dallner,G. (1976) in *Biochemical Analysis of Membranes,* Maddy,A.H. (ed.), Chapman and Hall, London, UK, p. 79.
8. Touster,O., Aronson,N.N., Dalaney,J.J. and Hendrickson,H. (1970) *J. Cell. Biol.,* **47**, 604.
9. Fleischer,S., McIntyre,J.O. and Vidal,J.C. (1979) *Methods Enzymol.,* **55**, 32.
10. Maunsbach,A.B. (1974) *Methods Enzymol.,* **31**, 330.

11. Trouet,A. (1974) *Methods Enzymol.,* **31**, 323.
12. Wattiaux,R., Wattiaux-De Coninck,S. and Ronveaux-Dupal,M.F. (1971) *Eur. J. Biochem.,* **22**, 31.
13. Collot,M., Wattiaux-De Coninck,S. and Wattiaux,R. (1976) in *Biological Separations in Iodinated Density Gradient Media,* Rickwood,D. (ed.), Information Retrieval Ltd., London, p. 89.
14. Fleischer,B. and Fleischer,S. (1970) *Biochim. Biophys. Acta,* **219**, 301.
15. Wattiaux,R., Wattiaux-De Coninck,S. and Vandenberghe,A. (1982) *Biol. Cell,* **45**, 475.
16. Wattiaux,R., Wattiaux-De Coninck,S., Ronveaux-Dupal,M.F. and Dubois,F. (1978) *J. Cell Biol.,* **78**, 349.
17. Pertoft,H. and Laurent,T.C. (1977) in *Methods of Cell Separations,* Catsimpoolas,C. (ed.), Plenum Press, NY, p. 25.
18. Reinhardt,P.H., Taylor,W.M. and Bygrave,F.L. (1982) *Biochem. J.,* **204**, 731.
19. Pertoft,H., Warmegard,B. and Hook,M. (1978) *Biochem. J.,* **174**, 309.
20. Appelkvist,E.L., Brunk,U. and Dallner,G. (1981) *J. Biochem. Biophys. Methods,* **5**, 203.

CHAPTER 6

Centrifugal Separations of
Mammalian Cells

A. BROUWER, R.J. BARELDS and D.L. KNOOK

1. INTRODUCTION

Detailed analysis of the biochemistry, function and behaviour of a specific cell population can be ascertained only on relatively homogeneous cell populations. However, cell preparations obtained from blood or tissues represent a mixture of different cell types. Many techniques are currently available for fractionating heterogeneous cell populations and for isolating fractions that are enriched for a specific type of cell even if it constitutes only a minor component of the original population.

The cellular properties used as the basis for cell separation or purification include cell size, density, electrical charge, (antigenic) surface properties and light-scattering characteristics. Since cell separations on the basis of parameters other than cell size and density cannot be accomplished by centrifugation techniques, they will not be considered in this chapter.

In practice, centrifugation proves to be a very powerful and flexible tool for the separation of cell populations. Examples of separations of cells by centrifugation for both analytical and preparative purposes will be described in this chapter. The major techniques available, namely, differential centrifugation, isopycnic density centrifugation, rate-zonal centrifugation (velocity sedimentation) and centrifugal elutriation, will be described in detail with a brief introduction outlining their theoretical background. For a more extensive introduction to the basic principles of centrifugation and sedimentation of particles in general, the reader is referred to other chapters in this volume and to other reviews (1 − 4).

2. USE OF CENTRIFUGATION FOR CELL SEPARATIONS

Centrifugation can be applied to separate cells solely on the basis of differences in cell density as well as differences in sedimentation rate, which reflects both cell size and density. The methods available are summarised and compared with respect to their essential characteristics in *Table 1*. More detailed descriptions of the methods summarised in *Table 1* are given in the following sections (Sections 2.1 and 2.2).

Table 1. General Characteristics of the Various Methods for the Separation of Cells by Centrifugation.

Property used	Nomenclature	Centrifugal force	Gradient shape	Specific equipment	Limitations	Major advantages
Density (ϱ)	Isopycnic equilibrium density centrifugation	Relatively high (100 – 30 000 g)	Continuous/discontinuous (ϱgradient $\cong$ ϱcells)	Swing-out, fixed angle or zonal rotor	Overlapping density profiles; high centrifugal forces; capacity	Least sensitive to cell aggregation
Sedimentation rate (D and ϱ)	Differential pelleting	1 – 300 g	No gradient	–	Low resolution	Rapid; easy
	Unit gravity velocity sedimentation	1 g	Continuous gradient	Special separation chamber	Capacity $\pm$ 50 x 10^6 cells; special equipment; long duration	Simple; inexpensive
	Velocity sedimentation or rate-zonal centrifugation	20 – 1000 g	Continuous gradient: (linear/isokinetic (ϱgradient $<<$ ϱcells)	Swing-out or zonal rotor	Low resolution or special equipment; capacity	Large capacity (zonal); simple (swing-out)
	Centrifugal elutriation	100 – 1000 g	No gradient	Elutriator rotor and Beckman J-21 or J-6B centrifuge	Special equipment	Rapid; high viability; high resolution and capacity

D: cell diameter; ϱ: density.

2.1 Separation on the Basis of Density

For density separations, cells are included in, or layered on top of, a continuous or discontinuous density gradient. The density range of the gradient should cover the densities of the cells to be separated. After centrifugation to equilibrium, each cell type will be found at a position in the gradient which corresponds to its own density, irrespective of its size. Equilibrium is generally achieved only if the cells are subjected to centrifugal forces much higher than those used for velocity sedimentation or centrifugal elutriation. Isopycnic density centrifugation can be used for both preparative and analytical purposes. For preparative separations, the applications are limited, since, even within a homogeneous population of one type of cell, there is heterogeneity in cell density (3). This often results in a serious overlap in the density profiles of different cell types (see Section 5.2.1). Isopycnic density centrifugation may be successfully applied, however, in combination with other separation techniques which separate cells on the basis of other criteria.

Cells can be introduced into the centrifuge tube in two ways, either by layering a (concentrated) suspension of cells on top of a prepared continuous or discontinuous density gradient or by distributing the cells throughout the gradient. In the latter case, cells are initially scattered throughout the density gradient and some (lighter) cells will find their equilibrium position by moving centripetally (flotation) while others sediment.

The choice of using continuous or discontinuous gradients depends on the aims of the experiment. Discontinuous gradients are occasionally useful for preparative purposes, for example, for the separation of red blood cells from white blood cells by a single-step density separation in Ficoll-metrizoate (see Section 5.4.5) or from non-parenchymal liver cells using metrizamide (see Section 5.2.2). Discontinuous gradients are generally not recommended because of the higher risks of artifacts (selection) at interfaces and the generally poor results obtained (3). Continuous gradients are often used analytically to correlate differences in cell density within populations of cells with differences in functional or biochemical parameters (see Section 5.2.1).

2.2 Separation According to Sedimentation Rate

For separations on the basis of sedimentation rate, cell size is the major determinant but cell density is also important. This is reflected by the equation describing the ideal behaviour of cells in suspension subjected to a centrifugal field. The sedimentation velocity of a cell (dr/dt) can be expressed as:

$$\frac{dr}{dt} = \frac{r_p^2 (\varrho_p - \varrho_m)\, \omega^2\, r}{K\eta}$$

where r_p is the cell radius, ϱ_p is the density of the cell, ϱ_m is the density of the medium, ω is the angular velocity of centrifugation, r is the distance to the centre of rotation, η is the local viscosity of the medium and K is a constant. Several techniques can be employed to separate cells on the basis of sedimentation rate and these are described in the following sections.

2.2.1 *Differential Pelleting*

This term refers to a procedure by which cells can be separated in the absence of a density gradient. A homogeneous solution containing a mixture of cells is centrifuged or left to sediment at unit gravity. The more rapidly sedimenting cells that pellet at the bottom of the tube are separated from the supernatant containing the less rapidly sedimenting cells. The resolution of this method is very poor and only few examples of satisfactory results can be found in the literature. Even for the separation of parenchymal liver cells from the much smaller non-parenchymal cells, the result is generally unsatisfactory. However, since differential pelleting is a very quick and simple procedure, it is often used as the first step in a purification procedure. The resolution of differential pelleting may be increased in some instances by use of a supporting column of a Percoll solution such as described by Pertoft and Laurent (4).

2.2.2 *Unit Gravity Velocity Sedimentation*

This technique does not involve centrifugation. A thin layer of cell suspension is placed on the top of a preformed density gradient and left to sediment into the gradient for one to several hours depending on the cells being separated. The shallow gradient has a maximum density less than that of the cells to be separated and is used only to stabilise the position of bands in the column against convection currents. The collection of fractions is started before the first cells reach the bottom of the gradient. Because the cells are sedimented only by the Earth's gravitational field, this technique is relatively slow. The application of unit gravity velocity sedimentation is discussed in Section 5.3.

2.2.3 *Velocity Sedimentation by Rate-zonal Density Gradient Centrifugation*

This method is essentially the same as that for unit gravity separations, but employs centrifugation to decrease the time required to achieve the separation. Velocity sedimentation can be performed either in a normal swing-out rotor or in a reorienting zonal rotor. For small numbers of cells ($1 - 50$ x 10^6) and for the separation of cells with large differences in sedimentation rate, centrifugation in swing-out rotors may give sufficiently good separations. For high cell numbers ($<10^9$ cells), large capacity reorienting zonal rotors can be used. Zonal rotors require more skill and care, especially when working under sterile conditions.

A rather specialised method for the velocity sedimentation of cells, with a high resolution, has been developed by Tulp *et al.* (5). This method employs a special separation chamber that can be used in an adapted swing-out rotor of a low-speed centrifuge. However, the capacity of the chamber (<50 x 10^6 cells) is rather low for preparative purposes.

In summary, velocity sedimentation can be a very simple technique when working with cells that are of very different sizes. However, for cells with relatively small differences in sedimentation rate and for larger numbers of cells, special equipment is necessary. The use of velocity sedimentation is discussed further in Section 5.3.

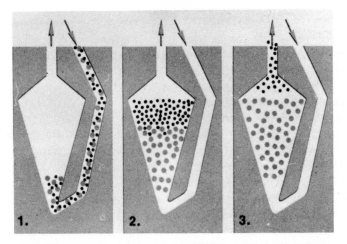

1. Sample suspended in a medium enters chamber

2. Sedimentation tendency of particles balanced by counterflow

3. Flow increased—slow-sedimenting particles elutriated out of chamber

Figure 1. A diagrammatic illustration of the behaviour of cells in the standard Beckman separation chamber during centrifugal elutriation (Reproduced, with permission, from the Beckman instruction manual for the JE-6B elutriator rotor).

2.2.4 *Centrifugal Elutriation or Countercurrent Centrifugation*

The term centrifugal elutriation describes a technique which involves the balance between a centrifugal force and a centripetal flow of liquid within a separation chamber (*Figure 1*). Because of the conical shape of the separation chamber, a gradient of liquid flow rate which opposes the centrifugal force is generated. Cells present in the separation chamber will be found in a position at which the two forces acting on them are at equilibrium. The position of each cell is determined by its size, shape and density. No pelleting occurs and the fractions can be harvested by increasing the flow rate or decreasing the centrifugal force allowing each fraction to be eluted from the chamber separately (*Figure 1*). The advantages of centrifugal elutriation include:

(i)   one can use any medium in which cells sediment;
(ii)   the viability of cells remains high;
(iii) the rotor has a large capacity range;
(iv)   the rotor is capable of high resolution and
(v)   only a short time is needed for separation.

On the other hand, the equipment is relatively expensive and requires an experienced operator. Application of this technique is described in detail in Section 5.4.

2.3 Gradients for Cell Separations

The range of media that can be used for preparing density gradients for cell separations is rather limited. As described in Section 5.1 of Chapter 1, there are several criteria that must be fulfilled by gradient media namely:
(i)   the medium should be non-toxic to cells;
(ii)   it should not permeate into cells;
(iii)   it should form low viscosity solutions and
(iv)   it should be possible to form isotonic gradients of sufficient density.

The list of media that can be used successfully for separating cells now includes Ficoll, Percoll, serum albumin, metrizamide and Nycodenz. A comparison of the properties of these media has already been given in Section 5 of Chapter 1. In this chapter, examples are given for the use of Percoll (Section 5.4.2) and metrizamide (Sections 5.2.1 − 5.2.3). Nycodenz has properties similar to those of metrizamide, but has some practical advantages (e.g., autoclavability) that make it particularly useful for separating cells (7).

The shape of density gradients (which can be determined by plotting the density of each fraction of the gradient) can be varied. The shape of a gradient can be discontinuous, or continuous; in the latter case either linear or non-linear gradients can be generated. Linear and non-linear gradients can be generated easily using simple gradient makers (see Section 5.2.1 of this Chapter and Section 5.1 of Chapter 2). In addition, one can use isokinetic gradients that are designed in such a way that the increase in the density and viscosity of the gradient solution balances the increase in centrifugal force along the length of the centrifuge tube. The latter is specifically developed for rate-zonal separations.

When separating cells on density gradients special care must be taken to make sure that the gradients are isotonic and that the osmolarity varies by less than 10% throughout the gradient. Otherwise, a cell travelling through the gradient may shrink or swell during the separation altering its sedimentation behaviour and ending up at a position reflecting a density and size other than that of the original cell. With gradients formed by a gradient maker, homogeneous osmolarity is easily achieved by starting with two isotonic solutions. In the case of discontinuous gradients and continuous ones generated by the diffusion of discontinuous gradients, there is a risk of local changes in osmolarity due to differences in the diffusion properties of the solute forming the density gradient and that used as the osmotic balancer, usually NaCl. One advantage of using colloidal media (e.g., Ludox or Percoll) is that they are osmotically inert and hence stable iso-osmotic gradients can be prepared simply by adding NaCl to a final concentration of 0.9% NaCl prior to forming the gradients by centrifugation.

2.4 Problems and Artifacts of Cell Separations

Various problems and artifacts may result in suboptimal or misleading separations. The significance of these artifacts and the means available to circumvent them are briefly discussed below. The major artifacts include the following.

2.4.1 *Cell Clumping*

A major problem encountered in cell separation is cell clumping, since clumped cells will behave differently during centrifugation. Serious clumping will prevent sufficient purification and result in a reduction of the cell yield obtained. The usual measures that can be taken to reduce cell clumping include working at low cell concentrations and the addition of components such as albumin, serum, DNase (to reduce gelation), protease or EDTA to the medium (6). In some instances, working at 4°C may reduce clumping (3). Although in the authors' experience with sinusoidal liver cells, the opposite has been found in that the clumping of cells during the initial steps of the isolation procedure can be largely prevented by keeping the temperature above 20°C.

2.4.2 *Overloading of Gradients with Cells*

For the optimal separation of cells by centrifugation in a density gradient, the number of cells applied should not exceed the capacity of the gradient. The gradient capacity can be defined as the number of specific cells that can be sedimented in a given experimental situation without observing any deviation from the ideal sedimentation pattern (3). Overloading a gradient with cells will result in the broadening of bands and an apparently faster sedimentation of cells. The gradient capacity must be estimated experimentally for any given situation, since no accurate theoretical descriptions are available (3).

2.4.3 *Wall Effects*

Wall effects are the consequence of the fact that the direction of sedimentation within a normal centrifuge tube is not parallel to the wall of the centrifuge tube even in swing-out rotors (*Figure 2*). This leads to concentration of cells at the periphery of the tube, thus increasing the risk of clumping and the loss of cells sedimented onto the wall of the centrifuge tube. In addition, the local density of the gradient may increase, resulting in a disturbance in the gradient stability. The loss of cells due to wall effects can be minimised by reducing the distance that cells have to travel through a gradient and by increasing the distance of the tube from the axis of rotation. Centrifugation of cells in a fixed-angle rotor, as is sometimes used for isopycnic separations, generally results in serious wall effects.

2.4.4 *Swirling*

Swirling or vortexing of the tube contents during acceleration and deceleration can induce serious disturbances in the gradient and the separated bands. The changes in the angular velocity of the rotor are accompanied by an angular momentum termed Coriolis's forces which cause rotational movement of the fluid. Swirling can be reduced by increasing the distance from the axis of rotation and by slow acceleration and deceleration. Sometimes, it is possible to introduce a special anti-vortex baffle into the separation chamber (5). The effects of swirling can also be partially counteracted by using steeper density gradients.

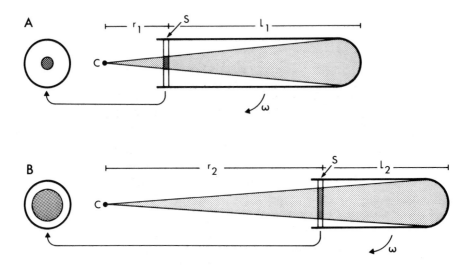

Figure 2. A schematic illustration of wall effects during centrifugation. In **A**, where the distance to the axis of rotation (C) is short (r_1) and the tube length is long (l_1) the radial direction of the centrifugal force allows only a minor proportion of the cells in the sample layer (S) to reach the bottom of the tube without coming into contact with the wall and hence wall effects are worse. The view of the sample layer surface given on the left indicates that only a small percentage of cells (indicated by the shaded area) will escape interaction with the wall of the tube under these conditions. In **B**, where there is an increased radius (r_2) and a decreased tube length (l_2), wall effects are much less pronounced.

2.4.5 Streaming (Droplet Formation)

When a suspension of cells is layered on top of a density gradient and left to stand without centrifugation, small droplets that fall into the gradient may be formed at the interface; this can occur only after a few minutes. This phenomenon appears to result from the diffusion of the gradient medium into the sample layer (30). It may be circumvented by reducing the cell concentration in the sample, by increasing the steepness of the gradient at the sample/gradient interface or by centrifuging the gradients immediately after loading the sample.

2.4.6 Centrifugal Force

Depending on the type of cell, all cells are affected to some extent by centrifugal forces. This may result in the loss of cells and reduced cell viability. In general, isopycnic centrifugation is carried out using higher centrifugal forces than for velocity sedimentation, since equilibrium must be achieved. The centrifugal forces can be reduced by shortening the distance that cells have to travel through a gradient. In addition, the use of low viscosity media (see Section 2.3) is recommended to ensure the optimal sedimentation velocities of cells. Centrifugal elutriation has the advantage that it can be carried out using relatively low centrifugal forces.

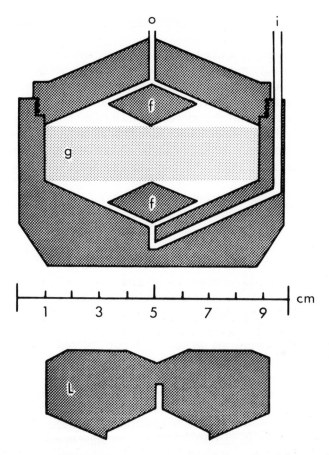

Figure 3. Special separation chamber for cell separation by centrifugation. This separation chamber, developed by Tulp *et al.* (5), can be installed in an adapted centrifuge and allows separation of cells and particles with high resolution using moderate centrifugal forces. The Perspex (darkly-shaded) chamber can be loaded and unloaded without disturbance to the gradient, since flow deflectors (f) are present. The lightly shaded area within the chamber (g) indicates the position of the density gradient during centrifugation. L, one of the vanes of the antivortex cross; i, inlet; o, outlet.

2.4.7 Osmolarity Effects

A change in osmolarity of the medium surrounding a cell will induce a change in cell volume. Swelling or shrinkage of cells due to hypo- or hyper-osmolarity, respectively, will affect their sedimentation properties. Therefore, the osmolarity of the medium must be adjusted to the same as that of the cells. In the authors' experience with different types of rat-liver cells, media used for isolation and suspension of cells can have an osmolarity exceeding the physiological osmolarity of rat plasma (~ 285 mOsm) by up to about 10% without affecting the morphology or viability of cells. Hypo-osmolarity, leading to swelling, appears to be more critical in causing changes in ultrastructural morphology and cell viability. Special precautions should be taken with regard to osmolarity when working with density gradients (see Section 2.3).

Most of the problems described in this section are not encountered when using centrifugal elutriation, which is described in detail in Section 5.4. A rather sophisticated alternative is offered by special separation chambers which resemble those used for unit gravity sedimentation but that can be inserted into a centrifuge (*Figure 3*). This chamber combines a relatively large effective surface and a short sedimentation distance for cells with special devices to prevent swirling and disturbances during layering and collecting of samples. The chamber is used in a large rotor of a low-speed centrifuge which maximises the distance from the centre of rotation. In this arrangement, artifacts due to swirling, wall effects and centrifugal forces are minimised. The equipment can be used for both isopycnic and rate-zonal (see Section 5.3) separations. Disadvantages are the relatively high price and the limited capacity ($<50 \times 10^6$ cells).

3. CHARACTERISATION OF CELLS AND ANALYSIS OF RESULTS

Before the separation of a heterogeneous population of cells can be attempted, methods must be devised for distinguishing between the various cell types that are present. Therefore, the selection of appropriate markers for the characterisation of the cells is very critical. The markers that can be selected depend, of course, not only on the cells to be purified, but also on the nature of the contaminating cells and on the preference of the investigator.

Generally, microscopy forms the basis of most methods for the identification of cells. Routine light microscopic examination is sometimes sufficient to identify cell types, for example, in the separation of large parenchymal and the smaller non-parenchymal liver cells. In most other instances, the cells must be stained before they can be identified. This can be done by routine histological staining of cell smears (e.g., May-Grünwald, Giemsa) such as are used for the morphological identification of various types of white blood cells as well as by immunochemical or cytochemical staining, by which the presence of a certain antigenic property or enzymatic activity can be visualised. These microscopic methods allow an assessment of the actual percentage of cells in the sample which are associated with the chosen marker. A different type of information can be obtained by biochemical analysis of purified cells. A positive marker for the desired cell type is of less value here, since variations in the levels of a marker from one cell to another can prevent its accurate use for the determination of cell purity. A negative marker which is absent from the target cell but present in all cell types that are possible contaminants is clearly more useful.

For the different types of liver cells that are used for most of the experimental examples given in this chapter, there are several independent markers, both biochemical and morphological, that can be used to distinguish each liver cell type from the other ones (6). These methods include light and electron microscopy and biochemical analysis (8 – 10). In this chapter, the discussion will be limited to light-microscopy techniques combined with enzyme cytochemistry. Parenchymal cells can be recognised directly using a light microscope because of their large size (diameter $\cong 20$ μm). Fat-storing cells also have a very characteristic morphology associated with the presence of numerous large lipid droplets, which also allows direct recognition by light microscopy. Kupffer cells, endothelial cells and lymphocytes can be distinguished from each other by light microscopy only after they have been stained histologically or cytochemically. Here, peroxidase and non-specific esterase will be described as examples of cytochemical markers. In the rat, peroxidase is found exclusively in Kupffer cells while, although esterase activity is present in all liver cells, lymphocytes do not exhibit this activity (11). The value of these markers has been confirmed in previous studies, by comparision with other, independent, markers such as ultrastructural morphology (8 – 13), histochemical staining, other enzyme markers (12) and specific vitamin A fluorescence (fat-storing cells) (13,14).

Apart from purity, the viability of a cell preparation must also be assessed. A rapid method for assessing cell 'viability' which is frequently used is based on the capacity of cells to exclude trypan blue. This gives an indication only of the integrity of the cellular membrane. In practice, cells which exclude trypan blue may in fact prove to be non-viable when assessed for functional or morphological properties (6). Therefore, each method for cell separation should be checked for its influence on cell morphology and function. The methods presented in this chapter have been extensively studied for their effects on these parameters and were found to be satisfactory with regard to ultrastructural morphology (8,9,13), biochemistry (10,12,14) and the ability of cells to survive in culture (15 – 17). Most of these studies employed rather laborious techniques which will not be included in the experimental protocols presented here.

4. GUIDELINES FOR DEVISING A METHOD FOR CELL SEPARATIONS

This section is designed to give guidelines for deciding on the best procedure to be used for the purification of a particular type of cell ('A' cells) from a heterogeneous suspension.

(i) As a prerequisite, a method to distinguish 'A' cells from all other cell types present in the suspension should be selected. This marker will be used to determine the percentage of 'A' cells present in the suspension and in any cell sample obtained after fractionation. In addition, criteria for cell viability should be defined to evaluate the effect of the separation procedure on the cells.

(ii) Determine the density profiles of 'A' cells and contaminating cells, by isopycnic density centrifugation in a continuous gradient using various density media (see Section 2.3) in isotonic solutions. Compare the density profiles obtained using different media, check the recovery and viability of the cells. Select the best gradient medium. Prepare a density gradient with a density range that includes that of all cells present in the suspension and determine the density profile of the cells. Any of the following results might be obtained: (a) complete purification of 'A' cells; (b) considerable enrichment of 'A' cells although not sufficiently pure. In this case, density separation in combination with other techniques may prove useful; or (c) there may be no enrichment of 'A' cells.

(iii) Determine the size distribution of 'A' cells and contaminating cells by light microscopy. The extent of overlap will give a first indication as to whether a separation on the basis of sedimentation rate is likely to be successful.

A combination of the results from steps (ii) and (iii) will indicate whether there is a good possibility of purifying 'A' cells on the basis of sedimentation rate by rate-zonal or elutriation techniques. A theoretical estimation of the sedimentation rate of 'A' cells can be calculated from the average size and density of the cells. The homogeneity of 'A' cells with respect to these parameters and the extent of overlap with contaminating cells will indicate whether high resolution methods such as centrifugal elutriation or those employing special separation chambers are necessary. The estimated sedimentation rate of 'A' cells can be used to predict the experimental conditions necessary for purification by rate-zonal centrifugation or centrifugal elutriation. A trial separation can be carried out, for instance, with centrifugal elutriation. The experimental conditions can be selected by using the rotor speed and flow-rate nomogram that is included in the Beckman instruction manual (see Appendix A). A fixed rotor speed can be selected and the suspension of cells fractionated by applying a series of flow rates centred around the predicted flow rate necessary for the elution of 'A' cells. The results may be used to establish the conditions necessary for the preparative purification of 'A' cells.

If neither of these methods alone leads to full separation from the contaminating cells, subsequent application of two methods which select for differences in density and sedimentation velocity may give better results. If this approach is also unsuccessful, attempts can be made to apply methods of cell separation that select for other criteria, such as mentioned in Section 1.

5. EXPERIMENTAL PROTOCOLS FOR CELL SEPARATIONS

5.1 Isolation of Rat-liver Cells

The experimental examples described in this chapter mainly concern the separation and purification of the various cell types present in the rat liver. Although the liver appears to be a relatively homogeneous tissue, it contains a variety of cell types (11). The main type of cell is the parenchymal cell, which accounts for over 90% of the total liver volume and about 60% of all liver cells (11). The other cells (non-parenchymal cells) include primarily sinusoidal cells, of which the Kupffer, endothelial and fat-storing cells are the most abundant (11). In total, the relatively small sinusoidal cells account for about 35% of all liver cells but only for a small percentage of the liver volume (11).

There are several methods for the preparation of cell suspensions from the rat liver (6). Generally they involve either perfusion of the liver followed by incubation with collagenase resulting in a suspension that contains all types of liver cell, namely parenchymal cells and all the types of sinusoidal cells (6,18).

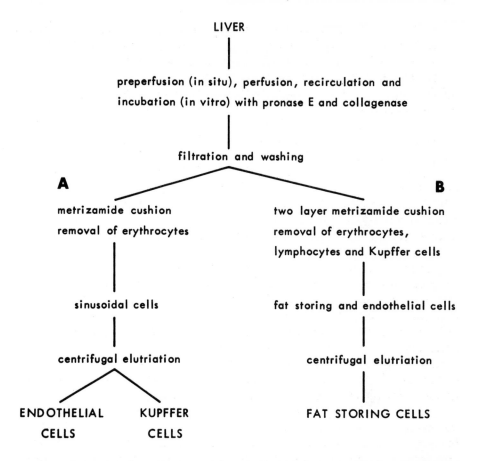

Figure 4. A schematic illustration of the methods for the isolation of sinusoidal liver cells (Section 5.1) and the subsequent purification of endothelial, Kupffer and fat-storing cells. The preparative metrizamide density gradients is described in detail in Sections 5.2.2. and 5.2.3 for **A** and **B**, respectively. Purification by means of centrifugal elutriation are described in Sections 5.4.3. (**A**) and 5.4.4. (**B**).

Alternatively, the liver is perfused and incubated with pronase E, which selectively destroys parenchymal cells, resulting in a suspension of non-parenchymal liver cells consisting almost exclusively of sinusoidal cells (6,8). In order to optimise the digestion of the extracellular matrix and obtain a higher yield, especially of fat-storing cells, collagenase can also be included in the incubation medium during pronase treatment (*Figure 4*) (13).

For the experiments described below, the following two methods for liver cell isolation are used.

(i) Parenchymal liver cells are obtained from 3-month-old female WAG/Rij rats by perfusion and incubation of the rat liver with collagenase-containing medium (18) as described in Appendix B. The parenchymal cells are harvested from the liver digest by differential centrifugation at 50 *g* for 3 min.

(ii) Sinusoidal cells are obtained from the liver of female BN/BiRij rats by perfusion and incubation of the liver with medium containing pronase E and collagenase (*Figure 4*). This method is described in detail in Appendix B. After incubation, the liver digest is centrifuged for 10 min at 300 *g* and the pellet represents the sinusoidal cell mixture that is used for the separation experiments.

5.2 Isopycnic Centrifugation of Cells

5.2.1 *Continuous Metrizaminde Gradient Fractionation of Sinusoidal Cells*

Freshly isolated sinusoidal cells are separated according to differences in density and the density profiles of both Kupffer and endothelial cells are analysed.

Equipment, chemicals and solutions

A simple gradient maker with magnetic spinbar and stirrer;
peristaltic pump, silicon tubing (i.d., 1.3 mm; o.d., 3.0 mm) and glass capillary extensions (15 cm);
Abbé refractometer;
centrifuge with a swing-out rotor;
centrifuge tubes (Corex, 15 ml);
medium for sinusoidal cells, Gey's balanced salt solution (GBSS); NaCl (8000 mg/l); KCl (370 mg/l); $MgSO_4.7H_2O$ (70 mg/l); $NaH_2PO_4.2H_2O$ (150 mg/l); $CaCl_2.2H_2O$ (220 mg/1); $NaHCO_3$ (227 mg/l); KH_2PO_4 (30 mg/l); $MgCl_2.6H_2O$ (210 mg/l); glucose (1000 mg/l); adjusted to pH 7.4; osmolarity: 275-285 mOsm. Sterilise by filtration and store at 4°C;
30% (w/v) metrizamide (Nyegaard & Co., Oslo) in GBSS without NaCl; osmolarity, 285 mOsm; this solution is pH 7.4;
sinusoidal liver cells (see Section 5.1);
0.5% w/v trypan blue in physiological saline (0.9% NaCl in water).

Peroxidase staining (8):

conical tubes (15 ml);
centrifuge with a swing-out rotor;
Bürker counting chamber;
microscope;
water bath (37°C);
incubation medium: prepare the following mixture: dissolve 15 mg of 3,3-diaminobenzidine tetrahydrochloride.$2H_2O$ (DAB) (Merck) in 15 ml of 0.05 M Tris-HCl (pH 7.4) containing 7% w/v sucrose (300 mOsm). Add 10 μl of 30% H_2O_2. Use within 30 min.
CAUTION: DAB is suspected of being carcinogenic.

Experimental protocol. Freshly isolated sinusoidal cells (200 – 300 x 10^6 cells) from one rat liver (see Section 5.1) are used. Wash the cells 3 times with 15 ml of GBSS and collect them by centrifugation for 5 min at 300 *g* at room temperature. This is done to remove most of the parenchymal cell debris and to reduce the risk of cell clumping. The final pellet is suspended in about 6 ml of GBSS.

Dilute the 30% (w/v) metrizamide stock solution with GBSS to a final concentration of 20% (w/v). This 20% metrizamide solution (solution A) is used to make a continuous gradient from 20% to 8%. The gradient is prepared using a simple gradient maker with two vessels of equal dimensions (20 – 50 ml).

Fill the first vessel of the gradient maker with 8.5 ml of solution A. Mix 5.4 ml of cell suspension with 3.1 ml of solution A and place this into the second vessel. If clumping occurs add 0.34 mg of DNase to the cell suspension. The second vessel of the gradient maker is connected to a two-channel peristaltic pump. Each channel is connected to a 15 cm long glass capillary, the outlet of which is placed at the bottom of a Corex centrifuge tube. Adjust the pump speed to 1 ml/min (per channel). Keep the gradients at 4°C. After the gradients are made, the capillaries are carefully removed from the tubes.

Centrifuge the gradient for 30 min at 2000 g at 4°C with slow acceleration and deceleration. To avoid cell clumping, the run time should not exceed 45 min. After centrifugation, the gradients are fractionated preferably by upward displacement (see Section 7.3 of Chapter 2) alternatively it is possible to use a reversal of the loading procedure. For this latter method the glass capillary is inserted with its end a few millimetres above the pellet. Collect about 25 fractions of equal volume (~0.4 ml). The pellet is collected separately.

Analysis. Determine the cell number and their viability in each fraction. For cell viability assessment, mix equal volumes of the cell suspension and 0.5% trypan blue (± 25 μl) and immediately determine the percentage of cells stained by the dye. The stained cells represent non-viable cells. Determine the density of each fraction from the refractive index using an Abbé refractometer.

Peroxidase staining reaction: add a few drops of each gradient fraction (±0.5 x 10^6 cells) to about 2 ml of incubation medium. Incubate for 30 min at 37°C. Centrifuge the suspension for 3 min at 300 g. Remove the supernatant and resuspend the cells in about 200 μl of GBSS. Determine the percentage of positively staining cells in a haemocytometer. Erythrocytes will be stained almost homogeneously black, while Kupffer cells will have a brown to black colour. The erythrocytes and Kupffer cells can be easily distinguished on the basis of cell size. Endothelial and white blood cells will remain unstained.

The results obtained in a typical experiment are shown in *Figure 5*. Almost all of the erythrocytes present in the original preparation are recovered in the pellet (~30 x 10^6 cells). The peroxidase-positive Kupffer cells and the negative endothelial cells and lymphocytes all form broad bands ranging between 1.060 g/cm^3 and 1.098 g/cm^3. There is a large overlap between the different cell types.

As a result, the maximum purity of Kupffer cells that can be obtained using this technique amounts to 25% in fraction 10. This fraction represents only a small percentage of the total number of Kupffer cells present. For endothelial cells, the maximum yield and purity is much higher. Pooling of fractions 14 – 24 would result in an endothelial cell preparation of 90% purity which contains about 50% of the endothelial cells in the gradient. Such results show

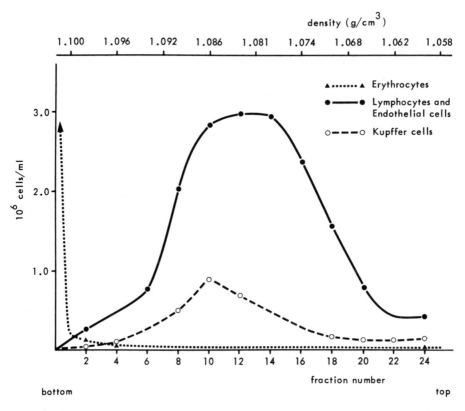

Figure 5. Fractionation of sinusoidal liver cells on a continuous metrizamide gradient. Sinusoidal liver cells were centrifuged, fractionated and analysed as described in Section 5.2.1. The viability of all fractions throughout the gradient was at least 90%.

that Kupffer and endothelial cells of the normal rat liver cannot be completely purified by isopycnic density centrifugation alone. Previous results from the authors' laboratory have shown that the density separation of these cell types can be improved by preloading of the Kupffer cell population with Jectofer® *in vivo*. To do this rats are injected intramuscularly (biceps femoris) with 0.2 ml of Jectofer® per 150 g of body weight 4 days and 1 day before isolation of the cells as described in Appendix B. Jectofer® is an iron-sorbitol-citric acid complex (Astra, Södertälje, Sweden) containing 50 mg iron per ml. After this treatment the density distribution of Kupffer cells is shifted towards higher densities and relatively pure fractions of each cell type can be obtained (9).

5.2.2 Single-Step Discontinuous Metrizamide Gradients for the Preparative Purification of Sinusoidal Liver Cells

This procedure is routinely used in the authors' laboratory to eliminate red blood cells and parenchymal cell debris from the sinusoidal cell suspension obtained after treatment of the liver with pronase and collagenase (10,16). A single-step density separation between denser erythrocytes, which sediment to the bottom, and the floating sinusoidal cells is achieved.

Equipment, chemicals and solutions

Centrifuge with a swing-out rotor;
Gey's balanced salt solution (GBSS) (see Section 5.2.1);
30% (w/v) metrizamide in GBSS without NaCl (see Section 5.2);
15 ml plastic centrifuge tubes;
sinusoidal liver cells (see Section 5.1);
trypan blue staining (see Section 5.2.1);
peroxidase staining (see Section 5.2.1).

Esterase staining:

Equipment as for peroxidase staining (see Section 5.2.1.);
physiological saline (0.9% NaCl);
fixative: 2.5% glutaraldehyde in 0.1 M sodium cacodylate buffer (pH 7.4). **CAUTION: Cacodylate is highly toxic** and may be replaced by 0.1 M phosphate.
Incubation medium: mix (in a fume cupboard) the following solutions: 9.5 ml of 0.1 M potassium phosphate buffer (pH 7.4), 0.5 ml ethylglycol-monomethylether containing 11 mg of 1-naphthylacetate (keep under a nitrogen atmosphere), 0.5 ml 1.0 M NaOH, 0.5 ml of a mixture of equal parts of 4% *p*-rosanilin in 2.0 M HCl and 4% sodium nitrite in distilled water. Adjust the mixture to pH 7.8; filter through cellulose paper and use within 30 min.

Experimental protocol. Dilute a suspension of $100-400 \times 10^6$ sinusoidal cells to 10 ml with GBSS. Add 14 ml of 30% (w/v) metrizamide and mix thoroughly but gently by pipetting. Divide this mixture equally between two centrifuge tubes. Carefully layer $1-2$ ml of GBSS on top of each tube and centrifuge at room temperature for 15 min at 400 g and allow the rotor to coast to a halt. Collect the sinusoidal cells by slowly aspirating the coherent top layer between the GBSS and metrizamide solution into a Pasteur pipette. The whole procedure is carried out at room temperature to reduce cell clumping. The sinusoidal cell suspension can be analysed directly and used as such or fractionated further into Kupffer and endothelial cells by centrifugal elutriation (see Section 5.4.3).

Analysis. Determine the cell number and viability of the sinusoidal cell suspension. Determine the percentage of peroxidase-positive cells as described in Section 5.2.1.

Esterase staining reaction: incubate a few drops of the cell suspension ($\sim 0.5 \times 10^6$ cells) with $1-2$ ml of fixative for 7 min at 4°C. Centrifuge the cells for 3 min at 700 g. Wash the pellet twice with 0.9% NaCl. Resuspend the cells in about 200 μl of 0.9% NaCl. Mix equal parts of this cell suspension and the incubation medium; incubate the mixture for 10 min at 37°C. Determine the percentage of positive-staining cells in a haemocytometer. Kupffer, endothelial and parenchymal cells will be stained red; erythrocytes and

Table 2. Preparation of Sinusoidal Liver Cells by use of a Single-step Discontinuous Metrizamide Gradient.

Parameter	Composition
Cell yield	44 x 10^6 cells/g liver
Viability (trypan blue)	99%
Cytochemical staining:	
peroxidase-positive cells	27%
esterase-positive cells	80%
Composition:	
Kupffer cells	26%
endothelial cells	54%
erythrocytes	1%
fat-storing cells	1%
parenchymal cells	1%
other cells (e.g., lymphocytes)	18%

Sinusoidal liver cells isolated from 3-month-old female BN/BiRij rats were centrifuged and fractionated as described in Section 5.2.2. The cells were stained and analysed by light microscopy for peroxidase and esterase activity.

leukocytes will remain unstained. Fat-storing cells will be faintly stained, but can be recognised by their typical appearance.

The results of a typical experiment are given in *Table 2*. The viability of the sinusoidal cell preparation, as judged by trypan blue exclusion, is 99%. This first indication of cellular integrity can be confirmed by electron microscopical studies (8 – 13), biochemical analysis (10,12) and the behaviour of the cells in maintenance culture (15,16,19). Kupffer and endothelial cells represent the majority of the cells in this preparation (80%). The esterase-negative lymphocytes are the major contaminants, erythrocytes being virtually absent.

5.2.3 *Two-step Discontinuous Metrizamide Gradients for the Preparative Partial Purification of Fat-storing Cells*

This procedure is used as the initial step in the purification of fat-storing cells from a sinusoidal cell suspension (13).

Equipment and chemicals. These are the same as described in Section 5.2.2.

Experimental protocol. Sinusoidal liver cells isolated from 12-month-old female BN/BiRij rats are used (100 – 400 x 10^6 cells).

Dilute the 30% (w/v) metrizamide stock solution with GBSS to 18% (w/v). Place 5 ml of this 18% metrizamide solution in each of two centrifuge tubes. Dilute the cell suspension to 12 ml with GBSS and mix this with 8 ml of the 30% metrizamide solution to give a final concentration of 12% metrizamide. Carefully layer half of the suspension into each tube on top of the 18% metrizamide solution. Carefully layer 1 – 2 ml of GBSS on top of the 12% metrizamide solution. Keep the tubes at 20°C and centrifuge for 17 min at

Table 3. Isopycnic Separation of Sinusoidal Liver Cells on a Two-step Discontinuous Metrizamide Gradient.

Cell type	Composition			
	Low density fraction		High density fraction	
	Cell number (10⁶)	% of total	Cell number (10⁶)	% of total
Fat-storing cells	20.1	15.8	2.6	1.7
Kupffer cells	6.7	5.3	37.6	25.5
Endothelial cells	98.2	77.0	107.5	72.8
Other cells (lymphocytes)	2.5	2.0		
Total	127.5	100.1	147.7	100

Sinusoidal cells isolated from 12-month-old female BN/BiRij rats were centrifuged in the metrizamide gradient and fractionated as described in Section 5.2.3. The low density (<1.069 g/cm³, on top of 12% metrizamide) and the high density (<1.098 g/cm³, on top of 18% metrizamide) fractions were stained for esterase (low density fraction only) and peroxidase activity.

1400 *g* without braking. Collect the cell layers found at the two interfaces in each tube.

Analysis. The two fractions can be characterised with regard to yield, viability and composition as described in Section 5.2.2. The results are given in *Table 3*. The low density fraction contains about 45% of the total cells recovered. This fraction is relatively enriched in fat-storing cells and contains about 90% of the fat-storing cells recovered. The major contaminants of this fraction are the endothelial cells, while the degree of contamination by lymphocytes, which are difficult to remove by subsequent centrifugal elutriation, is not very significant. As a result, this fraction can be used for the preparation of purified fat-storing cells by centrifugal elutriation (see Section 5.4.4). The high density fraction contains the majority of the Kupffer cells and lymphocytes as well as a portion of the endothelial cells.

5.2.4 *Other Applications of Isopycnic Centrifugation for the Separation of Cells*

As indicated earlier, isopycnic density centrifugation is widely used for the fractionation of heterogeneous samples of cells from various sources. In the previous sections, representative examples of isopycnic separations have been described. These examples essentially reflect the major experimental variations of isopycnic density centrifugation and these protocols can be easily modified for the separation of other cell types. The additional information on the densities and other details of specific cell types can be extracted from the literature. A useful overview by Pertoft and Laurent of the density ranges of a large variety of cell types has been published (4). This list includes all types of blood cells and cells from a wide range of tissues.

Next to, and often in combination with, velocity sedimentation, separation on the basis of density has played an important role in the separation and characterisation of bone marrow-derived cells (20). This extensive field, which involves many different cell types representing all stages of differentiation between stem cells and various types of mature differentiated cells in blood and tissues, has been reviewed by Williams (20).

5.3 Velocity Sedimentation of Cells

The principles of velocity sedimentation have been described in Sections 2.2.2 and 2.2.3. Several examples of cell separation by velocity sedimentation and the major experimental conditions are given in *Table 4*. Unless one uses rather specialised equipment, rate-zonal sedimentation is often limited in both its degree of resolution and the number of cells that can be separated in a single step. The specialised equipment that is used includes both large capacity reorienting zonal rotors (1) and special separation chambers (see Section 2.2.3). The use of zonal rotors for separations is described in Chapter 7 and a detailed protocol for the separation of white blood cells by rate-zonal sedimentation in a zonal rotor is given in Section 8.1 of Chapter 7. With regard to the separation of sinusoidal liver cells, the only method using velocity sedimentation that has been reported to give satisfactory results employs a special separation chamber and centrifuge (5). With this arrangement, the separation of sinusoidal cells into Kupffer and endothelial cells can be accomplished (5) with a resolution that is comparable with centrifugal elutriation (cf., Section 5.4.3). It has also been applied to the separation of dividing leukaemia cells and white blood cells (5). Other examples of velocity sedimentation can be found in studies on the fractionation of various types of colony-forming cells in the bone marrow (20,22), which is often performed at unit gravity (22). No detailed experimental protocols using rate-zonal sedimentation will be presented here.

5.4 Centrifugal Elutriation of Cells

The concept of applying centrifugal elutriation to the separation of cells was described by Lindahl as early as 1948 under the name of 'counterstreaming centrifugation' (23). Lindahl's idea was taken up and developed by Beckman in the mid 1960s, resulting in a relatively simple elutriator rotor which has become commercially available. The method uses the principle of counterflow centrifugation in which the movement of particles of different sedimentation rate is opposed by a controlled centripetal flow of liquid (elutriation) (*Figure 1*). This method does not need a density gradient, and pelleting, which can damage cells, does not occur (see also Section 2.2.4). The elutriator rotor is designed to separate and/or purify suspensions of cells or particles over the size range $5 - 50$ μm in diameter, mainly according to size (see Appendix A). It can also be used to provide cell-free serum or media.

5.4.1 Description of the Elutriator Rotor System

The original model of the elutriation rotor, the JE-6B, consists of a single separation chamber plus a by-pass chamber as counter balance. The JE-6B elutriator rotor is used either in J-21 series or in modified J-6B series cen-

Table 4. Examples of Cell Separation by Velocity Sedimentation.

Cell sample	Cell type(s) separated	Gradient	Run time/cen-trifugal forces	Equipment	Results Purity	Yield	Viability*	Reference
White blood cells	Lymphocytes(a); monocytes (b)	BSA (1.5 – 6.5%)	9 min/ 20 g	Special separation chamber (*Figure 3*)	(a) > 90% (b) 90%	(a) > 90% (b) 67%	99%	5
White blood cells	Lymphocytes (a); monocytes (b); basophils (c)	Ficoll (2 – 4%)	2 h/ 1 g	Unit gravity separa-tion chamber	(a) – (b) 69 – 77% (c) –	(a) – (b) 28% (c) –	>98%	22
White blood cells	Lymphocytes (a); monocytes (b)	Ficoll (3.4 – 16.9%)	45 min/ 2000 r.p.m.	Zonal rotor (MSE AXII)	–	–	–	22
Pre-fractionated white blood cells	Lymphocytes (a); basophils (b)	Ficoll (4.5 – 9.5%)	15 min/ 85 g	100 ml tubes; swing-out rotor	(a) > 95% (b) 62 – 72%	(a) ±90% (b) 30 – 50%	–	21
Human bone mar-row cells	CFU-C	Ficoll (2 – 4%)	2 h/ 1 g	Unit gravity separa-tion chamber	–	–	–	(See Chapter 7)
Pre-fractionated canine gastric cells	Parietal cells (a); chief cells (b)	Ficoll 2 – 4%	50 min/ 1 g	Unity gravity separation chamber	(a) < 60% (b) 85%	(a) ±50% (b) ±30%	–	22
Sinusoidal liver cells	Kupffer cells (a); endothelial cells (b)	Percoll (3.5 – 18%)	8 min/ 16 g	Special separation chamber (*Figure 3*)	(a) 98% (b) 97%	(a) 68% (b) 86%	>95%	5

*Yield represents the percentage of cells of one type that is recovered in the purified fraction.

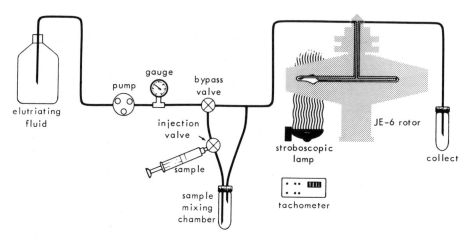

Figure 6. Scheme of the elutriator system, showing the positioning of accessories within and outside the centrifuge (Reproduced, with permission, from the Beckman instruction manual).

trifuges. A larger elutriator rotor, the JE-10x is also available. The JE-10x rotor can be used for the same applications as the standard JE-6B rotor. The separation chamber of the JE-10x has a 10 times greater volume (40 ml) than the JE-6B rotor chamber. Between 10^7 and 10^{12} cells can be separated in a single run (cf., Appendix A for the standard rotor). The JE-10x rotor can only be used in the J-6 series centrifuges. Elutriator rotors require several accessories, namely a stroboscope unit, a peristaltic pump for liquid delivery and removal (*Figure 6*) and a modified centrifuge door with a window.

In this section, only the smaller JE-6B rotor with its accessories, in combination with a J-21 centrifuge will be described. The later metal JE-6B elutriator rotor has a slightly different design from the original plastic rotor and has different caps on the rotor chambers. In our experience, both types of this rotor fractionate cells equally well. However, the metal JE-6B rotor is much more practical and can be assembled more easily without risks of leakage, or wrong connections.

Rotor chambers. The elutriator rotor has two epoxy-resin chambers. One of these chambers, the so-called by-pass chamber, is always installed in the by-pass position. One of the two types of separation chamber, the standard or Sanderson chamber (*Figure 7*), is always installed in the elute position of the rotor. The Sanderson chamber has a slightly higher resolving power (see Appendix A), which makes it especially useful for the separation of cells with only small differences in sedimentation rate. In cases where there is a great tendency for the cells to form clumps, the Sanderson chamber inlet (*Figure 7*) may have an advantage over the standard chamber in reducing the formation of clumps.

Use of the Sanderson chamber (Beckman order no. 354691) which is supplied for the metal JE-6B rotor may result in a poor separation or purification because of its smaller internal volume and height. To avoid this problem, the Sanderson chamber (order no. 335206) from the plastic elutriator rotor can

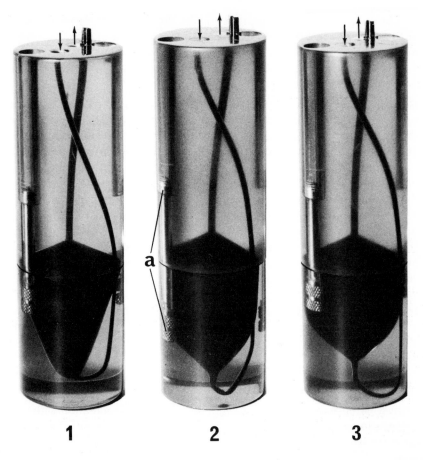

Figure 7. Separation chambers for centrifugal elutriation (in the JE-6B elutriator rotor). **(1)** Standard Beckman chamber. **(2)** and **(3)** Sanderson chamber. Note the difference in internal chamber volume and height between the original Sanderson chamber **(2)** and the later one **(3)**. The original chamber **(2)** can be combined with the cap of the later model only if the screws (a) are modified (see also Section 5.4.1).

be used in combination with the cap of the Sanderson chamber for the metal JE-6B rotor. The screws to connect this new type cap to the old chamber need to be longer than the standard ones; these can usually be made by most departmental workshops (see *Figure 7*).

Installation of the centrifugal elutriation system. Setting up the elutriator system is done in two stages involving firstly the assembly of the elutriator rotor followed by the installation of the stroboscope and rotor in the centrifuge and connection of the rotor inlet and outlet to the flow system. An accurate and complete description of all the necessary steps and preparations is included in the Beckman instruction manual. In addition, it contains an extensive section on troubleshooting. Therefore, the installation of the system will not be described in further detail here. However, it is important to emphasise the lubrication of the 'O' rings, in particular the two which are in contact with the rotating seal. Insufficient lubrication and/or wearing of these 'O' rings will cause cross-leakage between the inflow and outflow channels within the rotor. In this case, the effective flow rate within the separation chamber is reduced and does not correspond to the measured flow rate.

Operation. The collection of separated cell fractions of increasing diameter during centrifugal elutriation of a heterogeneous cell sample can be accomplished by either stepwise increases in the flow rate or stepwise decreases in the rotor speed. The protocols described below all involve variation in flow rates at a fixed rotor speed, since, in our experience, it is much easier to predict and control the course of the separation. The desired flow rates must be checked in advance and can even be corrected during the run. The variations in rotor speed are more difficult to control, especially with the standard J-21 centrifuge and the original type of stroboscope unit which gives a reading of the rotor speed only after 1 min: with the later types of stroboscope units (Kit numbers 354700 and 354679 for the J2-21 and J-6B centrifuges, respectively), continuous monitoring of the rotor speed is possible. If, in addition, the relatively inaccurate standard rotor speed control is replaced by one that gives a finer regulation of speed then the system is better-suited for achieving separations by variation in rotor speed. Even then, one has to take into account that there will always be an overshoot in speed of the rotor when changing from one speed to another.

5.4.2 *Separation of Parenchymal Liver Cells into Ploidy Classes*

Rat-liver parenchymal cells consist of a heterogeneous population with regard to their ploidy (24). The relative proportions of the different ploidy classes within the liver are dependent on the strain of the rat. In addition, there is an age-related shift towards higher ploidy classes (24). *Table 5* shows the distribution pattern of ploidy classes in livers of 3-month-old female WAG/Rij rats. At this age, mononuclear tetraploids (4n) represent the majority of the parenchymal cell population. The following section describes the separation of diploid, tetraploid and octaploid cells by means of centrifugal elutriation (25).

Equipment, chemicals and solutions

J-21 (or J-6B) modified centrifuge (Beckman);
JE-6B elutriator rotor with a stroboscope unit (Beckman);
Sanderson chamber in combination with the by-pass chamber;
silicone tubing (i.d., 3.0 mm; o.d., 5.0 mm, LKB), silicone tubing (i.d., 3.18 mm; o.d., 6.35 mm, Beckman) (for construction of the flow sytem);
centrifuge with a swing-out rotor;
Bürker counting chamber;
microscope;
peristaltic pump type 2115 (LKB) or an equivalent low-pulse pump;
50 ml centrifuge tubes;
elutriation medium: L-glutamine (131 mg/l), L-aspartic acid (13.3 mg/l), L-threonine (23.8 mg/l), L-serine (31.5 mg/l), L-glycine (37.6 mg/l),

L-alanine (53.5 mg/l), L-glutamic acid (132.4 mg/l), KCl (223.7 mg/l), $NaH_2PO_4.H_2O$ (96.6 mg/l), $MgCl_2.6H_2O$ (101.7 mg/l), $NaHCO_3$ (2.01 g/l), glucose (3.60 g/l), fructose (3.60 g/l), sucrose (67.4 g/l), the solution is adjusted to pH 7.4, osmolarity: 308 mOsm, store at $-20°C$;
parenchymal liver cell suspension (see Section 5.1);
trypan blue staining (see Section 5.2.1).

Experimental protocol. Parenchymal cells isolated from 3-month-old WAG/Rij rats are used. Assemble the rotor as directed in the Beckman Instruction Manual. The Sanderson chamber should be used in preference to a standard chamber, since, in the authors' experience, it can fractionate a greater number of parenchymal cells, thus allowing more cells to be elutriated; 140×10^6 parenchymal cells can be loaded into this chamber. Prepare the centrifuge (type J-21 Beckman) and install the stroboscope and flow harness. Place the rotor into the centrifuge, connect the silastic tubing to the rotor in the correct manner and fill the entire system with elutriation medium. Remove all the air bubbles from the system. Adjust the rotor speed to 1350 r.p.m. (210 g) and check the pump speed settings required. The cells are loaded into the elutriation system at 4°C. Use an initial flow rate of 12 ml/min. Wait until all of the cells have entered the elutriation chamber; this will take about a minute.

Collect a total of three elutriated fractions which can be termed I, II and III, corresponding to three successive flow rates of 19, 32 and 41 ml/min, respectively. Usually fractions of 100 ml of eluate are collected in each case, with the exception of fraction II (32 ml/min) for which 150 ml is collected. This is because cells of fraction II are still being elutriated after the initial 100 ml of eluate has been collected. Keep the collected cells at 4°C until the run is completed.

Stop the pump after the last fraction has been collected. Stop the centrifuge, collect the pellet from the Sanderson chamber and cells remaining in the mixing chamber. This enables determination of the overall recovery of cells. Centrifuge all of the fractions in a swing-out rotor at 1000 g for 10 min. Resuspend each of the pellets in about 3 ml of elutriation medium. Store the fractions at 4°C until further use.

Analysis. Determine the cell concentration, sample volume and hence calculate the cell yield of each fraction. Determine the viability of cells using trypan

Table 5. Proportions of Diploid and Polyploid Parenchymal Cells in the Liver of 3-month-old Female WAG/RIJ rats.

Ploidy class nuclei	Diploid (2n)	Tetraploid (4n)		Octaploid (8n)		Decahexaploid (16n)
	MD	BD	MT	BT	MO	BO
% of total cell number	9	18	53	16	4	2

MD, mononuclear diploid; BD, binuclear diploid; MT mononuclear tetraploid; BT, binuclear tetraploid; MO, mononuclear octaploid; BO, binuclear octaploid (24).

Table 6. Centrifugal Elutriation of Parenchymal Liver Cells: Separation into Ploidy Classes.

Fraction	Flow rate (ml/min)	Cell number (10^6 cells)	Viability (%)	Composition
Original suspension		137.8	84	2n, 4n, 8n, 16n, aggregates, debris
I*	19	39.8	72	2n, 4n, debris
II	32	54.7	85	4n (90%)
III	41	15.5	83	4n, 8n, aggregates
Pellet		17.1	82	As original suspension
Mixing chamber		5.2	85	As original suspension
Total		127.0 (92%)	—	—

Parenchymal cells were isolated from 3-month-old female WAG/Rij rats (cf., Section 5.1) and subjected to centrifugal elutriation as described in Section 5.4.2. Total recovery of cells after elutriation was 92%. Cell fractions were analysed for their proportions of ploidy classes by TGZ-3 analysis (Section 5.4.2.).
*Fraction I was further purified by centrifugation in a Percoll density medium (Section 5.4.2.) resulting in a suspension of 32 x 10^6 cells with a viability of 90%. This preparation consisted of 50% diploid (2n) cells, contaminated mainly with tetraploid cells.

blue. Calculate the overall recovery. Prepare dried slide preparations of each fraction using a sedimentation chamber. Stain the cells with Feulgen to stain the nucleus clearly. The ploidy classes in each fraction are determined from photomicrographs (x 1000) by measurement of the nuclear diameter (e.g., with a TGZ-3 particle size analyser) and determining the number of nuclei per cell (24).

Typical results are shown in *Table 6*. Fraction I contains most of the diploid cells in addition to tetraploid cells, non-viable cells, debris and some sinusoidal cells. When fraction I is subjected to centrifugation (275 *g* for 10 min at 4°C) in an isotonic 27% (v/v) Percoll solution, a pellet consisting of 50% diploid cells with high viability is obtained (*Table 6*). Fraction II consists of a relatively homogeneous (90%) population of viable tetraploid cells. The octaploid cells cannot be separated from tetraploid cells and small aggregates (fraction III). The overall recovery is usually over 90%, with only small losses of cells in the pellet fraction and the mixing chamber.

5.4.3 *Purification of Kupffer and Endothelial Cells*

Sinusoidal cells prepared using a single-step discontinuous metrizamide gradient (Section 5.2.2), consists of a mixture of Kupffer and endothelial cells, with lymphocytes as the main contaminant. From this mixture, purified Kupffer and endothelial cells can be prepared by means of centrifugal elutriation.

Equipment, chemicals and solutions

Equipment as in Section 5.4.2;
elutriation medium; GBSS (see Section 5.2.1);
medium for peroxidase staining (see Section 5.2.1);
medium for esterase staining (see Section 5.2.2);

trypan blue staining (see Section 5.2.1);
sinusoidal liver cells (see Section 5.2.2).

Experimental protocol. Sinusoidal liver cells isolated from 3-month-old BNBi/Rij rats are used (see Section 5.1). The isolated cells are first centrifuged in discontinuous metrizamide gradients as described in Section 5.2.2. The cells thus prepared can be elutriated using either the Sanderson or the standard chamber. The elutriation system is installed as described in Sections 5.4.1 and 5.4.2.

When using the Sanderson chamber, the rotor speed is adjusted to 3250 r.p.m. and the rotor is equilibrated to 4°C. The sinusoidal cells ($100-500$ x 10^6 cells) are loaded into the elutriation system in elutriation medium with an initial flow rate of 18 ml/min. A total of three elutriated fractions are collected, which are termed L, E and K, corresponding to flow rates of 18, 32 and 48 ml/min, respectively. Collect 100 ml of eluate for the L fraction (18 ml/min). For the E and K fractions 150 ml of each are collected. All fractions must be kept at 4°C until the run is completed. Cells remaining in the Sanderson chamber are collected after the elutriation. Centrifuge each of the fractions at 450 g for 7 min and resuspend the pelleted cells in about 3 ml of GBSS.

If the standard chamber is used, the only necessary modifications are the rotor speed and the successive flow rates. Instead of 3250 r.p.m., the rotor speed used is 2550 r.p.m. The successive flow rates used are 13.5, 22.5 and 45.0 ml/min for the lymphocyte, endothelial and Kupffer cell fractions, respectively. The use of the standard chamber results in endothelial and Kupffer cell fractions of slightly less purity (results not shown) than are obtained when the Sanderson chamber is used.

Analysis. Determine the cell yield and viability of each fraction and calculate the overall recovery as described in Section 5.2.1.

Estimate the percentages of Kupffer and endothelial cells in the various cell fractions on the basis of the number of cells reacting positively or negatively after staining for peroxidase (see Section 5.2.1.) and esterase (see Section 5.2.2.) activities.

The typical results presented in *Table 7* show that both endothelial and Kupffer cells are obtained in separate fractions with a purity of 80% and 84%, respectively. Most ($\sim$70%) of the cells present in the original suspension are recovered in the purified fractions, with only a minor proportion of each cell type being lost in other fractions; the overall recovery of cells is usually about 90%. The viability of all cell fractions is at least 90%.

5.4.4 Purification of Rat-liver Fat-storing Cells

The following protocol represents the final step in the purification of fat-storing cells from rat liver (13).

Equipment, chemicals and solutions

Equipment as in Section 5.4.2;

Table 7. Purification of Rat-liver Kupffer and Endothelial Cells by Centrifugal Elutriation.

Fraction	Main type of cell	Total (10^6 cells)	Viability (%)	% of cells staining for		Composition (%)*			
				Peroxidase	Esterase	P	L	E	K
Starting sample	Sinusoidal cells	167.4	87	24.7	86.8	0.6	13.2	61.5	24.7
L	Lymphocytes	19.9	80	3.0	28.6	–	71.4	25.6	3
E	Endothelial cells	82.5	95	9.0	89.2	–	10.8	80.2	9
K	Kupffer cells	34.7	97	83.5	95.1	–	4.9	11.6	83.5
Pellet	Cell aggregates	9.8	95	35.4	96.0	16.0	4.0	44.6	35.4

*P: parenchymal cells; L: lymphocytes; E: endothelial cells; K: Kupffer cells.
Sinusoidal liver cells were isolated from 3-month-old rats (see Section 5.1) and separated from erythrocytes by single-step metrizamide density centrifugation (see Section 5.2.2). The sinusoidal cell suspension was subjected to centrifugal elutriation using the Sanderson chamber as described in Section 5.4.3.

Table 8. Purification of Rat-liver Fat-storing Cells by Centrifugal Elutriation.

Fraction	Main type of cell	Yield (10^6 cells)	Viability (%)	% of cells staining for		Composition (%)*			
				Peroxidase	Esterase	FSC	L	E	K
Starting sample	Endothelial and fat-storing cells	133	94	8	98	16	2	75	8
F1	Fat-storing cells	14	80	1	88	78	12	9	1
F2	Fat-storing cells	6	85	1	94	53	6	40	1
Pellet	Endothelial cells	108	93	11	99	7	1	81	11

*FSC: fat-storing cells; L: lymphocytes; E: endothelial cells; K: Kupffer cells.
Sinusoidal liver cells isolated from 12-month-old rats (see Section 5.1) were initially separated by a two-step metrizamide density centrifugation (see Section 5.2.3). The top layer fraction, designated as starting sample, was subjected to centrifugal elutriation as described in Section 5.4.4.

chemicals and solutions (see Section 5.4.3);
sinusoidal liver cell preparation (see Section 5.1).

Experimental protocol. Freshly isolated sinusoidal cells from a 12-month-old rat are first subjected to centrifugation in a two-step discontinuous metrizamide gradient as described in Section 5.2.3. The top layer or low density fraction thus obtained is further purified by centrifugal elutriation.

Install the elutriator system with the Sanderson chamber as described in Section 5.4.1. Set the rotor speed at 3250 r.p.m. and the initial flow rate at 16 ml/min, with the entire system equilibrated at 4°C.

Wash the cells from the top layer once with GBSS by centrifugation at 450 g for 7 min and thoroughly resuspend the pelleted cells in about 5 ml of GBSS. Introduce this cell sample (50 – 500 x 10^6 cells) into the mixing chamber and wash the cells into the rotor. Collect two fractions (F1 and F2) of 100 ml each at flow rates of 16 and 18 ml/min, respectively. Collect the cells remaining in the chamber after the run. Centrifuge all fractions at 450 g for 7 min and resuspend the cells in about 2 ml of GBSS.

Analysis. Determine the cell yield, viability and composition of each fraction as described in Section 5.4.3. Typical results are given in *Table 8*. The fat-storing cells, which constitute only a minor component of the starting suspension (16%), are primarily recovered in fraction F1, with a purity of about 80%. Both lymphocytes and endothelial cells contaminate this fraction to about 10% of the total; Kupffer cells are practically absent.

The second fraction, F2, contains additional fat-storing cells, but with a lower purity (53%) and mainly contaminated with endothelial cells. About 60% of the fat-storing cells present in the original sample are recovered in the two fractions together. The loss of cells into the pellet fraction is mainly the result of cell clumping.

5.4.5 *Separation of Mononuclear Leukocytes from Human Blood*

During the past 10 years, several groups have reported the separation of functionally distinct types of white blood cells by centrifugal elutriation (27). Various protocols that can be used for the purification of lymphocytes (27,28), monocytes (27,28) and granulocytes (27,29) have been developed. The separation of mononuclear leukocytes by centrifugal elutriation as described here is the same as that originally devised by Fogelman *et al.* (28).

Equipment, chemicals and solutions

Equipment for centrifugal elutriation (see Section 5.4.1) using the standard chamber.
Centrifuge with a swing-out rotor and tubes (50 ml);
elutriation medium: Krebs-Ringer phosphate buffer containing 25 mM glucose and 1% (w/v) bovine serum albumin (BSA);
equipment and chemicals for esterase staining (Section 5.2.2);
heparinised human blood (50 ml);
Ficoll-Paque (Pharmacia Fine Chemicals A.B) or Lymphoprep (Nyegaard

& Co., Oslo);
phosphate buffered saline (PBS) (50 mM; pH 7.4).

Experimental protocol. Place 15 ml of the Ficoll-Paque or Lymphoprep in each of six 50 ml centrifuge tubes. Dilute 50 ml of heparinised human blood by the addition of 150 ml of PBS and carefully layer 33 ml on top of the Ficoll solution of each tube. Centrifuge for 15 min at 1500 g at room temperature. Collect the cell layer on top of the Ficoll solution, which contains lymphocytes, monocytes, basophils and platelets. Dilute this suspension with elutriation medium to a final volume of about 5 ml. Prepare the elutriator system (see Section 5.4.1) filled with the Krebs-Ringer medium. Set the rotor speed at 2030 r.p.m. and the flow rate at 4.7 ml/min. Introduce the cell sample ($\sim$5 ml) into the mixing chamber ($\pm$100 x 10^6 cells) and collect nine elutriation fractions of 50 ml each, corresponding to flow rates of 4.7, 8.0, 10.0, 11.0, 12.0, 12.7, 13.5, 14.5 and 16.0 ml/min, respectively. Centrifuge the eluted fractions at 450 g for 7 min and resuspend the pellets in about 3 ml of elutriation medium. Collect the pellet from the separation chamber.

Analysis. Each fraction from the elutriator rotor is analysed for cell concentration and the percentage of cells positive for non-specific esterase activity (monocytes). Typical results are given in *Table 9*. The majority of cells are eluted in fractions 3 − 6. These fractions contain primarily lymphocytes, in particular, fractions 3 − 5 contain populations of highly purified lymphocytes. The larger esterase-positive monocytes are eluted only in fraction 9 or remain in the separation chamber. The pellet fraction represents a purified monocyte

Table 9. Separation of Mononuclear Leukocytes by Centrifugal Elutriation.

Fraction	Flow rate (ml/min)	Cell number (10^6 cells)	Esterase staining*	
			Negative (%)	Positive (%)
Original sample	−	104	85.6	14.4
1	4.7	0	−	−
2	8.0	3.0	100	0
3	10.0	17.8	99.7	0.3
4	11.0	25.0	99.3	0.7
5	12.0	25.0	98.0	1.9
6	12.7	10.8	95.2	4.8
7	13.5	1.2	95.3	4.7
8	14.5	0.8	84.5	15.5
9	16.0	2.8	58.7	41.3
Pellet	−	10.0	28.4	71.6

White blood cells were obtained by Ficoll-Paque centrifugation of human blood and subjected to centrifugal elutriation as described in Section 5.4.5. The cellular composition of fractions was determined by non-specific esterase staining and light microscopic examination.
*Esterase-positive cells represent monocytes; esterase-negative cells represent lymphocytes (mainly fractions 1 − 8) and granulocytes (fraction 9 and pellet).

preparation. The esterase-negative cells that contaminate this fraction are mainly granulocytes. The platelets present in the original sample are recovered mostly in fractions 1 and 2.

5.4.6 *Separation of Other Types of Cells*

Centrifugal elutriation has been extensively applied to separate a large variety of cells from different tissues and species. A complete list of all major publications employing this technique can be obtained from Dr. Sussmann of Beckman Instruments International SA (Geneva). This extensive list includes a brief abstract of each publication with the essential experimental details necessary for carrying out the separations. Another possibility for obtaining specialised information on elutriation techniques is from the Elutriator Users Group (Contact address: Dr A. Lodola, Biological Laboratory, University of Canterbury, Kent CT2 7NY, UK). The Users Club issues a newsletter containing articles on problems and new applications of centrifugal elutriation.

6. ACKNOWLEDGEMENTS

The authors wish to thank Ms. G.C.F. de Ruiter, Ms. E.Ch. Sleyster, Ms. A.M. Seffelaar and Mr. S.J. Bukvic for their participation in the experiments. Dr. A.C. Ford and Ms. L. Vermeer-Greven are thanked for their help in the preparation of the manuscript.

7. REFERENCES

1. Rickwood,D. (1978) in *Centrifugal Separations in Molecular and Cell Biology*, Birnie,G.D. and Rickwood,D. (eds.), Butterworth Publishers, London, p. 219.
2. Price,C.A., ed. (1982) *Centrifugation in Density Gradients*, Academic Press, NY.
3. Pretlow,T.G. and Pretlow,T.P. (1982) in *Cell Separation: Methods and Selected Applications*, Vol. 1, Pretlow,T.G. and Pretlow,T.P. (eds.), Academic Press, NY, p. 41.
4. Pertoft,H. and Laurent,T.C. (1982) in *Cell Separation: Methods and Selected Applications*, Vol. 1, Pretlow,T.G. and Pretlow,T.P. (eds.), Academic Press, NY, 115.
5. Tulp,A., Kooi,W., Kipp,J.B.A., Barnhoorn,M.G. and Polak,F. (1981) *Anal. Biochem.,* **117**, 354.
6. Brouwer,A., Leeuw,A.M.de, Praaning-van Dalen,D.P. and Knook,D.L. (1982) in *Sinusoidal Liver Cells,* Knook, D.L. and Wisse,E. (eds.), Elsevier Biomedical Press, Amsterdam, p. 509.
7. Ford,T.C. and Rickwood,D. (1982) *Anal. Biochem.,* **124**, 293.
8. Knook,D.L. and Sleyster,E.Ch. (1976) *Exp. Cell Res.,* **99**, 444.
9. Knook,D.L., Blansjaar,N. and Sleyster,E.Ch. (1977) *Exp. Cell Res.,* **109**, 317.
10. Knook,D.L. and Sleyster,E.Ch. (1980) *Biochem. Biophys. Res. Commun.,* **96**, 250.
11. Wisse,E. and Knook,D.L. (1979) in *Progress in Liver Diseases,* Vol. **VI**, Popper,H. and Schaffner,F. (eds.), Grune and Stratton, Inc., p. 153.
12. Wilson,P.D., Watson,R. and Knook,D.L. (1982) *Gerontology,* **28**, 32.
13. Knook,D.L., Seffelaar,A.M. and Leeuw,A.M.de (1982) *Exp. Cell Res.,* **139**, 468.
14. Knook,D.L. and Leeuw,A.M.de (1982) in *Sinusoidal Liver Cells,* Knook,D.L. and Wisse,E. (eds.), Elsevier Biomedical Press, Amsterdam, p. 45.
15. Leeuw,A.M.de, Brouwer,A., Barelds,R.J. and Knook,D.L. (1983) *Hepatology,* **3**, 497.
16. Leeuw,A.M.de, Barelds,R.J., Zanger,R.de and Knook,D.L. (1982) *Cell Tissue Res.,* **223**, 201.
17. Leeuw,A.M.de, Martindale,J.E. and Knook,D.L. (1982) in *Sinusoidal Liver Cells,* Knook,D.L. and Wisse,E. (eds.), Elsevier Biomedical Press, Amsterdam, p. 139.
18. Bezooijen,C.F.A.van, Grell,T. and Knook,D.L. (1977) *Mech. Ageing Dev.,* **6**, 293.
19. Barelds,R.J., Brouwer,A. and Knook,D.L. (1982) in *Sinusoidal Liver Cells,* Knook,D.L. and Wisse,E. (eds.), Elsevier Biomedical Press, Amsterdam, p. 449.

20. Williams,N. (1982) in *Cell Separation: Methods and Selected Applications,* Vol. **1**, Pretlow,T.G. and Pretlow,T.P. (eds.), Academic Press, NY, p. 85.
21. MacGlashan,D.W., Lichtenstein,L.M., Galli,S.J., Dvorak,A.M. and Dvorak,H.F. (1982) in *Cell Separation: Methods and Selected Applications,* Vol. **1**, Pretlow,T.G. and Pretlow,T.P. (eds.), Academic Press, NY, p. 301.
22. Wells,J.R. (1982) in *Cell Separation: Methods and Selected Applications,* Vol. **1**, Pretlow,T.G. and Pretlow,T.P. (eds.), Academic Press, NY, p. 169.
23. Lindahl,P.E. (1948) *Nature,* **161**, 648.
24. Bezooijen,C.F.A.van, Noord,M.J.van and Knook,D.L. (1974) *Mech. Ageing Dev.,* **3**, 107.
25. Bezooijen,C.F.A.van, Bukvic,S.J., Sleyster,E.Ch. and Knook,D.L. (1984) in *Pharmacological, Morphological and Physiological Aspects of Liver Aging,* Bezooijen,C.F.A.van, (ed.), Vol. **1**, p. 115.
26. Sleyster,E.Ch. and Knook,D.L. (1982) *Lab. Invest.,* **47**, 484.
27. Sanderson,R.J. (1982) in *Cell Separation: Methods and Selected Applications,* Vol. **1**, Pretlow,T.G. and Pretlow,T.P. (eds.), Academic Press, NY, p. 153.
28. Fogelman,A.M., Seager,J., Edwards,P.A., Kokom,M. and Popjak,G. (1977) *Biochem. Biophys. Res. Commun.,* **76**, 167.
29. Lionetti,F.J., Hunt,S.M. and Valeri,C.R. (1980) in *Methods of Cell Separation,* Vol. **3**, Catsimpoolas,N. (eds.), Plenum Publishing Co., p. 141.
30. Remenyik,C.J., Dombi,G.W. and Halsall,H.B. (1980) *Arch. Biochem. Biophys.,* **201**, 500.

APPENDIX A

Estimation of the Flow Rate and Rotor Speed Necessary to Separate Cells by Centrifugal Elutriation

The information given below has been extracted from the Beckman Instruction Manual for the JE-6B Elutriation System and Rotor.

Flow Rate and Rotor Speed

Figure 1 is provided for the operator who wishes to determine the approximate flow rate and rotor speed for the separation of specific particles. This nomogram has been generated from Equation 1:

$$F = XD^2 \left(\frac{\text{r.p.m.}}{1000} \right)^2 \qquad \text{Equation 1}$$

where F is the flow rate at the pump in ml/min; X is a constant which is equal to 0.0378 for the Sanderson chamber and 0.0511 for the standard chamber; D is the diameter of the particles in micrometres (μm) and r.p.m. is the rotor speed in revolutions per minute.

Equation 1 is an expression relating to the conditions existing at the elutriation boundary. It is derived by setting the velocity of a particle sedimenting in a gravitational force field (from Stoke's Law) equal to the flow velocity at the elutriation boundary, where flow velocity (V_f) is equal to flow rate (ml/min) divided by cross-sectional area (cm²).

Mathematically expressed as:

$$V_f = \frac{F}{A} = \frac{D^2 (\varrho_p - \varrho_m)}{18\eta} \omega^2 r \qquad \text{Equation 2}$$

214

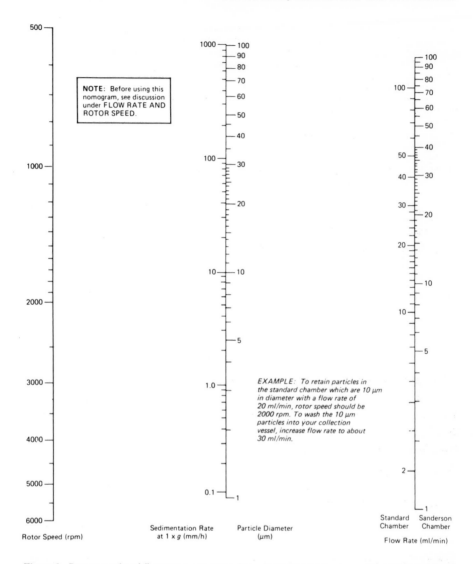

Figure 1. Rotor speed and flow rate nomogram. Use a straight edge to connect the rotor speed and the flow rate that correspond to the smallest, lightest particles you wish to retain in the rotor. Smaller, lighter particles will wash out. To elute a selected particle size, tilt one end of the straight edge counter-clockwise so that, according to *Table 1* and the centre scale of the nomogram, the desired fraction of cells will wash out of the rotor. Then decrease the rotor speed or increase the flow rate accordingly.

Solving Equation 2 for *F*, one arrives at Equation 1, where *X* is a constant which incorporates the cross-sectional area of the chamber at the elutriation boundary and the distance of the elutriation boundary from the axis of rotation, $\triangle\varrho$ is 0.05 g/cm³ (a reasonable density difference when working with cells), η is 1.002 mPa.s (the viscosity of pure water at 20°C), and factors which convert r.p.m. to radians per second.

Table 1. Specifications of the Sanderson and Standard Chambers.

Chamber	Separable particle size (min and max. approx.)	Particle diameter variation within an eluted population*		Number of particles to fill chamber		Chamber volume (ml)
		Population of small particles	Population of large particles	Min	Max	
Standard	5 – 50 μm	± 2.5 μm	± 5 μm	± 10^7	± 10^9	4.2
Sanderson	5 – 50 μm	± 1.5 μm	± 2.5 μm	± 10^5	± 10^7	5.9

*If particles are the same size, separation can be made using either chamber if population densities differ by at least 0.025 g/cm³.

Before using the nomogram, the reader should note that if the sedimentation rate of the particles has been determined in a medium other than the elutriation medium (or if the diameter was calculated in a different medium, since, with the indicated assumptions at unit gravity the sedimentation rate is proportional to D^2), a new experimental sedimentation rate (S_E) must be calculated before the nomogram is used. This can be calculated using the following equation:

$$S_E = S_P \left(\frac{\eta}{\triangle\varrho}\right) P \left(\frac{\triangle\varrho}{\eta}\right) E \qquad \text{Equation 3}$$

Where E is the elutriation medium and P is the medium previously used for measuring the sedimentation rate (S_P).

After using the nomogram, if viscosities other than that of pure water at 20°C (at 4°C, for example the viscosity of water is 1.567 mPa.s) and liquid/particle differences other than 0.05 g/cm³ are required, then the parameters read from the nomogram must be adjusted:

$$F_{new} = F_{nomogram} \left(\frac{\triangle\varrho}{0.05\ \eta}\right) new \qquad \text{Equation 4}$$

Once the magnitude of the adjustments required is known, use the nomogram as follows. Mark the known parameters on two of the scales in *Figure 1*. Then, using a ruler, draw a line through these two points so that the line intersects the third scale. Flow rate, particle size, and rotor speed are read where the ruler intersects these scales. With the ajdustments of Equations 3 and 4, these parameters should usefully approximate the elution parameters for the particles in which you are interested.

APPENDIX B
Isolation of Liver Cells

1. ISOLATION OF PARENCHYMAL LIVER CELLS

The isolation of parenchymal liver cells was carried out as described elsewhere (1).

1.1 Media

The standard medium used for the isolation of parenchymal cells consists of 0.1 mM L-aspartic acid, 0.2 mM L-threonine, 0.3 mM L-serine, 0.5 mM glycine, 0.6 mM L-alanine, 0.9 mM L-glutamic acid, 0.9 mM L-glutamine, 3 mM KCl, 0.5 mM NaH_2PO_4, 0.5 mM $MgCl_2$, 24 mM $NaHCO_3$, 20 mM glucose, 20 mM fructose, 25 mM Hepes, 150 mM sucrose; adjusted to pH 7.4 at 37°C; osmolarity 300 mOsm/l (2). The dissociation medium used during some steps of the isolation procedure consists of the standard medium containing, in addition, 75×10^2 units of collagenase (Sigma) and 46×10^3 units of hyaluronidase (Sigma) per 100 ml medium.

1.2 Preparation of the Parenchymal Cell Suspension

The method for the isolation of parenchymal cells given here is a modification of an earlier procedure (3). Several steps in this isolation procedure have been modified and the method finally used for the preparation of the liver parenchymal cells is as follows. Under ether anesthesia, the liver, portal vein and inferior *vena cava* are dissected free and left *in situ*. A cannula (Braunüle No. OG20) is inserted into the portal vein. An initial first perfusion of the liver is done for 60 sec using 15 ml of dissociation medium, this blanches the organ evenly. The medium used for the first perfusion is passed through the liver and, via an incision in the abdominal aorta, to the outside of the body, where it is discarded. During the first perfusion, a cannula is inserted into the thoracic portion of the inferior *vena cava* to transport the effluent and the abdominal aorta is then clamped off. A second perfusion is done using the standard medium at a rate of 15 ml/min for 15 min to eliminate calcium ions from the liver; this dissociates the desmosomes. The medium is kept at 37°C in a reservoir and a mixture of 95% oxygen and 5% carbon dioxide is bubbled through the medium at a flow rate of 200 ml of gas/min. The perfusion pressure should be 40 Torr. After the second perfusion, a third perfusion is done with dissociation medium at a rate of 15 ml/min for 20 min. The perfusion conditions are the same as for the second perfusion, except that the perfusate is returned to the medium reservoir. After this recirculating perfusion, the liver is excised and cut into pieces as small as possible using razor blades. The pieces of tissue are incubated for 3 min at 37°C with 50 ml of dissociation medium, in which an oxygen tension of at least 50×10^3 Pa is maintained. The cell suspension obtained is filtered through five layers of gauze and through a 50 μm Nylon filter. Cellular debris and non-parenchymal cells are removed by centrifugation at 50 g for 5 min. The final pellet consists almost entirely of parenchymal cells. The yield usually ranges from 30 to 40 x 10^6 isolated cells per gram wet weight of liver. After isolation the cells should be used immediately.

2. PREPARATION OF SINUSOIDAL LIVER CELL SUSPENSIONS

2.1 **Media**

The composition of the isolation medium (GBSS) is given in Section 5.2.1.

2.2 **Preparation of Sinusoidal Cell Suspensions**

Sinusoidal liver cells are isolated from 3-month-old female BN/BiRij rats (4). The liver is first perfused *in situ* through the portal vein with Gey's balanced salt solution (GBSS) for 5 min. This is followed by a post-perfusion of 6 min with 0.2% pronase E (Merck) dissolved in GBSS. During this post-perfusion process, the liver is carefully excised and placed on a sieve. The excised liver is then connected to a circulation perfusion system containing 0.05% collagenase type I (Sigma) and 0.05% pronase E (Merck) dissolved in GBSS and perfused for 30 min. All media should be kept at 37°C and perfused at a flow rate of 10 ml/min.

Following perfusion, Glisson's capsule should be removed and the main vessels sectioned. The paste-like substance is then stirred in 100 ml GBSS containing 0.05% collagenase type I and 0.02% pronase E at 37°C for 30 min. The pH of this mixture is kept at pH 7.4 by the addition of 1 M NaOH. The suspension is filtered through Nylon gauze. The resulting filtrate is then centrifuged at 300 g for 5 min to pellet the sinusoidal cells.

3. REFERENCES

1. Van Bezooijen,C.F.A., Grell,T. and Knook,D.L. (1977) *Mech. Ageing Dev.,* **6**, 293.
2. Seglen,P.O. (1974) *Biochim. Biophys. Acta,* **338**, 317.
3. Van Bezooijen,C.F.A., Van Noord,M.J. and Knook,D.L. (1974) *Mech. Ageing Dev.,* **3**, 107.
4. De Leeuw,A.M., Barelds,R.J., De Zanger,R. and Knook,D.L. (1982) *Cell Tissue Res.,* **223**, 201

CHAPTER 7

Separations in Zonal Rotors

J. GRAHAM

1. INTRODUCTION

There are two types of zonal rotor, batch-type and continuous-flow. Because continuous-flow rotors are less widely used in the laboratory, these will be mentioned only briefly. Of the batch-type rotors there are those that are loaded and unloaded while spinning (i.e., dynamically) and those (reorienting rotors) which are loaded and unloaded while the rotor is stationary. This chapter will concentrate mainly on the dynamically-loaded and unloaded rotors since they are the most widely available. Although the operation of reorienting rotors will be considered in detail, no specific methods for the isolation of particles in these rotors will be presented. Any particular gradient and centrifugation system used with a dynamically-loaded rotor can usually also be used with the appropriate reorienting rotor. Dynamically-loaded batch-type rotors were originally categorised on the basis of size and speed, for example, the AXII is a low-speed rotor (up to 5000 r.p.m.) and the BXIV and BXV are ultracentrifuge rotors (up to 48 000 and 35 000 r.p.m., respectively, in their titanium forms). Appendix III lists the range of zonal rotors currently available commercially. Although the BXIV rotor, as produced by MSE, is no longer commercially available, it remains probably the most widely used zonal rotor in laboratories within the UK. The design and operation of a zonal rotor will thus be described with reference to the MSE BXIV rotor; comparisons with the MSE AXII rotor will be made where appropriate. The operation of the Kontron TZT 48 and TZT 32 zonal rotors (the modern equivalents of BXIV and BXV rotors) will then be discussed, since the design of these is significantly different to that of the MSE BXIV. Finally, the operation of reorienting rotors will be presented with reference to the Sorvall TZ 28 rotor.

2. MSE BXIV AND AXII ROTORS

2.1 General Design

The body of the BXIV rotor is in two parts which fit together by a screw-thread and are sealed by an 'O'-ring: the bottom part supports a central core which also carries the septum assembly (see *Figures 1* and *2*). The septum assembly, made of Noryl, comprises a cylinder carrying four septa which fits over the core; the four septa extend to the wall of the rotor effectively dividing it into four sectors.

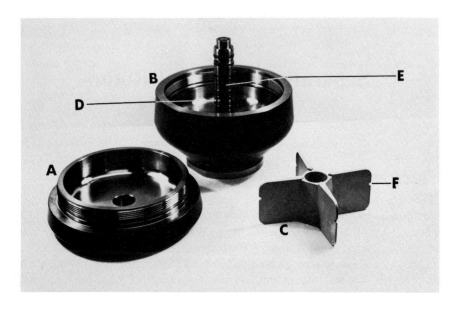

Figure 1. Disassembled BXIV rotor. **A**, top; **B**, bottom-supporting rotor core; **C**, septa assembly; **D**, exit of annular channel at core surface; **E**, exit of central channel at core surface; **F**, septa channel connecting **D** to the edge of the rotor.

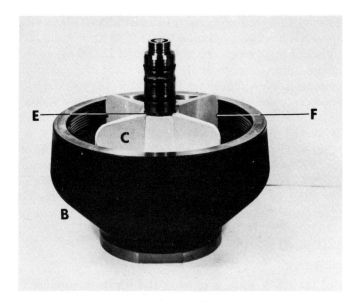

Figure 2. The BXIV rotor − partially assembled, see Figure 1 for key.

The gradient fills the entire enclosed space and it is introduced while the rotor is spinning at low speed. To provide access to the top of the gradient (core of the rotor) a central channel within the core exits at the surface of the core just above the septa assembly. To provide access to the bottom of the gradient (wall or edge of the rotor) an annulus in the core exits in a groove at the core surface within the vane-assembly. Radial channels from the edge of each vane open onto the groove: a continuous path is thus established from the annulus to the edge of the rotor.

The MSE AXII rotor consists of two thick Perspex discs separated by a Perspex septa assembly and sealed in an aluminium collar and secured by 10 locking bolts; the core is separate and locates in a central hole in the discs. The fluid lines to the centre of this rotor are identical to that of the BXIV; while the channels from the annulus to the edge of the rotor are formed between the bottom of the vanes and a Perspex disc which supports the vanes.

Since these rotors are designed for loading while the rotor is spinning at low speed, the normal safety mechanisms which prevent activation of the rotor drive while the lid of the chamber is open must be overridden. This is achieved by the depression of a microswitch by the guard tray which covers the open chamber during zonal operation.

Access from the outside to the interior of the spinning rotor is provided by the fluid seal (feed head), which facilitates the transfer from a static to a

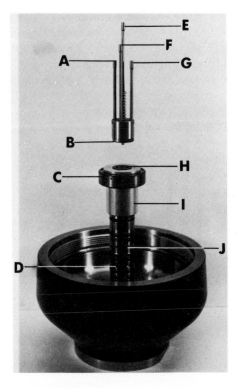

Figure 3. The BXIV rotor and feed head to show fluid channels. **A**, cooling water inlet; **B**, stainless steel static seal; **C**, bearing; **D**, annular exit (edge line); **E**, channel to centre; **F**, channel to edge; **G**, cooling water outlet; **H**, Rulon rotating seal; **I**, metal collar extending downwards over core from bearing; **J**, centre channel exit.

rotating system. The rotating seal is made from Rulon (a filled fluorocarbon): it carries a central channel and an annulus and is seated on top of the core of the spinning rotor (*Figure 3*). It is housed within a bearing assembly which allows the seal to rotate with the core. A static stainless steel seal is in contact, under pressure, with the Rulon seal. The face of this seal also contains a central channel and an annulus which exit distally as two tubes (*Figure 3*); two further tubes supply cooling water to the face of the seal.

In modern forms of the BXIV (post-mid-1970) the Rulon seal forms part of the detachable feed head and the fluid seal is completely enclosed. The feed head casing consists of two threaded parts (*Figure 4*): the lower part carries the bearing assembly itself (plus Rulon seal); the upper part carries the stainless steel seal. A metal collar extends downwards (*Figures 3* and *4*) from the inner aspect of the bearing to fit over the core. The feed head casing is restrained from rotation by three bayonet pins projecting from the guard tray (*Figure 5*). Older models of the BXIV (pre-mid-1970) incorporate the Rulon seal into the top of the rotor core. The detachable part of the feed head comprises the bearing and metal collar, which extends over the core, and the static seal. In both types, a spring between the stainless steel seal and the top of the feed head casing supplies the requisite pressure at the fluid seal. The AXII rotor is similar to the earlier BXIV rotors in that the Rulon seal is part of the rotor core. A metal sleeve which extends upwards from the core (*Figure 6*) supports the bearing assembly, and the static stainless steel seal which is housed within the bearing assembly is restrained from rotation by a metal bar.

In the AXII rotor, the feed head remains in place during all phases of the

Figure 4. The BXIV rotor feed head. **A**, top of feed head casing; **B**, groove for bayonet pin of guard tray; **C**, centre channel; **D**, edge channel; **E**, stainless steel seal; **F**, bearing; **G**, bottom of feed head casing.

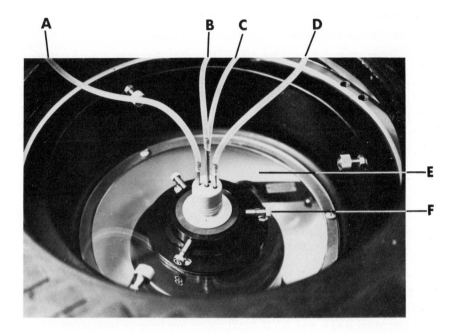

Figure 5. The BXIV rotor and feed head assembled for operation in an MSE Superspeed 65 centrifuge. **A**, cooling water inlet; **B**, centre channel; **C**, edge channel; **D**, cooling water outlet; **E**, guard tray; **F**, bayonet pin for securing feed head.

Figure 6. The AXII rotor positioned in MSE Mistral 6L centrifuge. **A**, aluminium casing, **B**, upper Perspex disc; **C**, static seal; **D**, metal sleeve supporting bearing; **E**, restraining bar; **F**, Perspex septum.

zonal operation. Because the BXIV operates at high speed under vacuum the feed head is only in place during loading and unloading of the rotor at 2000 – 3000 r.p.m.: after loading the feed head is detached and the rotor capped off prior to acceleration to high speed; the feed head is only replaced when the rotor decelerates to 2000 – 3000 r.p.m. prior to unloading.

2.2 **Operation**

2.2.1 *Ancillary Equipment*

(i)   Silicone or polypropylene tubing (internal diameter 2.5 – 3.0 mm) for connections to feed head.

(ii)   Peristaltic pump capable of delivering 10 – 100 ml/min (e.g., Hilo flow pump).

(iii) Plastic tubing connectors.

(iv)   Spencer-Wells artery forceps (haemostats) for clamping lines to the feed-head.

(v)   50 ml plastic syringes.

(vi)   Recording u.v. spectrophotometer with continuous-flow cell to monitor the absorption of the effluent from the rotor. The system marketed by ISCO is excellent for this task − it includes a special zonal flow cell, whose internal diameter is nowhere less than 3 mm and a recorder with considerable absorbance expansion and contraction facilities.

2.2.2 *Practical Details*

The assembly and operation of zonal rotors is covered adequately in most instruction manuals and only a few practical details will be described here.

(i)   The MSE BXIV and AXII zonal rotors are initially loaded via the edge with the least dense gradient first. Successively denser parts of the gradient then displace the lighter gradient solutions centripetally which will eventually emerge from the central channel. The fluid flow is then reversed and the sample is fed to the core of the rotor followed by some low-density solution (overlay) to push the sample into the centrifugal field; alternatively, in isopycnic separations, the sample may be fed to the edge of the rotor mixed with the gradient. The AXII and BXIV rotors are designed only for centre unloading, that is, they must be unloaded by pumping dense liquid to the wall of the rotor and collecting the gradient from the centre. These rotors are not suitable for edge unloading (by pumping water to the centre) since an enormous loss of resolution would occur as material reaching the wall of the rotor which would be distributed over the entire edge surface of the rotor and must then leave via a single channel in the mid-point of each vane. One rotor which overcomes this problem is the BXXIX (not generally available commercially) in which the wall and vanes are tapered.

(ii)   Several arrangements for the flow connections to the feed head of varying degrees of sophistication are suggested in the instruction manuals. The simplest system which minimises the length of tubing and hence mixing of the gradient is, however, the most desirable. *Figure 7* describes a system which allows for the greatest flexibility. For loading the gradient, the fluid line to the wall (annulus) passes through a tubing connector B and a peristaltic pump to either a gradient former (continuous gradient) or to a vessel containing the lightest gradient material (discontinuous gradient). In the former case, it is convenient to include a tubing connector

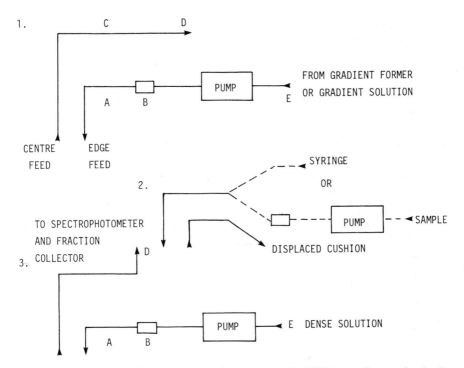

Figure 7. Simple feed head connections for operation of a BXIV rotor. See text for details.

between the pump and the gradient former. Although the tubing lengths should be minimal, they should not be so short as to impose any sideways pressure on the feed head nor render manipulation of the tubing connector difficult. The tubing from the centre channel should be of sufficient length to reach the tubing connector B or allow easy application of a syringe and long enough to dip below the surface of about 50 ml water in a beaker placed on top of the guard tray: a steady stream of bubbles emerging from the tube implies that the gradient is entering the rotor correctly.

(iii) During passage of the first low density solution into the rotor, the pump should be stopped and the tubing to the gradient former disconnected or removed from the gradient solution. There should be no movement of solution into the rotor: the spinning rotor will tend to draw liquid into itself unless sufficient pressure is applied by the peristaltic pump to the tubing passing through it. This problem is obviously particularly important when a discontinuous gradient is used.

(iv) Referring to *Figure 7*, any air bubbles drawn into the tubing can be removed either by inclusion of a bubble trap in the line between the pump and the rotor or by clamping at A; disconnecting B; pumping to expel the air bubble; finally reconnecting B and unclamping at A.

(v) When the light gradient emerges at D, the pump is stopped and the lines are clamped at A and C. The sample can be applied either by a syringe at

D or through the pump: the edge line is disconnected at B; the line B − E flushed out with water; filled with sample and then B connected to D (*Figure 7*).

(vi) The only major problem which may be encountered during loading or unloading is cross-leakage at the fluid seal, either from the annulus outwards or from the annulus to the centre channel. The former is not necessarily detrimental to the gradient and only becomes important if excessive (i.e., >0.3% of the total rotor capacity). Leakage inwards is more problematical: during loading, small columns of liquid will be expelled from the centre channel by the displaced air within the rotor; during unloading small columns of the dense unloading medium will mix incompletely with the emerging low density gradient and a streaming effect within the liquid path will cause oscillations on the absorbance tracing. Cross-leakage may be overcome by decreasing the flow rate; if this is unsuccessful it may be necessary to remove the feed head and clean the fluid seal. In the modern forms of the BXIV this is relatively easy since both the static and rotating seals are part of the detachable feed head assembly; the Rulon seal may even be re-polished if necessary. In the older BXIV rotors only the static seal can be removed; in this case the Rulon seal may be partially cleaned by gently applying a moistened tissue to its surface as it rotates. There is no way that the AXII feed head can be detached and replaced.

(vii) Prior to unloading the rotor by passing dense sucrose to the wall of the rotor, it is advisable to remove any air trapped in the edge line. The latter (*Figure 7*) is disconnected at B and the length of tubing B − E primed with unloading solution. Water is fed from a syringe at D until the cushion emerges from the tube beyond A. After clamping at A, B is reconnected, D is connected to the flow cell of the recording spectrophotometer and A is unclamped and the pump started.

These and other important points regarding cleaning and maintenance of the rotor and seal have been described elsewhere (1).

3. BECKMAN BATCH-TYPE ZONAL ROTORS

Beckman manufacture a range of zonal rotors basically similar in design to the MSE BXIV and BXV for use in ultracentrifuges; in addition, there is also available the Z-60 for very high speed work (capacity 330 ml, maximum speed 60 000 r.p.m.) and the JCF-Z which operates in medium-speed centrifuges. These rotors have the advantage of possessing interchangeable cores which permit their use in either the normal dynamic mode (i.e., unloading by passing dense medium to the edge of the rotor and collecting from the centre) or the BXXIX type dynamic mode which permits unloading by feeding water to the centre of the rotor and collecting from a tapered edge or used in a static reorienting mode (see Section 5). The dynamic mode of the BXXIX type is advantageous for the harvesting of a single band of material which occurs close to the edge of the rotor (for a more detailed description of this type of rotor see reference 1).

4. KONTRON ZONAL SYSTEM

4.1 General Design

The Kontron TZT 48 and TZT 32 rotors have been developed to simplify the operation of zonal rotors in that no bearing assembly is required. The general design of these rotors is very similar to that of the MSE BXIV or BXV, the major difference is that the rotating seal on top of the rotor core which is made of stainless steel is part of the core and is not detachable, while the stationary seal which forms part of a removable feed head is made of Teflon. Indeed, this arrangement is rather similar to that of the earlier MSE rotor systems in which the stainless steel stationary seal was also part of a removable feed head. The Kontron stationary Teflon seal (see *Figure 8*) is surrounded by a metal jacket which contains circulating cooling water; the four external tubes are thus the standard central and annular lines plus two tubes for the cooling water. As with all zonal rotor systems, the seal assembly is secured on top of a Perspex guard tray which surrounds the top of the rotor and effectively seals the chamber from the atmosphere, and pressure is maintained at the seal interface by spring discs between the top of the stationary seal and the underside of the feed head assembly. Unlike all other dynamically-loaded rotors, the bearing assembly around the seal has been dispensed with and this eliminates much of the vibration which normally arises from the ball-race, especially if this becomes contaminated with spilled gradient medium. Such vibrations increase the possibility of leakage across the seal; in the Kontron system the chance of leakage is considerably reduced. Any peripheral leakage from the seal is collected in a channel in the collar of the guard tray and a tube from the floor of the channel which exits at the surface of the guard tray enables any leaked solution to be removed using a syringe.

Within the rotor, the vane assembly is now manufactured from Delrin although in some of the earlier models it was made of titanium. In the most recent models the vane assembly and core are a single unit made from Delrin and

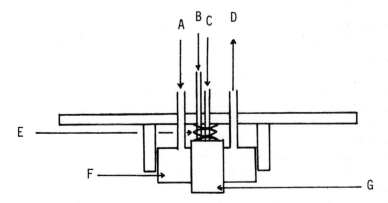

Figure 8. Kontron TZT 48 stationary seal assembly (vertical section-diagrammatic representation). **A**, cooling water in; **B**, annular channel; **C**, core channel; **D**, cooling water out; **E**, spring discs; **F**, cooling jacket; **G**, stationary seal.

the rotating metal seal is inserted separately on top of the core. Unlike the MSE BXIV, the line to the edge of the rotor exits at the bottom of the vane rather than at its mid-point.

A series of inserts are available which permit the reduction in the rotor volume (normally ~650 ml) to 325 ml, 160 ml and 140 ml. With the volume reduced to 140 ml it can be used essentially as a Beaufay-type rotor. A modified core and feed head enables the conversion of the TZT 48 or 32 to a reorienting rotor (see Section 5.2).

4.2 Operation

The reduced tendency of the seal to leak means that the rotor can be filled at 40 − 45 ml/min: thus the rotor can be filled in 15 − 17 min. As long as the rotor is cold at the start of the run, this means that loading can be carried out without chamber refrigeration and therefore condensation within the chamber is reduced, if not eliminated entirely. Otherwise, the operation of the rotor, together with the arrangement of fluid lines and ancillary equipment, is essentially identical to that of the MSE BXIV rotor. There is a gantry device which can be fixed to the side of a Kontron centrifuge which permits the collection of the tubes from the feed head into a single manifold. Although this device facilitates the manipulations involved in the loading and unloading of the rotor, the extra lengths of tubing required between the gradient pump and the rotor tend to promote mixing within the fluid flow.

5. REORIENTING ZONAL ROTORS

5.1 General Design

Reorienting (Reograd) rotors can be loaded with the gradient while stationary and then accelerated very slowly and smoothly to about 1000 r.p.m. before acceleration to the desired running speed. During this early acceleration (up to 1000 r.p.m.), the gradient reorients (2) from a vertical to a horizontal position (*Figure 9*). Likewise, the rotor must be decelerated very slowly and smoothly below 1000 r.p.m. to permit reorientation back to a vertical gradient for unloading in a stationary mode. Excellent linear gradients can be recovered from such rotors in spite of these two gradient reorientations. Moreover, these rotors do not require any sophisticated seals for their operation. During reorientation, the maximum shear occurs where the greatest changes in area occur (3), that is, those zones which are at the top and bottom of the stationary rotor which find themselves at the centre and edge of the spinning rotor. Shearing forces decrease towards the centre of the gradient. It is therefore advantageous to include relatively large amounts of overlay and cushion in these rotors to minimise the amount of shear experienced by the sample. A more thorough description of the design of these rotors has been given elsewhere (1).

Models which are available commercially are the TZ 28 (Du Pont-Sorvall) which is designed to operate in either high-speed or ultra-speed Sorvall centrifuges equipped with a slow acceleration facility and the Kontron TZT 48 and TZT 32 rotors which can be modified to operate in the reorienting mode.

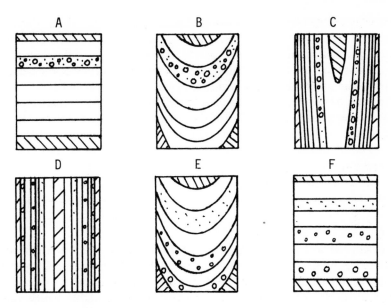

Figure 9. Schematic diagram of fractionation in a reorienting gradient (Reograd) zonal rotor. **A**, rotor at rest; **B**, accelerating (early phase); **C**, accelerating (late phase); **D**, at speed; **E**, decelerating; **F**, at rest.

5.2 Operation of the Sorvall TZ 28 Rotor

The TZ 28 rotor can be loaded statically or dynamically, but it must be unloaded statically. A detachable distributor which is located centrally at the top of the rotor core contains a central and an annular channel. The central channel passes down the rotor core and exits at the port at the bottom of each of the six radial vanes. In a static mode the gradient may be introduced into the rotor low density end first through silicone tubing and a plastic connector which fits into the central space of the distributor.

Unlike a dynamically-loaded rotor with a fluid seal, the rotor chamber is not completely filled with the gradient prior to introduction of sample: instead it is filled to a volume which will permit the subsequent introduction of the sample. The gradient is then allowed to reorient by controlled slow acceleration to 1000 r.p.m. over a period of at least 10 min.

To load the rotor dynamically with gradient the annular channel of the distributor is used. This channel connects with six exit ports which occur at the surface of the central rotor core and immediately adjacent to each vane. Liquid introduced into the channel whilst the rotor is spinning at 1000 r.p.m. is drawn into the rotor by centrifugal force and is directed along the surface of each vane from the core to the wall of the rotor. In this mode, therefore, the gradient is introduced high density end first. The sample is generally loaded into the annulus while the rotor is spinning, although it could precede the loading of the gradient in a static mode.

Prior to unloading the rotor, deceleration from 1000 r.p.m. must be accomplished slowly and smoothly to permit reorientation of the gradient. This is best achieved with the drive on by fine adjustment of the rate control knob. Alternatively the rotor may be permitted to coast to a stop without the brake. The only method of unloading the rotor is to connect a line to the central space of the distributor (as in static loading) and to reverse the flow of the peristaltic pump so that the gradient is drawn out of the rotor dense-end first.

The main advantage of the rotor is the elimination of a complicated fluid seal and consequently elimination of the attendant leakage problems. Its main disadvantage is that unloading, dense-end first, is not ideal, in that mixing tends to occur due to an inverse gradient in the channel to the bottom of the vane and the vertical part of the tubing to the peristaltic pump.

In the reorienting mode, the Kontron TZT 48 and TZT 32 rotors are loaded and unloaded statically; such modified rotors cannot be loaded dynamically.

6. CONTINUOUS-FLOW ROTORS

The high-speed continuous-flow rotors of the BIX-type which were manufactured by Beckman and which required a special centrifuge for their operation will not be considered here and the reader is referred elsewhere for a description of them (1). There are three rotors currently available which can be adapted for continuous flow operation in standard centrifuges, the Sorvall TZ 28 and the Beckman CF-32Ti and JCF-Z rotors. The operation of these rotors will now be considered briefly.

6.1 **Sorvall TZ 28 Rotor**

This is designed primarily for the harvesting of particles from large volumes of culture medium by sedimentation onto the wall of the spinning rotor. This rotor can only be used in the RC-5 series centrifuges. As shown in *Figure 10* the input sample is fed to the bottom of the rotor chamber via the channels in the septa. As the fluid is displaced upwards, particles sediment out towards the wall of the rotor and the clarified fluid eventually passes out of the rotor core. Flow rates up to 1400 ml/min can be used and this rotor is normally used to harvest relatively large particles such as mitochondria, protozoa, bacteria and, at low flow rates, some of the larger viruses.

6.2 **Beckman CF-32Ti and JCF-Z Rotors**

These rotors can be used to band material such as a virus within a short density gradient between the surface of an extended rotor core and the wall of the rotor (see *Figure 11*). The fluid seal is essentially the same as that of a batch-type rotor, that is, it consists of a central channel and an annulus, but unlike batch-type rotors it remains connected to the rotor during all phases of the centrifugation run. Moreover, the channels within the rotor are rather different. The central channel exits at the bottom of the tapered core surface; the annulus channel is the common path for two channels within the rotor, one of which connects with the top of the rotor core and one which connects with the top of the wall of the rotor. Fill the rotor with gradient, normally a simple discontinuous one, in the usual way, that is, while the rotor is spinning at

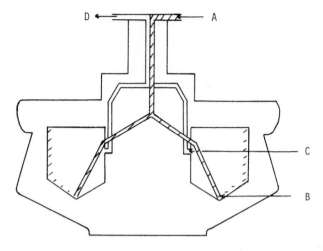

Figure 10. Continuous-flow operation of the Sorvall TZ 28 (vertical section-diagrammatic representation). Sample enters at **A**; exits from the rotor chamber at **B**. Material from the input sediments to the wall of the spinning rotor and the clarified fluid leaves the chamber at **C** to emerge from the rotor at **D**.

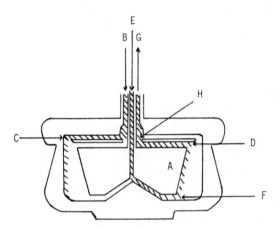

Figure 11. Continuous-flow operation of Beckman CF-32Ti (vertical section-diagrammatic representation). **A**, extended rotor core. The annular channel (**B**) forms the common external path to the edge of the rotor (**C**) and the top of the core (**D**). Gradient is fed to the wall of the rotor via **B** and **C**. Sample passes down the core channel (**E**), passes up over the surface of the core between **F** and **D** to exit at **G**. To unload the rotor an air-block at **H** ensures the passage of dense sucrose to the wall from **B** to **C** and the effluent is collected at **E**.

2000 r.p.m., introduce the gradient, low density first, via the wall of the rotor (see *Figure 11*). Then feed buffer down the central channel, this passes up over the rotor core and exits in the annulus. Accelerate the rotor to the running speed; during this time buffer is continuously fed to the bottom of the core. When the running speed is attained, feed the virus-containing material to the bottom of the core via the central channel. When all of the material has been fed into the rotor, maintain the loading speed for a further period of time to permit banding of the last virus particles.

To unload the rotor, decelerate it to 2000 r.p.m. and pump dense sucrose to the wall of the rotor via the annular channel. Since this channel is part of the common external channel to the top of the core, the latter branch must be blocked off to prevent access of the dense sucrose to the core surface. This is achieved by introducing a small bubble of air ahead of the sucrose unloading solution. The bubble of air remains at the centre of the core effectively blocking off the core surface channel. The rotor contents exit via the bottom of the core and central channel.

6.3 Practical Considerations

The efficiency of a continuous-flow separation should be assessed in terms of the 'percentage clean-out', that is, the total amount of virus recovered as a percentage of the total virus input. Values in excess of 90% should be aimed for. The percentage clean-out is a function of the centrifugal force experienced by the virus suspension and the time taken for the virus suspension to pass over the core surface. The centrifugal force experienced by the sample depends on the diameter of the rotor core and the speed of the rotor. The time spent by the virus suspension travelling over the core depends on the flow-rate through the rotor. Maximisation of the percentage clean-out is thus a function of increasing rotor speed and decreasing flow rate. A more thorough evaluation of these problems is given elsewhere (1).

7. GRADIENT DESIGN

Since the gradient within a zonal rotor fills the space of an enclosed cylinder, in the spinning rotor equal volumes of gradient occupy a decreasing radial thickness the further they are from the core. The volume/radius relationship for the MSE BXIV is shown in *Figure 12*. Thus, unlike tube gradients, a zonal rotor gradient which is linear with volume will not be linear with radius. To achieve a gradient linear with radius in a zonal rotor, such a gradient must be convex with volume. In the absence of a gradient mixer that can be programmed, the most convenient way of achieving such a gradient is to use a constant-mixing-volume gradient former. An inexpensive device for preparing convex gradients is shown in *Figure 13*, this was designed by Birnie and Harvey (4). In this closed mixing system the volume of gradient removed from the mixing chamber by the pump is replaced by an equal volume of dense solution delivered into the mixing chamber from the reservoir. The form of the gradient can be modified by altering the ratio of the volume of the dense medium in the reservoir and the light medium in the mixing chamber. The gradient profile shown in *Figure 14* was recovered from an MSE BXIV rotor using 400 ml of 45% (w/v) sucrose in the mixing chamber and 400 ml of 5% (w/v) sucrose in the reservoir.

A more convenient system is now marketed by Kontron (*Figure 15*). In this design the lower mixing chamber receives dense solution from the upper chamber through a central hole in the separating disc. This connection is closed during filling with a tapered metal rod. The volume of the reservoirs can be

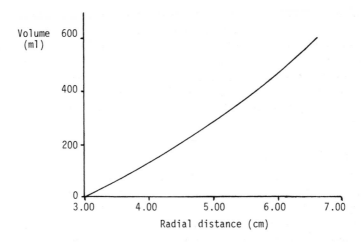

Figure 12. Volume-radius relationship from a 600 ml gradient starting 3.0 cm from centre of an MSE BXIV zonal rotor.

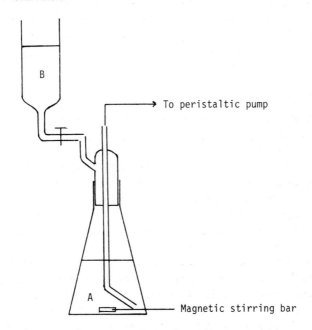

To peristaltic pump

Magnetic stirring bar

Figure 13. Simple constant mixing-volume gradient mixer (4). Low density solution is placed in the closed vessel (**A**); high density solution in the reservoir (**B**).

adjusted using plastic inserts. Moreover, a stick-on transfer is available which can be applied to the upper chamber to calibrate this chamber in terms of both volume and rotor radius.

Steensgaard and Roth (5) have produced a number of data processing programs for ultracentrifugation which are available from Kontron. These include a radius-volume relative centrifugal force (RCF) program for zonal

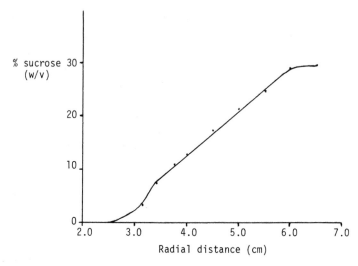

Figure 14. A 'linear-with-radius' sucrose gradient recovered from an MSE BXIV zonal rotor. For details see text.

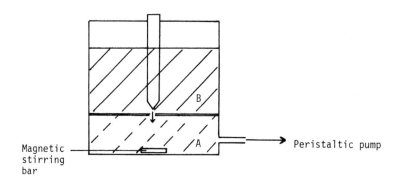

Figure 15. Kontron constant mixing volume gradient mixer **A**, mixing chamber; **B**, upper reservoir.

rotors and one for the calculation of sedimentation coefficients from centrifugation runs in zonal rotors. The quantitative aspects of zonal centrifugation have been discussed in detail by Steensgaard *et al.* (6).

8. EXAMPLES OF SEPARATIONS USING ZONAL ROTORS

8.1 Separation of Human Blood Cells

8.1.1. *Introduction*

Since cells have different sizes and/or shapes and/or densities it is possible to separate them on the basis of either sedimentation rate or density (see Chapter 6). For most cell separations it is necessary to devise gradients which are more or less isotonic throughout so as to minimise cell volume changes due to differences in tonicity between the cell and its surrounding medium. Consequently, it has become common to use gradients of high molecular weight solutes such as Ficoll (Pharmacia Fine Chemicals A.B.) or dextran or, more recently, metrizamide or Nycodenz (Nyegaard & Co.), which have lower osmotic activity in comparison with sucrose solutions. Shallow gradients of Ficoll dissolved in isotonic salt solutions which have a nearly uniform osmolarity can even separate large and small cells of the same type (8,9). At low concentrations, polymers such as dextran or Ficoll contribute little to the total tonicity of the gradient medium. Moreover, by judicious selection of gradient solutes whose density/osmolarity relationships are different, it is possible to maintain a constant controlled tonicity throughout. For example, Probst and Maisenbacher (10) used a linear sucrose gradient of $2-31\%$, an inverse gradient of $0.76-0\%$ saline and a constant Hanks salt solution to separate Ehrlich ascites cells on the basis of size. Alternatively, an iso-osmotic gradient of density $1.07-1.19$ g/ml can be generated from a linear metrizoate gradient and an inverse linear sucrose gradient.

The zonal system used to fractionate the cells from human blood described here was developed from a method for resolving lymphocytes in swing-out rotors (11). These gradients were designed so that the osmolarity throughout was $280-300$ mOsm using an increasing Ficoll gradient and inverse gradients of both metrizoate and Krebs-Ringer-Tris.

8.1.2 *Equipment and Chemicals*

MSE Coolspin or MSE Mistral 6L;
MSE AXII zonal rotor and ancillary equipment;
MSE simple gradient maker;
Ficoll, sodium metrizoate, Tris;
Krebs-Ringer-Tris buffer (125 mM NaCl, 5 mM KCl, 1.2 mM $MgSO_4$, 35 mM Tris, 1 mM NaH_2PO_4, pH 7.4);
Leishman's stain.

8.1.3 *Experimental Protocol*

(i) *Preparation of sample.* It is inconvenient to use whole blood for this experiment since the gradient becomes rapidly overloaded with red cells. The blood is therefore subjected to several low speed centrifugations to remove the majority of the red cells.

Centrifuge freshly collected human blood (400 ml) in the 6 x 250 ml swing-out rotor (Coolspin) at 1000 r.p.m. (500 g) for 10 min. Aspirate the plasma together with the 'buffy coat' from above the loosely-packed red cell layer; aspiration of some of the red cells is unavoidable. Dilute the remaining red cell layer to the original volume with Krebs-Ringer-Tris; mix and re-centrifuge. Aspirate the superantant and 'buffy coat' as before and repeat the whole procedure twice more. Dilute the combined supernatants with an equal volume of buffer; centrifuge at 1500 r.p.m. (800 g) and carefully aspirate the majority of the supernatant to leave approximately 60 ml, in which to resuspend the loosely-sedimented cells.

(ii) *Gradient solutions.* The following solutions are required:
(a) Low density gradient solution: 34.3 g Ficoll, 94.2 ml of 10% (w/v) metrizoate, 116.5 ml of 0.175 M Tris-HCl (pH 7.4), 163 ml of Krebs-Ringer-Tris made up to 900 ml.
(b) High density gradient solution: 169.3 g Ficoll, 76.6 ml of 10% (w/v) metrizoate, 122.8 ml of 0.175 M Tris-HCl (pH 7.4) made up to 900 ml.
(c) Cushions: (i) 25 g Ficoll, 5 ml of 10% (w/v) metrizoate, 15 ml of 0.175 M Tris-HCl (pH 7.4) made up to 100 ml. (ii) 400 ml of 35% (w/v) sucrose.

(iii) *Centrifugation.* Carry out all operations at 4°C. With the A-rotor spinning at 600 r.p.m., feed to the wall of the rotor a linear gradient generated from 450 ml each of the low and high density solutions; follow this with the 25% (w/v) Ficoll and sucrose cushions to fill the rotor. When the low density solution emerges from the centre of the rotor, reverse the flow and feed the sample to the centre followed by approximately 30 ml of overlay (Krebs-Ringer-Tris solution).

Increase the speed of the rotor slowly to 2000 r.p.m. while maintaining the centre channel open to a reservoir of Krebs-Ringer-Tris, approximately 30 ml of which is taken up by the expanding rotor. After 45 min, slowly decelerate the rotor and unload by pumping 40% (w/v) sucrose to the wall of the rotor.

Collect 50 ml fractions; dilute each with phosphate-buffered saline (PBS) and sediment the cells at 1500 r.p.m. for 10 min. Aspirate the majority of the supernatant and resuspend the pellet in the remaining supernatant.

8.1.4 *Analysis*

Apply one drop of each sample to a microscope slide; smear and air-dry. Stain the film with Leishman's stain (2 min) and differentiate by the addition of PBS (2 min). The first 100 – 150 ml of the gradient recovered from the rotor includes the original sample band and contains serum and small fragments of cells. Lymphocytes band broadly in the 300 – 500 ml fractions. The 600 – 900 ml fractions comprise monocytes in the lighter region which overlap the polymorphonuclear leukocytes in the heavier region. It may be possible to discern an enrichment of neutrophils in the densest part of the polymorphonuclear leukocyte band.

8.2. Fractionation of Membranes from a Rat-liver Nuclear Pellet

8.2.1 *Introduction*

The method (12) depends for its efficacy on the recovery of large (often paired) sheets of plasma membrane from the lateral (apposed) surfaces of rat-liver parenchymal cells when the liver is disrupted by gentle Dounce homogenisation in 1 mM $NaHCO_3$. These sheets of membrane sediment at 1000 g for 10 min, above the nuclear pellet and below the heavy mitochondria. Various methods (13,14) have been devised to optimise the recovery of these membranes from the low speed (1000 g for 10 min) pellet of the homogenate by differential centrifugation but they are unsatisfactory inasmuch as it is necesary to wash the pellet many times and as the membrane-containing zone is rather ill-defined, losses are generally quite substantial. Several workers (15 – 17) have therefore used gradients in an A-type zonal rotor using either the complete low speed pellet or sometimes the whole homogenate as the source of membranes. In addition to improving the recovery of membranes, the use of the zonal rotor permits the processing of as much as 100 g of rat liver at a time (16).

Discontinuous (non-linear) gradients have been devised on an empirical basis to resolve the nuclei, sheets of plasma membrane, heavy mitochondria and any low density microsomal material trapped in the low speed pellet.

8.2.2 *Equipment and Materials*

MSE Coolspin or MSE Mistral 6L;
MSE AXII Zonal rotor and ancillary equipment;
simple gradient former (200 – 300 ml total volume);
simple gradient former (600 – 700 ml total volume);
loose-fitting Dounce homogeniser;
muslin (cheese-cloth or nylon screening);
sodium bicarbonate, sucrose, Tris.

8.2.3 *Experimental Protocol*

Kill the rat by cervical dislocation; expose the liver and perfuse it with cold isotonic saline solution to remove the majority of the blood. After excision, transfer the liver to a chilled beaker; chop finely with scissors and homogenise in 1 mM $NaHCO_3$ (~40 ml) using 10 strokes of the pestle of a loose-fitting Dounce homogeniser. Observe the suspension under a phase-contrast microscope to check the degree of cell rupture, which should be at least 95%, and then filter the homogenate through two layers of muslin, cheesecloth or Nylon screening to remove connective tissue, etc. Centrifuge the filtered homogenate at 1000 *g* for 10 min; resuspend the pellet in 1 mM $NaHCO_3$ using two or three gentle strokes in the Dounce homogeniser. This forms the nuclear fraction for the A-rotor separation. Care is required because the nuclear pellet may appear to be bi- or tri-partite and the boundary between the supernatant and pellet may be difficult to see. The entire pellet is required for subsequent fractionation; if in doubt do not aspirate all the supernatant.

The method is based on that of Evans (16): introduce the following discontinuous gradient into the A-rotor via the wall: 300 ml of 6% (w/v) sucrose in 1 mM $NaHCO_3$, 100 ml of 24% (w/v) sucrose, a gradient of 100 ml of 24% (w/v) sucrose mixed with 100 ml of 36% (w/v) sucrose, a gradient of 300 ml of 36% (w/v) sucrose mixed with 300 ml of 54% (w/v) sucrose and 100 ml of 54% (w/v) sucrose. When the low density medium emerges from the centre of the rotor reverse the flow and apply the sample to the top of the gradient followed by 40 ml of 1 mM $NaHCO_3$. Increase the rotor speed to 3900 r.p.m. while the centre line is kept open to a reservoir of 1 mM $NaHCO_3$. After 40 – 50 min decrease the speed to 600 r.p.m. and unload the rotor by pumping 60% (w/v) sucrose to the wall of the rotor. Pass the effluent from the centre of

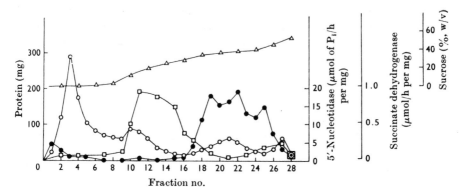

Figure 16. Fractionation of a nuclear pellet from rat liver in the AXII rotor (see text for experimental details). △—△ sucrose (%w/v); ○—○ protein; ●—● 5′-nucleotidase; □—□ succinate dehydrogenase. Reproduced from Evans (16) with permission.

the rotor through a flow cell and monitor the absorption at 280 nm continuously.

Dilute fractions from the gradient with an equal volume of 5 mM Tris-HCl (pH 8.0) and centrifuge at 20 000 g for 30 min; wash once by resuspension and recentrifugation in 0.25 M sucrose, 5 mM Tris-HCl (pH 8.0) and resuspend in this medium prior to analysis of enzyme activities.

8.2.4 *Analysis of Results*

The fractions from the gradient are characterised by enzymic analysis. Succinate dehydrogenase (succinate-cytochrome c reductase) is used as a mitochondrial marker and 5′-nucleotidase as a plasma membrane marker (see Appendix V). *Figure 16* shows the distribution of these enzymes and protein in the gradient. The nuclei (not shown) band at the edge of the rotor. This can be avoided by the inclusion of a cushion of approximately 100 ml of 60% (w/w) sucrose. The first peak of material contains small fragments of membranes and debris which just enter the gradient from the sample zone.

8.3 Fractionation of a Tissue Culture Cell Post-nuclear Supernatant

8.3.1 *Introduction*

The isolation of plasma membrane from tissue culture cell homogenates poses a number of problems which relate primarily to the homogenisation process. Unlike organised tissues such as liver or intestinal mucosa, the plasma membrane of tissue culture cells is not generally stabilised against vesiculation during cell disruption by intercellular junctional complexes or by the presence of an extensive cytoskeleton. It is crucial for the efficient separation of plasma membrane from smooth endoplasmic reticulum that the size of plasma membrane fragments should be maximised: to achieve this, the gentlest possible liquid shear conditions should be used and these are going to vary with different tissue culture cells. To facilitate this it is necessary to swell the cells osmotically whilst causing the minimal amount of damage to internal organelles such as nuclei and mitochondria. The following method has been designed specifically for Lettre cells and the homogenisation process should be regarded only as a guide.

8.3.2 *Equipment and Chemicals*

Refrigerated low speed centrifuge;
BXIV-type zonal rotor (MSE BXIV, Kontron TZT 48 or Beckman Ti-14) and the appropriate ultracentrifuge;
Dounce homogeniser (30 ml capacity), e.g., Model S from F.T. Scientific Instruments;
$NaHCO_3$, $MgCl_2$ and sucrose;
PBS (0.2 g KCl, 8.0 g NaCl, 0.2 g KH_2PO_4, 1.15 g Na_2HPO_4 in 1000 ml).

8.3.3 *Experimental Protocol*

(i) *Homogenisation.* Carry out all operations at $0-4°C$. Suspend approximately 3 x 10^9 cells (passaged $3-6$ times) in PBS ($30-40$ ml) and centrifuge at 500 g for 5 min. Repeat this washing step twice more. For the final washing step increase the centrifugation speed to give 800 g so that as much of the supernatant as possible can be removed by aspiration. Resuspend the cell pellet in 25 ml of 1 mM $NaHCO_3$, 0.2 mM $MgCl_2$ and after 5 min, homogenise using $10-15$ strokes of the pestle of the Dounce homogeniser. Check cell disruption by phase-contrast microscopy and then add 2 M sucrose, 1 mM $NaHCO_3$, 0.2 mM $MgCl_2$ to make the final sucrose concentration 0.25 M. Centrifuge the homogenate at 1000 g for 10 min to remove the nuclei, and keep the supernatant until required.

The reader should note that other cell lines or even later passage Lettre cells will almost certainly require different homogenisation conditions for efficient and reproducible cell disruption. Other hypotonic media which can be used are 10 mM Tris, 10 mM imidazole or 10 mM Hepes supplemented either with $MgCl_2$ ($0.2-1.0$ mM), $CaCl_2$ ($0.2-1.0$ mM) or KCl ($1-10$ mM) either singly or in combination, at a pH between 7.4 and 8.0. It is advisable to maintain the highest possible ratio of cells to volume of homogenising medium and to use the minimum number of passes of the pestle. For a more thorough discussion of homogenisation problems see Graham (18).

(ii) *Fractionation.* Introduce sequentially the following sucrose solutions into the BXIV rotor (at 2500 r.p.m.): 50 ml each of 15%, 20%, 25% and 30%, 100 ml of 35%, 50 ml of 40%, 100 ml of 50% and 200 ml of 55% (w/w) sucrose containing 1 mM $NaHCO_3$, 0.2 mM $MgCl_2$. When the rotor is full, feed the post-nuclear supernatant to the centre of the rotor and overlayer the sample with 30 ml of 0.1 M sucrose, 1 mM $NaHCO_3$, 0.2 mM $MgCl_2$. After centrifugation at 16 000 g for 60 min unload the gradient by feeding 60% (w/w) sucrose to the edge of the rotor.

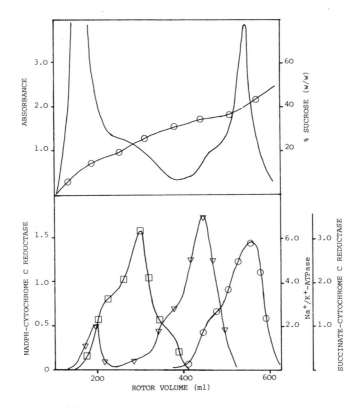

Figure 17. Fractionation of a Lettre cell post-nuclear supernatant. **Top panel**: ——— 280 nm absorbance, ⊖–⊖ density in % sucrose (w/w). **Lower panel**: ⊟–⊟ NADPH-cytochrome c reductase; △–△ Na$^+$/K$^+$-ATPase; ⊖–⊖ succinate-cytochrome c reductase. All activities are in μmol substrate reacted/h/mg protein. For experimental details see text. Reproduced from ref. 19 with permission.

8.3.4 *Analysis of Results*

The enzyme profile of the zonal effluent is shown in *Figure 17*. Plasma membrane fragments (containing Na$^+$/K$^+$-ATPase activity) sediment well ahead of the endoplasmic reticulum (containing NADPH-cytochrome c reductase activity) but just behind the mitochondria (containing succinate-cytochrome c reductase activity). Membrane fragments from later passage cells tend to be smaller and run closer to the endoplasmic reticulum (19). The general problems of fractionation and analysis of tissue culture cell membranes are also discussed elsewhere (20).

8.4 Harvesting of Virus from Tissue Culture Fluid

8.4.1 *Introduction*

The harvesting of virus from large volumes of tissue culture fluid presents the virologist with a major problem as the sedimentation of the particles into a pellet frequently causes inactivation of the virus. At the same time, simple centrifugation of tissue culture fluid will not resolve the virus from heavier membrane contaminants, and the volumes of culture fluid normally processed make it impracticable to purify the virus using sucrose gradients in swing-out rotors. The extra capacity available with the batch-type rotors makes the harvesting of virus from up to 1.1 litres (with the BXIV rotor) and up to 3 litres (with the BXV) a practical proposition. The process of harvesting virus is considered in greater detail elsewhere (1).

It is convenient to consider the process in two stages, firstly concentration and partial purification followed by purification and additional concentration. To purify 1.1 litres using the BXIV rotor the sequence of events would be: (a) concentrate approximately 550 ml onto a cushion of dense sucrose; (b) unload and recover the virus; (c) reload the second 550 ml and concentrate as in (a); (d) unload as in (b); and (e) the rotor is finally loaded with the combined virus concentrates, a sucrose gradient and a cushion of dense sucrose and recentrifuged to band the virus isopycnically. Steps (a) and (c) concentrate and partially purify the virus which will be contaminated by some soluble protein and any co-sedimenting membrane. In step (e) the gradient is chosen so that any membranous material will band away from the virus and soluble protein will remain in the sample region.

As an example of the double zonal centrifugation approach the concentration and purification of Sendai virus from allantoic fluid will be described.

8.4.2 *Equipment and Chemicals*

Refrigerated low-speed centrifuge;
BXIV-Type zonal rotor (MSE BXIV, Kontron TZT 48 or Beckman Ti-14) and the appropriate ultracentrifuge;
sucrose;
PBS containing 0.2 g KCl, 8.0 g NaCl, 0.2 g KH_2PO_4, 1.15 g Na_2HPO_4 in 1 litre.

8.4.3 *Experimental Protocol*

Clarify the allantoic fluid harvested from chick embryos 3 days after inoculation with Sendai virus by centrifugation at 3000 *g* for 20 min at 4°C. Feed the following solutions into the BXIV zonal rotor via the edge channel at 2000 r.p.m.: 25 ml of PBS, 450 ml of clarified allantoic fluid, 50 ml of 30% (w/w) sucrose in PBS and then fill the rotor with 55% (w/w) sucrose in PBS. After centrifugation at 45 000 r.p.m. (100 000 *g*) for 90 min at 4°C, displace the gradient from the centre by pumping 60% (w/w) sucrose to the edge. The absorption profile of the effluent at 280 nm is used to monitor the position of the virus (see *Figure 18*). Pool the major virus-containing fractions (normally contained in 80 − 90 ml) and dilute to between 20% and 25% (w/w) sucrose with PBS so that the total volume is 200 ml or less. Set up the second purification run in the BXIV zonal rotor using 50 ml of PBS, 200 ml of sample, 250 ml of 30% (w/w) sucrose in PBS and 55% (w/w) sucrose in PBS to fill the rotor; feed each of these solutions sequentially via the edge of the rotor. After

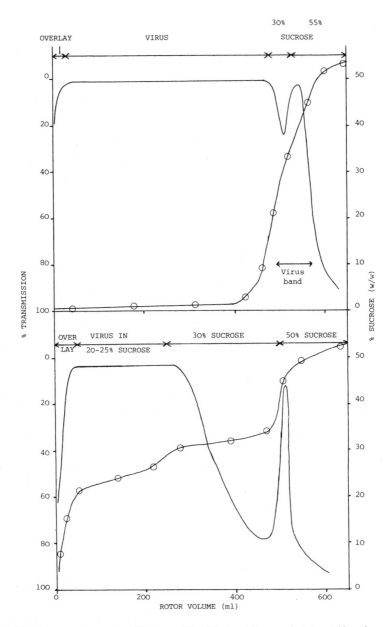

Figure 18. Concentration and purification of Sendai virus. **Top panel**: 1st centrifugation − concentration of virus from 450 ml allantoic fluid. **Lower panel**: 2nd centrifugation − concentration and purification of virus from 1st centrifugation. In both panels ——— % transmission at 280 nm, ⊖–⊖ density in % sucrose (w/w). For experimental details see text.

centrifugation at 45 000 r.p.m. for 2.5 h at 4°C, unload the rotor and monitor the effluent as before. The purified virus is normally harvested in less than 50 ml (*Figure 18*).

8.4.4 *Analysis of Results*

This method achieves excellent recovery (at least 90%) of intact virus and the purification measured in terms of increase in infectivity titre over the original starting material is regularly of the order of 50- to 100-fold. Other viruses may require modification of the centrifugation conditions depending on their sedimentation properties. The aim should be to band the virus sharply in the steep gradient formed between two layers of sucrose [in this case that between the 30% and 55% (w/w) sucrose] and to include in the final purification gradient the largest possible volume of virus-free sucrose between the leading edge of the sample and the banding position to maximise the separation from soluble protein and any slowly-sedimenting membranes.

8.5 **Separation of 40S and 60S Ribosomal Subunits**

8.5.1 *Introduction*

For the separation of ribosomal subunits on the basis of their sedimentation rate it is important that the radial distance occupied by the sample layer on top of the gradient is as small as possible in order to maximise the resolution of the gradients. For this reason it is inconvenient to use swing-out rotors for bulk preparations of subunits. In the method described here, 35 ml of sample can be fractionated and with the overlay recommended (65 ml), the radial distance occupied by this volume is 0.3 cm. To achieve a similar situation using the largest tube (38 ml) of the Beckman SW28 swing-out rotor, 24 tubes would be required. A zonal rotor clearly provides a big advantage for this purpose, both in terms of time-saving and consistent quality of the separation.

Rat-liver polysomes are prepared as described by Schreier and Staehelin (21) and incubated *in vitro* under conditions optimal for protein synthesis, in order to allow completion of nascent peptide chains and detachment of ribosomes from endogenous mRNA. Those ribosomes which have released their nascent chains will dissociate into '40S' and '60S' subunits in 0.5 M KCl with full retention of biological activity on subsequent lowering of the salt concentration.

8.5.2 *Equipment and Chemicals*

BXIV-Type zonal rotor (MSE BXIV, Kontron TZT 48 or Beckman Ti-14) and the appropriate ultracentrifuge;
solution A: 30 mM Tris-HCl (pH 7.6), 100 mM NH_4Cl, 3.5 mM magnesium acetate, 1 mM dithiothreitol (DTT), 20 amino acids (50 μM each), 1 mM ATP, 0.6 mM GTP, 25 mM creatine phosphate, 4 units/ml creatine phosphokinase, freshly-prepared 'pH 5 enzymes' (16);
gradient solutions: 65 ml of an overlay solution containing 20 mM Tris-HCl (pH 7.6), 0.3 M KCl, 3 mM $MgCl_2$, 1 mM DTT; 250 ml of 15% (w/v) sucrose, 450 ml of 40% (w/v) sucrose, both solutions containing Tris-HCl, KCl, $MgCl_2$ and DTT at the same concentrations as in the overlay.

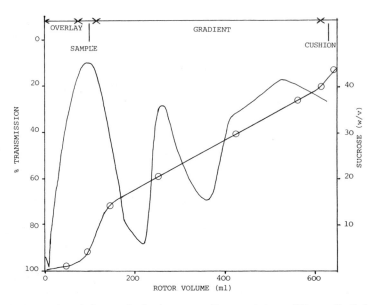

Figure 19. Separation of ribosomal subunits. ———— % transmission at 280 nm; ⊖—⊖ density in % sucrose (w/v). For experimental details see text.

8.5.3 *Experimental Protocol*

Incubate the polysome preparation in Solution A (~ 50 A_{260} units/ml); after 50 min at 30°C add KCl to a final concentration of 0.5 M. During the incubation load the zonal rotor. Introduce to the edge of the BXIV zonal rotor spinning at 2000 r.p.m., the following solutions: 500 ml of a $15-40\%$ (w/v) sucrose gradient (linear with volume), and 40% (w/v) sucrose to fill the rotor. Feed the sample (35 ml) into the rotor, overlayer it at the core of the rotor with overlay solution and centrifuge at 47 000 r.p.m. (115 000 g) for 4 h.

8.5.4 *Analysis of Results*

The primary aim of this particular separation is to isolate purified 40S subunits and, as shown in *Figure 19*, the resolution from both the soluble protein and 60S subunits is very good: the majority of the material is collected between 230 ml and 300 ml. This preparation is very active in binding initiation factors and methionyl-tRNA$_f$ (22). Clearly the 60S peak is heterogeneous and the leading edge of this material sediments almost to the wall of the rotor: this band probably also contains undissociated ribosomes and to purify this material a second gradient centrifugation would be necessary. The separation of ribosomes, polysomes and ribosomal units in zonal rotors has been discussed by Birnie *et al.* (23).

8.6 Separation of the F and HN Glycoproteins from Sendai Virus

8.6.1 *Introduction*

The F surface glycoprotein of Sendai virus is responsible for fusion of virus with host cells and for haemolysis; the HN glycoprotein possesses haemagglutination and neuraminidase activities (24). Efficient large scale separation of protein molecules depends on the positioning of the sample, in a small volume (15 ml), away from the core of the rotor by the use of a large volume (150 ml) of overlay. Under these conditions the sample layer occupies a radial distance of only 0.1 cm, maximising the resolution by sedimentation rate and is 1.4 cm from the core, thus ensuring that the sample experiences sufficient centrifugal force to sediment the proteins.

To detect the virus proteins within the gradient it is necessary to label them isotopically, since the gradient contains Triton X-100 which precludes the use of absorption at 280 nm to monitor the zonal effluent. A number of labelling methods can be used but the most convenient technique is to label the prepared virus with ^{125}I using Chloramine T: alternatively, a labelled amino acid precursor may be incorporated during the growing phase of the virus.

A linear gradient is required for the separation of the virus glycoprotein: this however cannot be generated using a two-chamber mixing device because of excessive frothing in the stirred compartment. The gradient must be formed from an initially discontinuous one by diffusion. Furthermore, the gradient loading speed must be reduced from the usual 20 ml/min because cross-leakage at the fluid seal occurs more easily with a detergent-containing solution.

8.6.2 *Equipment and Chemicals*

High-speed centrifuge (Sorvall RC5, MSE 21, etc.);
ultracentrifuge (MSE 65, etc.);
BXIV-Type zonal rotor (MSE BXIV, Kontron TZT 48 or Beckman Ti-14),
Spectrapor 2 dialysis tubing;
Chloramine T;
PBS;
Na ^{125}I (carrier-free);
tyrosine;
Nonidet P40 (NP-40);
Triton X-100;
sucrose.

8.6.3 *Experimental Protocol*

(i) *Virus labelling.* Incubate the following at room temperature for 2 min: 0.1 ml of virus (2 mg protein/ml in PBS), 0.1 ml of Chloramine T (3 mg/ml), 500 μCi (19 M Bq) of Na ^{125}I, 0.25 ml of PBS. Stop the reaction by addition of 1 ml saturated tyrosine solution and harvest the virus by centrifugation at 20 000 g for 15 min. To remove any unreacted isotope wash the pellet three times with PBS.

(ii) *Virus solubilisation.* The labelled virus is suspended in about 15 ml of PBS: make the solution 0.25% with respect to NP-40 and incubate it at room temperature for 15 min. Remove material which has not been solubilised by centrifugation at 100 000 g for 60 min (25). Dialyse the clear supernatant

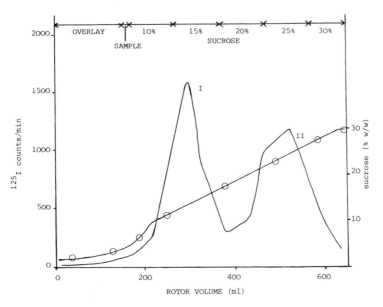

Figure 20. Fractionation of viral surface glycoproteins. ———— ^{125}I c.p.m.; ⊖—⊖ density in %
sucrose (w/w). For experimental details see text.

against multiple changes of PBS using Spectrapor 2 dialysis tubing for
2 − 3 days and then make the solution 1% with respect to Triton X-100.

(iii) *Gradient separation.* Sequentially feed the following solutions, all containing PBS and 1% Triton X-100 at 10 ml/min into a BXIV-Ti rotor at
2000 r.p.m.: 100 ml each of 10%, 15%, 20%, 25% (w/v) sucrose followed by
sufficient 30% (w/v) sucrose to fill the rotor. Remove the feed head; cap the
rotor and centrifuge it at 5000 r.p.m. for 2 − 3 h at 4°C. After deceleration to
2000 r.p.m., replace the feed head and introduce the sample (15 ml) and an
overlay of 150 ml of PBS containing 1% Triton (10 ml/min) to the core of the
rotor. Centrifuge the rotor at 45 000 r.p.m. (100 000 g) for 25 h and unload
the gradient by feeding 40% (w/v) sucrose to the wall of the rotor.

8.6.4 *Analysis of Results*

Two major bands of ^{125}I-labelled material are separated (*Figure 20*). The first
protein peak (I), which contains the F protein, migrates very slowly in the centrifugal field, indeed the trailing edge barely leaves the sample zone. On the
other hand the leading edge of the HN protein containing band (II) extends
close to the wall of the rotor. Very little cross-contamination between the
fusogenic and haemagglutination activities occurs between bands I and II.

8.7 Ultracentrifugal Analysis of Immune Complex Formation Between Mono-
clonal Antibodies and Human IgG

8.7.1 *Introduction*

The method for this investigation was developed by Steensgaard *et al.* (26).
These workers found that four different monoclonal antibodies differed quite

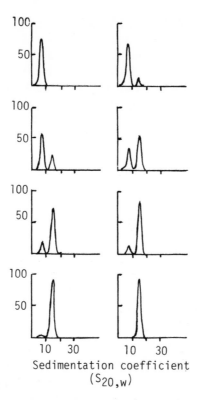

Sedimentation coefficient
$(S_{20,w})$

Figure 21. Centrifugation of immune complexes formed from the monoclonal antibody x3a8. The amount of antigen in the sample (2.2 ml) was 150 μg and the amounts of antibody as follows (a) 0 μg; (b) 160 μg; (c) 320 μg; (d) 640 μg; (e) 800 μg; (f) 960 μg; (g) 1120 μg and (h) 1280 μg. Reproduced from Steensgard *et al.* (26).

considerably in their immune complex formation with IgG. Their analytical zonal centrifugation system was able to estimate the number of antigenic determinants and antibody-binding sites. The isokinetic gradients (in which particles move at a constant speed through the gradient) used in this work were designed by Steensgaard and Jacobsen (27).

8.7.2 *Equipment and Chemicals*

Constant volume mixing chamber gradient former (the one made by Kontron Analytical is very suitable);
BXIV-type zonal rotor (MSE BXIV, Kontron TZT 48 or Beckman Ti-14) and the appropriate ultracentrifuge;
PBS;
5% (w/w) sucrose in PBS;
20% (w/w) sucrose in PBS.

8.7.3 *Experimental Protocol*

The monoclonal antibodies and human IgG antigen are prepared according to Steensgard *et al.* (26). Make the immune complexes by incubating antibody and antigen in a total volume of 2.2 ml at 37°C for 60 min; leave overnight at 4°C. Set up an isokinetic gradient in the zonal rotor with the 5% (w/w) sucrose solution in a 270 ml constant volume mixing chamber and 20% (w/w) sucrose in the upper chamber. Inject 2 ml of the sample to the centre of the rotor followed by 100 ml of PBS overlay. Centrifuge at 47 000 − 48 000 r.p.m. for 6 h. Unload the gradient by pumping 25% (w/w) sucrose to the wall and collect the gradient from the core into 10 ml fractions.

8.7.4 *Analysis of Results*

The sedimentation pattern obtained depends on the amounts of antigen and antibody and on the type of antibody. In the example shown in *Figure 21* the sedimentation of immune complexes formed from 150 μg antigen with increasing concentrations (up to 1280 μg) of monoclonal antibody x3a8 is shown. The first peak (7S) contains free antigen, the second (15S) contains the complex Ag_2Ab (26), there is, apparently, with this particular antibody, no AgAb complex. This was interpreted by the authors in the following manner: each binding site on its own is of low affinity, but once one site is occupied the second site is activated into a high affinity state: if there is insufficient antigen to form Ag_2Ab then AgAb dissociates.

9. ACKNOWLEDGEMENTS

I would like to thank the Cell Surface Research Fund for financial support, Dr. A. Wyke for providing the Sendai virus samples and Dr. M. Clemens and Miss V. Tilleray for the sub-ribosomal unit preparations. Also I am grateful to Dr. J. Steensgaard, for allowing me to cite his recent work on antibody-antigen complexes and for very valuable discussion on the Kontron TZT 48 zonal system.

10. REFERENCES

1. Graham,J.M. (1978) in *Centrifugal Separations in Molecular and Cell Biology,* Birnie,G.D. and Rickwood,D. (eds.), Butterworths, London, p. 63.
2. Hsu,H.W. and Anderson,N.G. (1969) *Biophys. J.,* **9**, 173.
3. Anderson,N.G., Price,C.A., Fisher,W.D., Canning,R.E. and Burger,C.L. (1964) *Anal. Biochem.,* **7**, 1.
4. Birnie,G.D. and Harvey,D.R. (1968) *Anal. Biochem.,* **22**, 171
5. Steensgaard,J. and Roth,H.E. *Data Processing Programs for Ultracentrifugation*, Kontron Analytic, Zurich, Switzerland.
6. Steensgaard,J., Moller,N.P.H. and Funding,L. (1978) in *Centrifugal Separations in Molecular and Cell Biology,* Birnie,G.D. and Rickwood,D. (eds.), Butterworths, London, p. 115.
7. Boone,C.W., Harell,G.S. and Bond,H.E. (1968) *J. Cell Biol.,* **36**, 309.
8. Pasternak,C.A. and Warmsley,A.M.H. (1973) in *Methodological Developments in Biochemistry,* Vol. 3, Reid,E. (ed.), Longmans, London, p. 249.
9. Graham,J.M., Sumner,M.C.B., Curtis,D.H. and Pasternak,C.A. (1973) *Nature,* **246**, 291.
10. Probst,H. and Maisenbacher,J. (1973) *Exp. Cell Res.,* **78**, 335.
11. Loos,J.A. and Roos,D. (1974) *Exp. Cell Res.,* **86**, 333.
12. Neville,D.M. (1960) *J. Biophys. Biochem. Cytol.,* **8**, 413.
13. Emmelot,P., Bos,C.J., Benedetti,E. and Rumke,Ph. (1964) *Biochim. Biophys. Acta,* **90**, 126.
14. Song,C.S., Rubin,W., Rifkind,A.B. and Kappa,A. (1969) *J. Cell Biol.,* **41**, 124.

15. El-Aaser,A.A., Fitzsimons,J.T.R., Hinton,R.H., Reid,E., Klucis,E. and Alexander,P. (1966) *Biochim. Biophys. Acta,* **127**, 553.
16. Evans,W.H. (1970) *Biochem. J.,* **166**, 833.
17. Hinton,R.H., Norris,K.A. and Reid,E. (1971) in *Separations with Zonal Rotors,* Reid,E. (ed.), University of Surrey Press, Guildford, UK, p. 52.1.
18. Graham,J.M. (1982) in *Cancer Cell Organelles,* Reid,E., Cook,G. and Morre,D.J. (eds.), Ellis Harwood Ltd., Chichester, UK, p. 342.
19. Graham,J.M. and Coffey,K.H.M. (1979) *Biochem. J.,* **182**, 173.
20. Graham,J.M. (1975) in *New Techniques in Biophysics and Cell Biology,* Vol. **2**, Pain,R.H., Smith,B.J. (eds.), Wiley & Sons, London, p. 1.
21. Schreier,M.H. and Staehelin,T. (1977) *J. Mol. Biol.,* **73**, 329.
22. Clemens,M.J., Echetebu,C.O., Tilleray,V.J., Pain,V.M. (1980) *Biochem. Biophys. Res. Commun.,* **92**, 60.
23. Birnie,G.D., Fox,S.M. and Harvey,D.R. (1972) in *Subcellular Components Preparations and Fractionation,* Birnie,G.D. (ed.), Butterworths, London, p. 235.
24. Scheid,A. and Choppin,P.W. (1974) *Virology,* **57**, 475.
25. Wyke,A.M., Impraim,C.C., Knutton,S. and Pasternak,C.A. (1980) *Biochem. J.,* **190**, 625.
26. Steensgaard,J., Jacobsen,C., Lowe,J., Ling,N.R. and Jefferis,R. (1982) *Immunology,* **46**, 751.
27. Steensgaard,J. and Jacobsen,C. (1979) *J. Immunol. Methods,* **29**, 173-183.

CHAPTER 8

Analytical Ultracentrifugation

R. EASON

1. INTRODUCTION

Analytical ultracentrifugation has had a profound effect on our understanding of the properties of biological macromolecules. Major experimental achievements such as the recognition of the discrete macromolecular nature of protein molecules, the association of proteins to form unique oligomers, the conformational changes that accompany many enzyme-substrate interactions, the demonstration of the semi-conservative replication of DNA and the description of the behaviour of supercoiled DNA are fully documented in fundamental textbooks of biochemistry and molecular biology.

Modern use is less extensive than in the 1950s and 1960s, when the analytical ultracentrifuge was one of the principal methods for molecular weight determination, since gel electrophoretic analysis of proteins and nucleic acids has now largely displaced the analytical ultracentrifuge. However, in certain cases, such as, for example, glycoproteins or highly charged proteins, detergent-protein binding interactions become atypical so that molecular weight determination by SDS-gel electrophoresis is less accurate. The analytical ultracentrifuge still remains a powerful tool to study interactions between macromolecules and also ligand-induced binding events. Active band centrifugation is an important method for analysing enzymes in their active conformation and when bound to other particles.

The biological situations that can be studied are very diverse. The same physical chemical principles apply to all components in the ultracentrifuge cell and the highly developed theoretical basis of analytical ultracentrifugation continues to be applied fruitfully to a wide variety of biological problems.

The nature of samples that can usefully be analysed requires careful consideration because, in most types of experiment with the analytical ultracentrifuge, purified samples are required. Contaminating components that contribute to the total absorbance or refractive index of the sample can lead to incorrect conclusions. However, a specific wavelength might be available where the contaminant does not absorb light. Alternatively, the contaminant might not interact with the molecules of interest so that it becomes separated during the experiment.

Experiments with the analytical ultracentrifuge involve a fairly small number of techniques and protocols. The range of equipment components used, that is cell assemblies and optical systems, is fairly limited. Correct use of these components demands great care and attention to detail. Practical in-

251

structions on the use of different cell assemblies are included here, but details of the arrangement of optical components and the theory of their operation are not fully discussed, because the basic systems are often modified by automatic and computerised devices and many are unique to a particular laboratory. Adjustment, alignment and focussing of optical components require extensive instructions found in the manuals associated with each particular installation and descriptions of the operation of schlieren and Rayleigh optical systems abound in textbooks and reviews.

An important consideration in using the analytical ultracentrifuge is the question of data evaluation and manipulation. Care in recording accurate data, making accurate measurements and applying the correct equations represents a major part of the practical work with the analytical ultracentrifuge. Basic determinations of sedimentation coefficients, diffusion coefficients and molecular weights can be made very simply without the use of complex apparatus, but extensive use is now made of automatic and computerised data recording and processing equipment. For more advanced analyses, interpretation of data often does rely solely on analytical manipulation. In addition, because of the complexity of the patterns, it is often necessary to simulate the behaviour of model systems and to match their behaviour with the experimental data in order to confirm an interpretation.

There are two major classes of experiment in analytical ultracentrifugation, sedimentation velocity analysis and sedimentation equilibrium analysis. In a velocity experiment, the components are driven by strong centrifugal forces and move towards an equilibrium position which they do not usually reach during the course of the experiment. The precise movement of the components is defined by flow equations and extensive theoretical treatments have been derived. Kinetic treatments provide a conceptually simpler picture of events when a particle is subjected to a strong centrifugal field. In equilibrium experiments, movement of the components is allowed to continue until an equilibrium situation has been established. The equations defining the equilibrium distributions of the components are usually simpler than flow equations for the same system.

2. SEDIMENTATION VELOCITY ANALYSIS

This experimental technique is a transport process which shares a common theory with electrophoretic techniques and gel chromatography. Flow of components in a centrifugal field depends on their size and shape which determine their sedimentation and diffusion characteristics.

At the beginning of this section, a simple two-component system, protein dissolved in dilute buffer [1.0 mg/ml bovine serum albumin (BSA) in 0.15 M NaCl], is considered. Many practical details encountered in analytical ultracentrifugation are covered in determining the sedimentation coefficient, the diffusion coefficient and hence the molecular weight of the protein. Conformational changes are then considered, followed by more complex experimental systems, including active band centrifugation and sedimentation analysis of interacting systems and polydisperse systems.

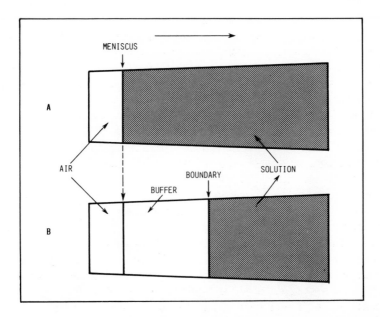

Figure 1. Moving boundary sedimentation analysis. **(A)** At the beginning of the run, there is an air space at the top of the cell and the protein molecules are uniformly distributed throughout the solution. **(B)** When the centrifugal force is applied (arrow), a boundary forms between the protein solution and the buffer due to the net migration of the macromolecules.

2.1 Sedimentation Coefficient of a Pure, Homogeneous Sample

A choice must be made between moving boundary centrifugation and band centrifugation. In moving boundary centrifugation the cell is filled with the solution to be analysed and this requires between 0.3 and 0.7 ml of sample, depending on the type of cell chosen. In band centrifugation, an aliquot of the solution is layered on top of a supporting gradient-forming solvent and this method requires between 5 and 50 μl of sample.

2.1.1 *Moving Boundary Analysis*

The overall nature of the experiment is shown in *Figure 1*. At the beginning of the experiment, the ultracentrifuge cell contains a homogeneous protein solution. When the centrifugal force is applied, the protein molecules move at a rate depending on their size and shape so that a boundary forms between the buffer solution and the protein solution. Measurement of the position of this boundary at intervals throughout the experiment allows calculation of the sedimentation coefficient of the protein.

(i) *Ultracentrifuge cell assemblies for sedimentation velocity analysis.* Typical ultracentrifuge cell assemblies that can be used for boundary analysis are shown in *Figure 2*. Single-sector centrepieces are identified by the angle subtended at the centre of rotation. In this figure, 2° and 4° single-sector centrepieces and a 2.5° double-sector centrepiece are shown. The exact components chosen to construct a cell assembly depend on the optical detection

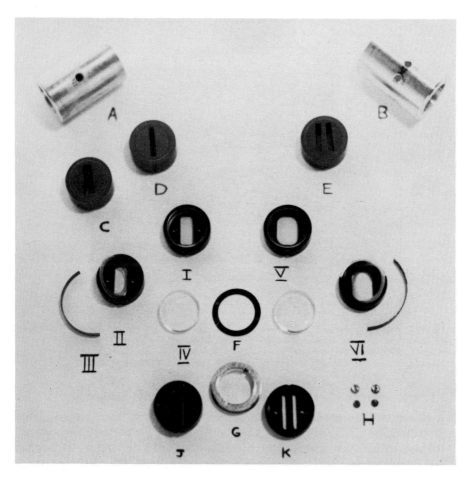

Figure 2. Components of cell assemblies. **(A)** Single-sector cell housing with hole at the top for filling the assembled cell. **(B)** Double-sector cell housing. **(C)** Single-sector 4° centrepiece, made of aluminium-filled epon. **(D)** Single-sector 2° centrepiece. **(E)** Double-sector 2.5° centrepiece. **(I)** Single-sector window assembly. **(II)** Single-sector window holder with plastic window liner. **(III)** Window gasket. **(IV)** Quartz window. **(V)** Double-sector window assembly. **(VI)** Components of double-sector window assembly. **(F)** Cell gasket. **(G)** Screw ring. **(H)** Housing plugs and gaskets. **(J)** Lower window holder with narrow slits for Rayleigh interference. **(K)** Upper window holder for Rayleigh analysis.

system to be used to detect the events in the cell during ultracentrifugation [see (v)].

(ii) *Assembly of cells.* Place a window gasket and window liner in the upper and lower window holder. Thoroughly clean two quartz or sapphire windows with distilled water using lens tissue for drying, and place in the window holder using the alignment marks to locate them correctly.

Stand the cell housing on end and insert the window assembly with the window facing upwards. Any debris or dust should be blown off with a gentle stream of air; an empty wash bottle is ideal for this purpose.

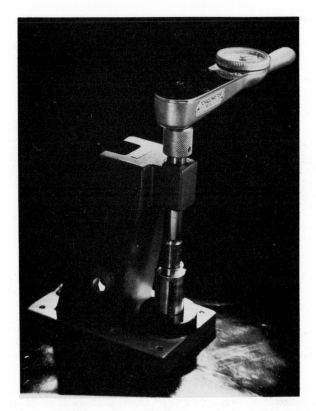

Figure 3. Torque wrench assembly. The cell is located by a bar at the base of the wrench and the screw ring is tightened by rotation of the wrench handle until the necessary torque is obtained.

Insert the centrepiece into the cell housing taking care to lower it evenly to avoid jamming or scratching of the components.

Insert the upper window assembly with the window pointing down. Centrepieces which are made of aluminium require a plastic gasket at both ends of the centrepiece in contact with each window.

Coat the screw-ring gasket lightly with lubricant, for example, Spinkote (Beckman-RIIC), a molybdenum disulphide-based lubricant, then insert into the cell assembly. Similarly, coat the screw ring and screw it carefully into the cell assembly with the inscribed letters pointing upwards.

Place the cell assembly into a torque wrench assembly *(Figure 3)* and tighten to 1.38 m.kg (120 inch-pounds), holding the torque for 15 sec to allow creepage of plastic in the components to occur.

(iii) *Filling the cell assembly.* Place the tightened cell assembly on its side in a wooden holder with the hole(s) in the cell housing facing upwards.

Carefully transfer the protein solution to a clean (sterile) glass or plastic 1 ml syringe fitted with a plastic cannula and avoid frothing.

For a 4° single-sector 12 mm centrepiece, 0.7 ml sample solution is needed.

For a 2° single-sector 12 mm centrepiece, 0.35 ml solution is needed.

Figure 4. Analytical ultracentrifuge rotors. The two-place titanium An-H rotor on the left contains a cell assembly and a counterbalance. The An-D aluminium rotor in the middle has two holes. When a double cell run is performed, the small holes in the rotor are used to define the reference edge. A six-place An-GTi rotor is on the right.

For a double-sector 12 mm centrepiece, insert 0.35 ml of protein solution into the right-hand sector (viewed through the screw ring end) and 0.36 ml of the corresponding buffer, or dialysate, into the left-hand sector.

Insert the plastic tube into the hole in the centrepiece. Push the tip of the tubing down to the bottom of the sector before expelling the liquid. The plastic tubing avoids the possibility of scratching the centrepiece material.

When the cell assembly is filled, remove any droplets with a lens tissue and carefully insert a plastic housing plug gasket into the hole(s) in the cell housing pushing them flat with a blunt probe. Insert the brass housing plugs into the holes and screw them finger-tight using a screwdriver.

(iv) *Insertion of cell assembly and centrepiece into rotor.* In this preliminary experiment, a two-place rotor *(Figure 4)*, for example, an An-D aluminium rotor or an An-H titanium rotor is used. A counterbalance is needed for the hole opposite the cell assembly.

Select the correct counterbalance depending on whether single- or double-sector cells are used and on the optical system chosen for the experiment. The edges of the holes in the counterbalance provide reference images at known distances from the centre of rotation. The counterbalance used for Rayleigh optics has an additional wire across the holes to provide sharp fringes.

Insert weights into the central hole in the counterbalance to match the weight of the cell assembly to within 0.5 g. The weight of the counterbalance should not exceed that of the cell assembly.

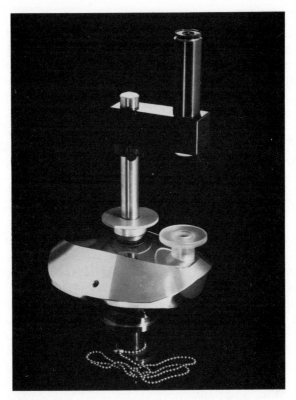

Figure 5. Cell alignment in the rotor. An aligning tool has been screwed onto the An-H rotor and the eyepiece is located over the cell. An aligning tool is on top of the cell to allow rotation so that the sector wall can be aligned with the hairlines in the eyepiece.

Place the counterbalance into the rotor and align it correctly using the scribe lines on the counterbalance and the rotor hole. Tighten the allen screw on top of the counterbalance to fix it firmly into the rotor and avoid movement when the cell assembly is being inserted.

Place the assembled cell into the rotor, screw ring uppermost and with the housing plug holes pointing towards the centre of the rotor.

Carefully push the cell into the hole using a small amount of lubricant if necessary.

Align the cell accurately in the rotor. This can be done using a wrench and the lines inscribed on the base of the cell housing and the rotor hole. Alternatively an aligning tool can be used *(Figure 5)*. Check the alignment of the crossed hair lines in the eyepiece using the lines on the alignment stand. Screw the alignment tool onto the rotor and locate the eyepiece over the cell. Place the plastic aligning tool on top of the cell and rotate it until the hairline in the eyepiece is co-linear with a sector wall. Remove the aligning tool.

Attach the filled rotor to the ultracentrifuge drive using a torque wrench for secure attachment.

257

(v) *Detection of events in the ultracentrifuge cell.* A wide range of optical systems and detecting systems are in use and it is not possible to deal with them in detail. Many of the optical systems are very sophisticated and recorders are often connected directly to a computer for accurate collection of data. In the simple preliminary experiment described here the discussion is restricted to the classical light sources and detection systems used to determine the sedimentation coefficient of the sample by boundary analysis.

The choice of optical system to detect and measure the position of the moving boundary depends on the nature of the sample under investigation. If the protein solution has a significant absorbance at an appropriate wavelength then either a u.v. photograph may be taken using a u.v. source of fixed wavelength, or a photoelectric scanner-monochromator can be used to produce a record of the absorbance at each position in the cell. Using this method, absorbance values of less than 0.05 can be recorded.

If the protein solution has no significant absorbance and if sufficiently concentrated solutions are available, then optical systems that detect changes in refractive index can be used. The schlieren system is useful for proteins if the concentration is above about 5 mg/ml. With Rayleigh optics, differences in refractive index between reference buffer and protein solution are measured by displacement of interference fringes. A displacement of one fringe corresponds to a concentration difference of about 0.25 mg/ml. In *Figure 6* comparative examples are given for a moving boundary experiment using different optical detection systems. The u.v. photograph is scanned using a microdensitometer to produce a result similar to that obtained for the scanner recording except that the direct relation to absorbance is not available. In the case of the photoelectric scanner, the wavelength is selected using a monochromator, usually a modified spectrophotometer or laser light source. The image of the cell is scanned by a photomultiplier with a narrow slit which moves from the bottom to the top. The difference between the absorbance of the reference buffer and the sample is drawn out as a continuous trace on a recorder, or input directly into an automatic data processing system. The image of the holes in the counterbalance are included in the recording to provide reference edges for measurement. When double-sector centrepieces are adopted with schlieren optics *(Figure 6d)*, the image of a baseline is superimposed on the photograph so that detection of minor components and areas under peaks is possible. The schlieren plate which usually has six photographs taken at intervals can be measured very simply by projection onto graph paper. More complex systems that allow automatic processing of schlieren images are also available.

Rayleigh interference photographs are measured in image enlarger-microcomparator devices with micrometer-driven stages. Again, automatic scanning devices have been described that simplify the collection of data.

(vi) *Determination of observed sedimentation coefficient.* During the ultracentrifuge run, the boundary between buffer and solution becomes more and more diffuse as the components in the system undergo diffusion. The sedimen-

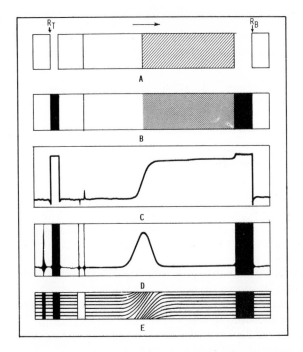

Figure 6. Results from a moving boundary experiment using various optical detection systems. (A) Diagram of an ultracentrifuge cell during a sedimentation velocity experiment. The boundary between buffer and solution is about midway down the cell. The images of the holes in the counterbalance are superimposed on the image of the cell and the top reference edge, R_T, and the bottom reference edge, R_B, are shown. The direction of the centrifugal force is indicated by the arrow. (B) Diagram of a u.v. light photograph of the system. Such a photograph is normally scanned in a densitometer so that the position of the boundary can be determined. (C) Diagram of a recording from the photoelectric scanner. This is effectively a direct recording of absorbance against distance. Usually a double peak is present at the meniscus because it is essential that the column of solution is slightly shorter than the column of reference buffer. The positive peak is the solution sector meniscus and the negative peak is the solvent sector meniscus. (D) Diagram of a schlieren photograph with a double-sector cell. A double meniscus is seen as above and a baseline is superimposed under the peak. The peak maximum corresponds to the midpoint of the boundary in the scanner trace. (E) Diagram of a Rayleigh interference photograph of the same system. The image of the wire in the upper reference hole in the interference counterbalance is present in **D** and **E**.

tation coefficient is defined as:

$$s = \frac{dr/dt}{\omega^2 r} = \frac{d \ln r}{dt} \cdot \frac{1}{\omega^2}$$

where the angular velocity, ω, in radians/sec is equal to $0.1047\,N$ (where N is the speed in revolutions per minute) and r (cm) is the distance of the boundary from the centre of rotation. When a graph of $\ln r$ against t (sec) is drawn, the slope is equal to $\omega^2 s$, thus the value of s can be found. The question is which point on the sedimenting boundary or schlieren peak should be chosen to measure the value of r for that measurement. Although the representative value is the second moment of the distribution (39), little or no error is introduced when the maximum of the schlieren peak or the 50% point on the

259

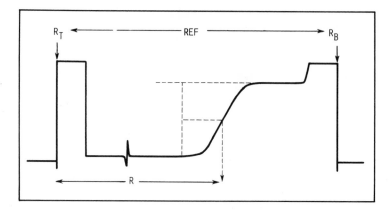

Figure 7. Determination of the distance of the 50% boundary point from the centre of rotation. The method used is described in the text.

bondary is taken. To evaluate the recordings for the moving boundary experiment (see *Figure 7*):

Measure the distance (REF) between the image of the reference edges (inner edges of the top and bottom holes in the counterbalance).

Divide by the actual distance between the edges to obtain the magnification factor for the recording (REF/1.7 = MAG).

For each record, measure the distance between the top reference edge and the chosen point on the recording (R cm). Divide by the magnification factor and add the real distance of the top reference edge from the centre of rotation using the relationship:

$$r = [(R/MAG) + 5.7]cm$$

Tabulate values of r (cm) against t (sec) (note that the absolute time is not required in this case). Determine the slope of this graph using the method of unweighted least squares and thereby calculate the sedimentation coefficient of the protein in units of S, where 1.00S, one Svedberg unit, is 1×10^{-13} sec.

(vii) *Conversion of observed s-value to standard conditions.* This observed sedimentation coefficient is a parameter that is dependent on the solution conditions used in the experiment. In the case of an 'uncharged', non-associating, monodisperse ('pure'), two-component ('H$_2$O-macromolecule') sample, the principal factors that influence the measured value of the sedimentation coefficient are the sample concentration and the temperature, viscosity and density of the solvent. These latter effects are corrected by converting the s-value to 'standard conditions', namely the value ($s_{20,w}$) expected in a solution having the density and viscosity of water at 20°C:

$$s_{20,w} = s_{T,solv} \left[\frac{\eta_{H_2O,T}}{\eta_{H_2O,20}} \right] \left(\frac{\eta_{solv}}{\eta_{H_2O}} \right)_{20} \frac{(1 - \bar{\nu}\varrho)_{20,w}}{(1 - \bar{\nu}\varrho)_{T,solv}}$$

where η is viscosity, $\bar{\nu}$ is the partial specific volume of the macromolecule and ϱ is the density of the solvent.

The effect of sample concentration can vary according to the nature of the sample, for example, the sedimentation coefficient may decrease with increasing concentration. An understanding of this behaviour can be found when the excluded volume effect in a thermodynamically non-ideal solution is considered. Ideally the entire volume of the system should be accessible to the macromolecule but, as the concentration increases, each molecule is excluded from a greater and greater volume of the system. This excluded volume is reflected in the value of the second virial coefficient B by the relationship:

$$s^0 = s_{obs} (1 + BMc)$$

where s^0 is the value at zero concentration. As the concentration becomes smaller, the influence of this term becomes zero and s_{obs} becomes identical to s^0.

To correct for concentration dependence, a number of runs at different initial concentrations are required. The range of concentration might require the use of more than one of the optical systems to give the sensitivity required at low concentrations. Different optical pathlengths through the sample can be used by adopting a variety of centrepieces of thickness varying from 1.5 mm to 30 mm. When analysing a number of samples at different concentrations it is convenient to use multiple cell rotors [see (viii)]. Sedimentation data at different concentrations are plotted against concentration and extrapolation to zero concentration gives $s^0_{20,w}$ from the relationship:

$$s^0_{20,w} = s_{20,w} (1 - k_s c)$$

where k_s is the concentration coefficient. Representative values of k_s are for globular proteins $1 - 2 \times 10^{-5}$ ml/μg, elongated proteins $1 - 3 \times 10^{-4}$ ml/μg and high molecular weight DNA $0.5 - 5 \times 10^{-1}$ ml/μg. For example, for a globular protein at a concentration of 0.5 mg/ml the value of the term $k_s c$ is about 5×10^{-3}, that is, the concentration-dependent effects are negligible.

In the case of high molecular weight DNA, rotor speed effects are observed. Material piles up and is rapidly lost from the plateau region in a boundary experiment and an abnormally high sedimentation coefficient is obtained. The effect is worse at higher concentrations of DNA. For example, in the case of DNA with a molecular weight of 100×10^6, the speed of centrifugation must be kept below 10 000 r.p.m.

(viii) *Use of multicell rotors.* A variety of rotors are available to analyse more than one sample in the same run. This is possible either by using wedge windows, which can deflect the image in the photographic plane, or a multiplexer than can electronically select the particular cell image to be recorded *(Figure 8)*.

2.1.2 Band Centrifugation Analysis

In preparative ultracentrifugation, it is usual to purify samples by zone centrifugation where the mixture is applied as a thin layer on a supporting gradient, using either preformed or self-generating gradients. The technique of band centrifugation is the corresponding procedure for the analytical ultra-

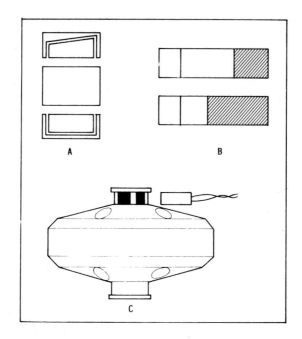

Figure 8. Multiple cell analyses. Two general types of system are used, wedge windows or a multiplexer system. **(A)** Diagram of an ultracentrifuge cell assembly containing a wedge window in the upper window assembly. The optical path is deflected so that the image from a wedge window cell assembly follows a different path from normal **(B)**. In the case of a two-place rotor containing one positive wedge window assembly and one normal cell assembly, two images are obtained, the upper image arising from the wedge cell assembly. The double image allows small differences in mobility to be more easily detected. **(C)** A multiple cell rotor may contain a number of normal cell assemblies. In this case the scanner system is used with a multiplexer which sorts the signals electronically. The rotor collar has a series of dark bands, one larger than the rest which are scanned by a photodiode assembly. These signals allow the scanner to determine which particular image is to be recorded.

centrifuge. Usually only shallow, self-generating density gradients are used to stabilise the bands. This method has the advantage that much smaller volumes of sample are required (50 μl or less) and the components, if there are not too many, can be visualised as discrete bands.

(i) *Band-forming centrepieces.* In order to achieve layering of the sample onto the supporting solution, special centrepieces, band-forming centrepieces, are required. Various designs are shown in *Figure 9*. Type I centrepieces are preferred when forming a layer on top of the supporting solution. Type III centrepieces are useful to underlayer the sample for, for example, flotation analysis of samples that are less dense than the main solution.

(ii) *Performing a band centrifugation run.* In this section a band sedimentation run of simian virus 40 (SV40) DNA is described.

Carefully clean the cell components and place the lower window assembly and centrepiece into the cell housing with the sample holes upwards.

Introduce 25 μl of sample buffer into the left-hand well and 25 μl of DNA

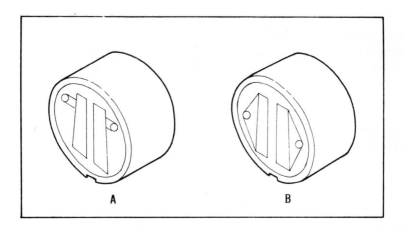

Figure 9. Band-forming centrepieces. (**A**) Type I double-sector band-forming centrepiece. The sample is introduced into the well next to the right-hand sector and reference buffer into the well at the left. The sectors are filled with supporting solvent. The sample is expelled from the well by the centrifugal force so that it layers on the top of the solvent to form a band. In some forms of this centrepiece, the two sectors communicate by an inscribed channel at the base of the sectors. This ensures that the levels in both sectors are identical. (**B**) Type III double-sector band-forming centrepiece. In this case, the sample is displaced from the well by solvent which migrates upwards from the bottom of the sector, so that a band is formed on top of the solvent as before. By adjusting the density of the sample and solvent systems, it is possible to arrange that the sample becomes underlayered and its centripetal movement can be followed.

solution (~ 10 μg/ml) into the right-hand well of a type I band-forming centrepiece. Avoid air locks by inserting the micropipette to the bottom of the well. Remove any droplets from the top surface of the centrepiece with a lens tissue. Quickly place the upper window assembly in the cell housing then complete the cell assembly and tightening procedure as before.

Place the assembled cell in a wooden holder and introduce 0.3 ml of CsCl solution, density 1.50 g/ml, in 0.01 M EDTA (pH 8.0) into both sectors. When filled, the level of the CsCl must not be higher than the scribe lines leading from the wells. Seal the cell assembly and place in the rotor as before. centrifuge at 40 000 r.p.m. and scan at 8 min intervals. When the speed of the centrifuge reaches 500 – 2000 r.p.m., transfer of the sample to the top of the buffer will occur. This can be checked by viewing the cell in the schlieren viewer. If the sample has layered as required, then a dark band will be seen at the top of the sector where refractive index differences between sample and buffer cause the light to become diffracted out of the optical track. The refractive index artifact will disappear within 10 – 15 min.

(iii) *Choice of solvent: band stability.* If a stable band is to exist throughout the centrifuge run, then the solvent must be sufficiently denser than the sample at every point down the cell (1). If the density difference is inadequate at any point then the band will spread forwards ('sinks') until a region of sufficient density difference is encountered. Such effects are usually clear on the recordings. When this difficulty is found, the experiment should be repeated using either a denser solution or a lower speed, unless diffusion problems arise.

Concentrated salt solutions are often used as the supporting solvent, for example, 0.5 or 1.0 M NaCl for proteins and CsCl solutions of density greater than 1.3 g/ml for DNA. If low ionic strength is required then solvents containing D_2O are often used instead.

To measure the sedimentation coefficient, plot the natural logarithm of the maximum in the concentration distribution against time. Even if the band profile is asymmetrical, only small errors arise in using the position of the peak maximum.

The conversion of the measured sedimentation coefficient in concentrated salt solutions to standard conditions requires special methods. To obtain a standard sedimentation coefficient, two approaches are possible (2). One can carry out a series of band sedimentation runs at different bulk solvent densities. Determine the observed (uncorrected) sedimentation coefficients, s_{obs}, from the movement of the band maxima. Calculate the product $s_{obs}\eta_r$, where η_r is the relative viscosity of the bulk solvent. Plot $s_{obs}\eta_r$ against the density of the bulk solvent. Extrapolate the value of $s_{obs}\eta_r$ to a density of 1.00 g/ml and hence obtain the standard sedimentation coefficient. Such graphs will not necessarily be linear. Alternatively, one can determine the buoyant density of the DNA [Section 3.4.3(i)]. At this point, the observed s-value is zero and $s_{obs}\eta_r$ is zero. Determine one value of $s_{obs}\eta_r$ as in the first method and construct a graph using these two points then extrapolate as before.

Concentration-dependent effects are expressed as variations in the shape of the sedimenting band. If the s-value decreases with concentration, then a band with a sharpened trailing edge and skewed leading edge will be present. If the diffusion coefficient, D, is large, this will tend to keep the band symmetrical but if D is small and concentration dependent, then front spreading and rear sharpening of the band will occur.

The technique of band sedimentation forms the basis for the important method of active band centrifugation (see Section 2.3).

2.2 Molecular Characterisation by Sedimentation/Diffusion

Methods to obtain the $s^0{}_{20,w}$ of a pure protein have now been presented. A simple kinetic derivation leads to the relationship:

$$s = \frac{M(1 - \bar{v}\varrho)}{Nf}$$

where M is the molecular weight, $\bar{v}$ is the partial specific volume, ϱ is the solvent density, N is Avogadro's number and f is the frictional coefficient of the particle. The s-value is a function of the size and the shape of the macromolecule. It can be used to determine the molecular weight of a macromolecule or to obtain information about the molecular conformation or changes in conformation. Boundary spreading during a moving boundary experiment reflects diffusion of the macromolecule. The diffusion coefficient, D, for an ideal solution is equal to RT/Nf, where R is the gas constant and T is the absolute temperature. Hence

$$\frac{s}{D} = \frac{M(1 - \bar{v}\varrho)}{RT}$$

If both D and s are known, then the molecular weight of the macromolecule can be calculated. If it is known that the protein is a simple globular protein, then instead of measuring D, an empirical relationship derived from s-values of proteins of known molecular weight can be adopted without further work.

$$s^0_{20,w} = 0.00242\ M^{0.67}$$

Similar equations for other macromolecules which represent a homologous series in terms of shape can be constructed (see Section 3.4 of Chapter 4).

2.2.1 Measurement of the Diffusion Coefficient

The flow of solute due to diffusion is given by the equation:

$$J = -\frac{c}{Nf} \cdot \frac{\partial\mu}{\partial r}$$

where μ is the chemical potential. It can be shown that:

$$J = -D \cdot \frac{\partial c}{\partial r}$$

where D, the diffusion coefficient, is given by:

$$D = \frac{RT}{Nf}[1 + BMc + ...]$$

where B is the second virial coefficient. The method for measuring D in the analytical ultracentrifuge is to form a sharp boundary between the buffer and protein solution. The spreading of the boundary is then followed at intervals of time. This can be achieved by using a synthetic boundary cell or, preferably, a synthetic boundary centrepiece *(Figure 10)*. To measure the diffusion coefficient, D, construct a cell assembly with a capillary-type synthetic boundary centrepiece. Introduce 0.15 ml of solution into the right-hand sector and 0.3 ml buffer into the left-hand sector. Seal the cell assembly. Use a heavy rotor, for example, an An-J or An-H rotor to minimise precession which will induce convection in the cell and centrifuge the rotor at 5000 r.p.m. When the centrifuge starts, the buffer layers on top of the solution to form a sharp boundary *(Figure 11)*. Schlieren photographs are a convenient method for recording the data. This is effectively a graph of $\partial c/\partial r$ against r. The flow equations relate flow to concentration by the relationship:

$$\frac{\partial c}{\partial t} = D\left(\frac{\partial^2 c}{\partial r^2}\right)$$

and a useful solution to this equation is:

$$\frac{dc}{dr} = \frac{c_0}{2(\pi Dt)^{\frac{1}{2}}} \cdot \exp[-r^2/4Dt]$$

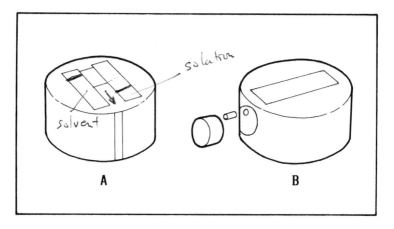

Figure 10. Synthetic boundary centrepieces. **(A)** Double-sector capillary-type synthetic boundary centrepiece. Two channels connect the sectors, one at the top and the other halfway up. A small amount of solution is introduced into the right-hand sector so that the level is below the middle channel. Reference buffer is introduced into the left-hand sector to almost fill the sector. The centrifugal force causes transfer of buffer through the middle channel. A layer of buffer is thereby superimposed on top of the solution to form a sharp boundary. **(B)** Valve-type synthetic boundary centrepiece. In this case, a hole in the side of the centrepiece accommodates a cup with a small hole in its base and a rubber valve which retains the liquid. A small amount of solution is introduced into the sector. The centrifugal force compresses the rubber valve and allows the buffer to drain from the cup to form a boundary on the solution.

where r is zero at the peak maximum this equation becomes:

$$H = \frac{c_0}{2(\pi Dt)^{1/2}} = \frac{A}{2(\pi Dt)^{1/2}}$$

where A is the area under the peak and H is the maximum height.
Thus $\left(\frac{A}{H}\right)^2 = 4\pi Dt$ and the slope of a graph of $\left(\frac{A}{H}\right)^2$ against t (sec) is $4\pi D$.
One convenient way to measure D is to cut out the peaks for weighing and match against a unit area cut from the photograph. Similarly, a digitiser can be employed for area and height determination. If a photoelectric scanner is used, then a derivative trace can be produced on the recorder along with the normal integral trace of c against r, which can also be numerically differentiated to produce data in the same form as the schlieren photograph. In sophisticated systems, direct processing of data by computer can be achieved. An analysis of boundary spreading in a low speed sedimentation run in the c against r traces can also be used to measure D (4).

2.2.2 Partial Specific Volume and Density Increments

To determine the molecular weight of the protein once s and D are known, it is necessary to evaluate $(1 - \bar{v}\varrho)$. The solvent density ϱ can be measured directly or obtained from tables (International Critical Tables). The partial specific volume, $\bar{v}$, is defined as the volume of water displaced by one gram of anhydrous protein. This can be determined pycnometrically if enough protein is available. Alternatively, the weight average value of the constituent amino

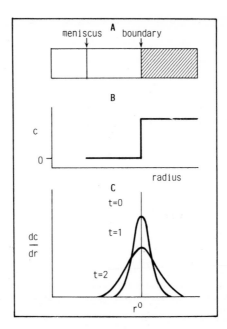

Figure 11. A synthetic boundary centrifugation experiment. **(A)** At the start of the experiment, the solvent is layered on top of the solution to form a sharp boundary. **(B)** The concentration distribution at the start of the experiment. **(C)** The derivative trace, such as is obtained from the schlieren photograph. At the start of the experiment, $t = 0$, a hypersharp peak is present at r^0. As time goes on, diffusion causes boundary spreading and the peaks become wider and shorter. The maximum height, H, and area, A, of the peaks are measured to determine the diffusion coefficient as described in the text.

acids, $\Sigma w_i \bar{\nu}_i$, is often used to determine $\bar{\nu}$. If the sample is not a simple protein but is instead a complex protein, for example, a glycoprotein, then it is often more reliable to determine the partial density increment $(\partial \varrho / \partial c)^0_\mu$ at dialysis equilibrium. In this case:

$$\frac{s}{D} = \frac{M}{RT} \cdot \left(\frac{\partial \varrho}{\partial c}\right)^0_\mu$$

The significance of this term is that it reflects all of the interactions between the macromolecule and other components in the solution and it is not necessary to know or even to try to define what these interactions might be. The molecular weight obtained refers to the molecular weight of the species at the defined concentration c.

To determine the density increment, dialyse the sample to equilibrium against buffer (usually 24 h with two changes of buffer). Measure the density of the dialysate buffer in a precision densitometer, for example, a DMA02C digital densitometer (Paar KG) and also the density of the sample solution. Determine the density difference $\Delta \varrho$, and thereby the density increment $\Delta \varrho / c$, which is the desired thermodynamic variable at constant chemical potential of diffusible components.

In a simple, two-component system, such as a pure protein in a physiological buffer, the terms $(1 - \bar{v}\varrho)$ and $(\partial\varrho/\partial c)_\mu$ are almost precisely equivalent and $(\frac{\partial\varrho}{\partial c})_\mu = (1 - \phi'\varrho)$ and so $\bar{v}$ is approximately equal to ϕ', the effective specific volume.

Another way to determine $\bar{v}$ is to obtain sedimentation equilibrium data (Section 3) in two solvents of different density (5). This is an important method for proteins combined with large amounts of lipid. The method is based on the change produced in sedimentation equilibrium concentration distribution when the density of the solution is increased by the use of D_2O.

$$\bar{v} = \frac{k - [(d\ln c/dr^2)_{D_2O}/(d\ln c/dr^2)_{H_2O}]}{\varrho_{D_2O} - \varrho_{H_2O} [(d\ln c/dr^2)_{D_2O}/(d\ln c/dr^2)_{H_2O}]}$$

where k is the ratio of the molecular weight of the protein in the deuterated solvent to that in the non-deuterated solvent. For simple proteins, the value of the term

$$[(d\ln c/dr^2)_{D_2O}/(d\ln c/dr^2)_{H_2O}]$$

is approximately 0.8. For a variety of proteins, the value of the constant k is 1.0155, as determined by deuterium exchange methods.

2.2.3 Detection of Conformational Changes

The sedimentation coefficient depends on the conformation of the macromolecule in solution as reflected by the frictional coefficient f. Stoke's law states that for a rigid sphere the frictional coefficient (f) is equal to $6\pi\eta r$ where η is the viscosity of the fluid and r is the radius of the sphere. As shown in the previous section, the term $(1 - \bar{v}\varrho)$ is equivalent to $\partial\varrho/\partial c$ and hence the original equation (Section 2.2) can be expressed as:

$$s = \frac{M\,(\partial\varrho/\partial c)}{N6\pi\eta R_s}$$

where R_s is the Stoke's radius of the rigid sphere. Thus, for example, when the volume of a spherical particle is doubled, the sedimentation coefficient will increase by a factor of 1.6. In the case of a less well-defined geometry, for example, a random coil, the Stoke's radius of the equivalent hydrodynamic sphere, R_s, is related to the radius of gyration, R_g, by the relationship $R_s = 0.665\,R_g$. The value of R_g is the distance of all particles from the centre of mass averaged over all conformations of the coil. In this way values of $s_{20,w}, M$ and $(1 - \bar{v}\varrho)$ or $(\partial\varrho/\partial c)_\mu$ are used to estimate particle dimensions. Frictional ratios can be calculated (6) from the equation:

$$f/f_{min} = \frac{(4/3)^{1/3}}{6\eta(\pi N)^{2/3}} \cdot \left(\frac{1 - \bar{v}\varrho}{\bar{v}^{1/3}}\right) \cdot \frac{M^{2/3}}{s}$$

where f_{min} is the frictional coefficient of the equivalent unsolvated sphere and

$$f_{min} = 6\pi\eta\,(3M\,\bar{v}/4\pi N)^{1/3}$$

so that axial ratios of ellipsoidal particles can be calculated.

Sedimentation velocity analysis can be used to detect changes in conformation caused by the binding of ligands. Two experimental procedures are con-

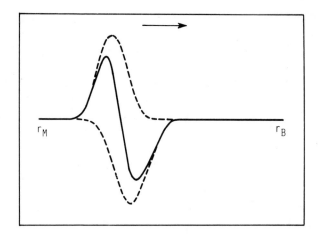

Figure 12. A paired sedimentation velocity experiment. This diagram is derived from a sedimentation velocity experiment using a Type I band-forming centrepiece with a communicating channel at the bottom of the sectors. An aliquot of one solution was introduced into the right-hand well and an aliquot of the other solution into the left-hand well. When the samples layer on the solvent, the sample in the right-hand sector gives rise to a positive peak on the scanner trace and the sample in the left-hand sector gives rise to a negative peak (dotted lines). If the solute molecules have different mobilities, then a difference curve (solid line) is obtained on the trace, where r_M and r_B are the positions of the meniscus and the cell bottom, respectively.

sidered here. The first method is the paired sedimentation velocity experiment. In this method, use two cells with the schlieren optical system. One cell assembly incorporates a 1° or 2° positive wedge window in the upper window holder to deflect the schlieren image vertically upwards (cf., *Figure 8*). Small differences in the mobility of the bands (~0.016s) can more readily be detected (7,8). In the second procedure, use one double-sector cell with the photoelectric scanner system. Load one sample into the left-hand sector and the other into the right-hand sector. The right-hand sector will give rise to a positive deflection on the recorder while the left-hand sector gives a negative deflection on the recorder *(Figure 12)*. Thus a difference plot is obtained and small differences in mobility will be apparent on the trace.

In the case of band sedimentation analysis of supercoiled DNA sedimenting through CsCl containing different amounts of the intercalating dye ethidium bromide, the conformation of the complex of ethidium bromide and DNA is a function of the amount of dye bound. The number of superhelical turns can be determined from the amount of dye bound when the sedimentation coefficient is a minimum (9).

2.3 Active Band Centrifugation

Active band centrifugation is an extension of the use of band centrifugation to study enzymes in their active form as active enzyme-substrate complexes using microgram amounts of enzyme (10). Two types of general analytical approach were originally described, the approximate method which readily gives rise to a value of the sedimentation coefficient of the enzyme, and the rigorous method

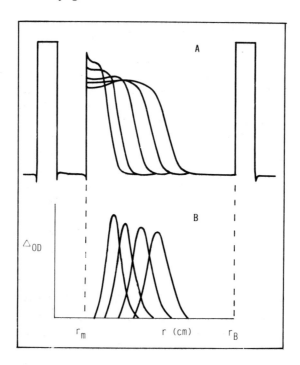

Figure 13. Active band centrifugation. **(A)** As the enzyme migrates through the substrate solution, the product is detected as an absorbing boundary and a series of such boundaries are super-imposed on the diagram. 'End effects' at the meniscus interfere with accurate determination of boundary position at early stages in the experiment but these effects disappear as the boundary migrates away from the meniscus. **(B)** Difference curves derived from the scanner traces shown above. The difference in absorbance between neighbouring scanner traces is determined and the peak maximum position measured. The sedimentation coefficient is determined using the 'approximate method' from the movement of the peak maximum.

which gives values for both s and D (11).

Insert the enzyme solution into the well of a type I band-forming centrepiece and fill both sectors with the buffer containing the substrate. For example, layer 10 μl of 3.3 μg/ml glutamate dehydrogenase onto 0.3 ml of substrate solution containing 0.15 M phosphate buffer (pH 7.5), 50 mM sodium glutamate, 2 mM NAD^+, 0.5 mM $NADP^+$ and centrifuge at 58 000 r.p.m. at 20°C with scans at 340 nm at 2 min intervals. As the enzyme migrates through the substrate solution, product is formed and is detected by an appropriate change in absorbance *(Figure 13a)*.

In the approximate method, difference curves are derived *(Figure 13b)* and the position of the optimum of each difference curve is determined. The standard graph of ln$r$ against t (sec) is drawn and from the slope, the sedimentation coefficient evaluated. In this procedure 'end effects' influence the data and until these effects become resolved in the course of the experiment, accurate s-values will not be obtained. This is observed in curvature of the graph of ln$r$ against time at the start of the experiment. When the data become linear, it is found that the value of s obtained is accurate. Interference of end effects depends on the speed chosen for centrifugation in relation to the size of the particle, that is, s and D. Under the correct conditions, sedimentation coefficients are obtainable with errors of less than 0.5%.

Soon after the introduction of this method, it was unclear how the value of s obtained related to the 'actual' s-value. This was clarified after the development of powerful simulation techniques which permitted careful scrutiny of the data obtained and the data was shown to be accurate and reliable. Simulation analysis also clarified uncertainties about substrate depletion in the course of the experiment.

The rigorous method, which allows evaluation of both s and D, is computationally very complex and requires computer processing of the data. Programs have been developed for such processing, but in other than the most simple cases, the computational demands become very great and interpretation becomes less certain. In this situation the preferred approach is to adopt the simulation of data by computer and match the experimental results with the data obtained for best fit (12).

2.4 Sedimentation Analysis of Interacting Systems and Polydisperse Systems

The behaviour of interacting macromolecules in sedimenting systems has been actively studied for more than 20 years. Where the biological system under study is unstable, so that lengthy methods of analysis are not possible, sedimentation velocity analysis can be an important technique, especially in view of advances in data analytical methods. The analytical ultracentrifuge is one of the few methods able to study interaction kinetics in non-equilibrium flow situations.

2.4.1 Polydisperse Systems: s-value Distributions

In band sedimentation experiments, the concentration profile gives a direct indication of the distribution of the sedimentation coefficients in the sample. For non-resolved mixtures, s_w, the weight average sedimentation coefficient is given by $\Sigma w_i s_i$, where w_i is the weight fraction of the ith component, sedimentation coefficient s_i and may be calculated in the case of band sedimentation from the movement of the centre of gravity of the entire distribution. By neglecting the effects of diffusion, data in the form of c against r can be changed to the sedimentation coefficient distribution of c against s by

$$s(r) = \frac{\ln(r/r^0)}{\omega^2 (t - t_0)}$$

In the case of a schlieren pattern which relates the concentration gradient, dc/dr, to the radial position, r, the distribution of sedimentation coefficients, $g(s)$, is given by:

$$g(s) = \left(\frac{dc/dr}{c_0}\right) \cdot \left(\frac{r}{r_m}\right)^2 \cdot 2 \, \omega^2 rt$$

where c_0 is the initial concentration, $(r/r_m)^2$ corrects the values of dc/dr for

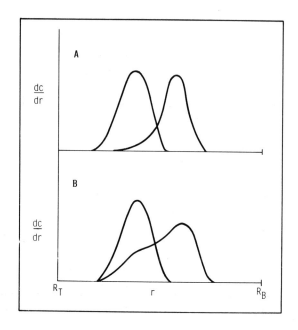

Figure 14. The effect of increasing association constant on an interacting macromolecular system. **(A)** Monomer-dimer associating system schlieren diagram. Only one distribution is seen in each case and the effect of increasing the association constant is to move the pattern towards the position where a pure dimer would be found and to sharpen the profile. **(B)** A monomer-trimer system. In this system, more than one peak is present as the stability of the oligomer increases.

radial dilution. The weight fraction of material with s-values between s and $s + ds$ is $g(s) \cdot ds$ (13).

2.4.2 *Interacting Systems*

When particles interact, their hydrodynamic properties are altered and this is reflected in the s-value. A consideration of the frictional properties of assemblies of subunits can lead to a prediction of the sedimentation behaviour of the oligomer, thus for an oligomer comprising n rigid spherical subunits (14) the following equation can be deduced:

$$\frac{s_n}{s_1} = 1 + \frac{1}{n} \Sigma_i \Sigma_j \, (1/d_{ij})$$

where d_{ij} is equal to R_{ij}/R and R_{ij} is the distance between the ith and jth subunit and s_1 the sedimentation coefficient of a monomer sphere of radius R. Interacting systems are indicated by changing sedimentation profiles. For example, when concentration is increased, additional components may appear. When a rapid equilibrium is involved, then the sedimentation profile will normally be independent of speed. If the system is a slowly associating one then the weight average sedimentation coefficient will vary with speed because, in this case, the oligomer is sufficiently stable to migrate significantly before

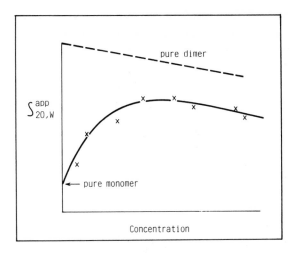

Figure 15. Sedimentation behaviour of a monomer-dimer system. The apparent sedimentation coefficient, s^{app}, is determined over a range of concentration. The value for the pure monomer is derived from the limiting value at zero concentration. The dashed line shows the sedimentation behaviour of the pure dimer. The experimental values are then compared with simulated curves derived using the equation:

$$s^{app} = [s_d^0 - 2(s_d^0 - s_m^0)/[(1 + 4Kc)^{1/2} + 1]]/(1 + k_s c)$$

where s_m^0 and s_d^0 are the sedimentation coefficients of the monomer and dimer at zero concentration and k_s is the concentration coefficient (see Section 2.1.1). The value of K that gives the best fit to the experimental data is then determined.

dissociating and the reacting mixture will be perturbed differently according to the speed of centrifugation.

A monomer-polymer system where all solutes achieve chemical equilibria instantaneously is effectively the same as a single non-ideal solute in ultracentrifuge experiments. The effect of increasing association constant is shown in *Figure 14* for a monomer-dimer and monomer-trimer system (16). In the case of the monomer-dimer system, only one boundary or peak is seen. If the rate of association is very slow, then two peaks will begin to separate out. To distinguish between heterogeneity and association interactions, fractionation experiments are essential to avoid ambiguity. Experiments can be performed at various initial concentrations of solute and a graph of the apparent $s_{20,w}$ against concentration constructed *(Figure 15)*. The graph is compared numerically with model equations to determine the association constant (15).

Complex sedimentation distributions, including those obtained in association reactions during active band centrifugation, may present too many difficulties for the calculation of interaction constants and the values of s and D of the components. An important practical choice then arises as to whether to attempt this or to assume particular models and simulate them for comparison with the real data. The latter approach is now well developed. The association reactions may be mediated by a ligand and predictive computer modelling methods are used to match experimental behaviour (17 – 19).

Pressure effects can play an important role in interacting systems. Thus, in an interacting system, any change in the partial specific volume as a result of association will affect the association constant of the reaction in a gradient of hydrostatic pressure, $\partial \varrho / \partial r = \varrho \omega^2 r$ because:

$$\left(\frac{d\ln K}{d\varrho}\right)_T = -\frac{\Delta V}{RT}$$

The hydrostatic pressure at any radius r in the gradient is related to the pressure at the meniscus, P_0, by the relationship:

$$P_r = \frac{\omega^2 \varrho}{2} (r^2 - r_0^2) \varrho_0$$

The equilibrium constant at radius r, K_r, is related to the equilibrium constant at the meniscus, K_0, by:

$$\ln K_r = \ln K_0 - \frac{1}{RT} \int \Delta V \cdot \frac{dP}{dr} \, dr$$

thus (20),

$$\ln K_r = \ln K_0 - \frac{\Delta V \varrho \omega^2}{2 RT} (r^2 - r_0)$$

If there is an increase in the partial specific volume for the above reaction then as the pressure increases, the amount of monomer will increase.

3. SEDIMENTATION EQUILIBRIUM ANALYSIS

In a sedimenting system all the components are moving towards a position of equilibrium under the influence of various potential gradients. When equilibrium is attained, the components become distributed throughout the solution and the total potential of any component is the same at all levels in the cell *(Figure 16)*. The total potential depends both on the chemical potential of each component and its potential energy in the centrifugal field. The situation can also be thought of as one in which the processes of sedimentation are balanced by those of diffusion.

Analysis of the distributions of components at sedimentation equilibrium offers an extremely powerful method to determine absolute molecular weights and study molecular interactions. The consideration of equilibrium experiments begins with analysis of a pure two-component system, for example, a pure protein in dilute buffer. Then experiments involving three or four-components systems will be considered followed by polydisperse and interacting systems.

3.1 Molecular Weight of a Pure, Homogeneous Sample

An experiment to determine the molecular weight of a protein is described, followed by a consideration of the factors that influence the choice of conditions for a sedimentation equilibrium run. A number of general points arise at several stages in this experiment about the sample and its behaviour at many stages in this experiment and these will be considered at each stage.

Dialyse the protein to equilibrium in the selected buffer. In general, the sam-

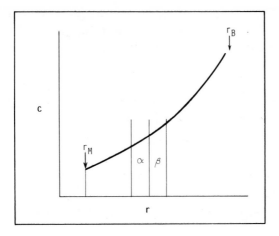

Figure 16. Diagram of a solute distribution at sedimentation equilibrium. Between the meniscus (at r_M) and the cell bottom (at r_B), the solute is distributed in a manner such that the total potential in region α is exactly balanced by the total potential in region β.

ple concentration dictates the method chosen for detection of events during centrifugation. Transfer 90 μl to the right-hand sector of a 12 mm double-sector cell assembly and 110 μl of dialysate to the left-hand reference sector then seal the assembly. The amount of sample dictates the length of the column of solution in the ultracentrifuge cell which in turn determines the time taken to reach equilibrium defined by the equation:

$$t_{0.1\%} = 0.7\,(b - a)^2/D$$

where b is the radius of the bottom of the solution and a is the radius of the meniscus.

It is essential that the image of the column of solution is enclosed within the column of reference buffer or dialysate. Often this is achieved using fluorocarbon or silicone oils to define the base of the columns. Care should be taken that no precipitate forms at the interface between the oil and solution. Centrifuge the sample for 30 h at 12 000 r.p.m. at 20°C then scan the cell at 280 nm with the photoelectric scanner. Increase the speed to 40 000 r.p.m. until the protein has sedimented away from the meniscus. Scan the cell again to determine the background absorbance that must be subtracted from the sample scan.

In general, when the rotor speed is chosen, either the 'low speed' or the 'high speed' method can be used (21). If the sample is homogeneous (i.e., containing only one macromolecular species) then the 'low speed' technique can be adopted using either optical system. If only very low concentrations of sample are involved, then the 'high speed' method can be used and this approach is also desirable if a mixture of components is present and the molecular weight of the smallest component is required. The objective in the low speed approach is to obtain concentration distributions so that the concentration at the bottom of the cell is three or four times greater than the concentration at the meniscus.

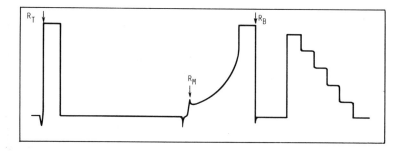

Figure 17. Scanner trace of a system at sedimentation equilibrium. The entire scan between the top of the cell (R_T) and the bottom of the cell (R_B) is shown and the position of the meniscus (R_M) is marked. At the right of the trace, calibration stairsteps are shown by means of which it is possible to convert chart height to absorbance. Each step corresponds to a change in absorbance of 0.2

In the high speed method, the recommended value of the effective reduced molecular weight $\sigma = \omega^2 M(1 - \bar{v}\varrho)/RT$ is 5 cm^{-2} for column lengths of about 3 mm. An estimate of the required speed, N, can be obtained from equations of the form:

> *low speed:*
> $\log (N) = 6.34 - 0.496 \log$ (mol. wt.)
> *high speed:*
> $\log(N) = 6.88 - 0.506 \log$ (mol. wt.)

Construct a graph from the calibration stairsteps *(Figure 17)* to relate chart deflection to absorbance. Convert the height of the graph minus the background to absorbance at chosen intervals across the cell. In a two-component system, the distribution of the macromolecule is given by the expression:

$$\frac{d\ln c}{dr^2} = \frac{M_2(1 - \bar{v}_2\varrho)\omega^2}{2RT\,[1 + BM_2 c + ...]}$$

where c_2 is the concentration of the macromolecule at radius r, $\bar{v}_2$ and M_2 are the partial specific volume and molecular weight of the macromolecule, respectively. Draw a graph of $\ln c$ against r^2 and determine the slope. If values for $\bar{v}_2$ or $(\partial\varrho/\partial c_2)^0_\mu$ are available, then the apparent weight average molecular weight of the protein can be found for the ideal case when the value of the virial coefficient, B, is zero. An alternative approach is to consider the concentration distribution throughout the cell, where the concentration, c_r, at radius r is given by:

$$c_r = c_0 \exp[\sigma(r^2/_2 - r_0^2/_2)]$$

where σ is given by:

$$\sigma = \frac{M(1 - \bar{v}\varrho)\omega^2}{RT}$$

and c_0 is the concentration at radius r_0. The best value of σ is determined by non-linear least squares curve fitting procedures.

If Rayleigh interference photographs are used to record the concentration

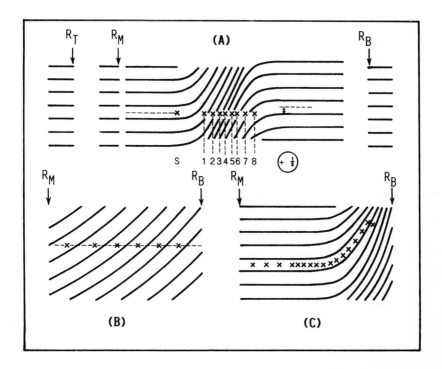

Figure 18. Measurement of Rayleigh interference photographs. **(A)** Diagram of a synthetic boundary cell some time after layering of buffer on top of the solution has occurred. The objective is to determine the number of fringes that correspond to the initial concentration c_0. Starting at the middle of a light fringe on the left, each fringe crossed is numbered until number 8 is reached. Then there is a fractional fringe increment to measure. This is the fractional vertical distance shown by the double-headed arrow and corresponds in this case to 0.5 fringe. Hence the fringe number in this case is 8.5. **(B)** Fringe pattern of a system at 'low speed' equilibrium. The dotted line allows the whole number of fringes to be determined and the fractional number at each end can be found by drawing a graph of distance *versus* fringe number and extrapolating to R_M and R_B. **(C)** Fringe pattern from a 'high-speed' experiment. In this case, the fringe displacement at constant increments in radial distance is measured.

distributions then an additional experiment is necessary. The initial concentration, c_0 must be found using a synthetic boundary cell. The results are evaluated as shown in *Figure 18* (22), where data from low-speed and high-speed runs are included. A photograph from the same cell assembly filled with water or a photograph of the cell which has been shaken after the original run and taken back up to speed is used as a blank to correct for distortion. The blank values are subtracted from the fringe displacements in the experiment. The region next to the meniscus must be horizontal before evaluation can begin. The natural log of the vertical displacement is plotted against r^2. In low-speed experiments, the interference plates can either be read along fringes (for lower concentrations) or 'across fringes' that is points of intersection of a horizontal line with successive fringes noted. Readings are extended as near the meniscus and cell bottom as possible.

3.1.1 *Concentration Dependence and Non-ideality*

When the graph of ln$c$ against r^2 is linear, the slope is constant, implying a constant molecular weight throughout the solution. However, the graph may curve downwards. The decreasing slope is principally due to increasing non-ideality effects (reflected through the second virial coefficient B) as concentration increases down the cell. The apparent molecular weight at any point, often determined from the slope of the tangent at that point, is related to the true molecular weight by an equation of the form:

$$M_{app} = M(1 + Bc + ...)$$

The non-ideal effects, reflected in the second virial coefficient B, can arise from a number of sources, for example, buffer-macromolecule interactions, and one approach is to repeat the experimental analysis using another buffer system by, for example, increasing the salt concentration to diminish the non-ideality effects. Another approach is to adopt an appropriate data handling method. Methods of handling the data from a sedimentation equilibrium experiment make it possible to derive a good estimate of the 'ideal' molecular weight of the macromolecule even when strong non-ideality is present (23).

Alternatively, the graph of ln$c$ against r^2 may slope upwards and this implies an increase in apparent molecular weight with concentration. Point by point molecular weight values can be derived from the gradient of the tangent to the graph at selected intervals, and this is the basis for the analysis of polydisperse and interacting systems as described in Section 3.2.

3.1.2 *Derivation of Apparent Molecular Weight and Concentration Data*

When analysing a series of samples at different dilutions, the same double-sector cell assembly can be used repeatedly. Dismantling is not always necessary and more concentrated samples can be inserted after rinsing with the appropriate solutions before use. Alternatively, multi-channel cells may be used. Six- or eight-channel centrepieces are commonly used when a number of samples at different initial concentrations must be studied *(Figure 19)*. Each channel of the six-channel centrepiece can hold a column of up to 3 mm in diameter and each channel of the eight-channel centrepiece can hold up to 1 mm diameter columns. With both types of cell assembly, adsorption of the solutes can be minimised by filling and emptying the centrepiece two or three times. The final filling must be completed rapidly to avoid loss from the small volumes by evaporation.

Fill the solvent side, that is, the left side, first. When the sample is introduced, place the more concentrated solution into the innermost hole first and the less concentrated sample into the outer hole. Dense FC 43 fluorocarbon (Maxidens, Nyegaard & Co.) is often used as a cushion and $5 - 10$ μl can be used in the multi-channel centrepiece, but care should be taken to ensure that no precipitation of sample occurs at the interface. If this happens then do not use a cushion of FC 43. With the six-channel centrepiece, up to 0.11 ml sample can be used with up to 0.12 ml dialysate in the reference side and with the eight-channel centrepiece, up to 0.02 ml sample and up to 0.025 ml of dialysate can

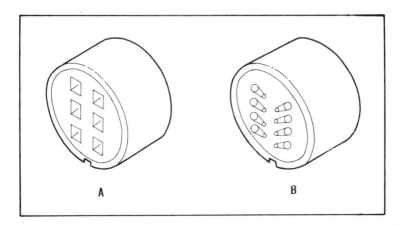

Figure 19. Multi-channel centrepieces. **(A)** A six-channel centrepiece and **(B)** an eight-channel centrepiece.

be added. Prior to use, the samples should be dialysed to equilibrium and it is often the practice to prepare diluted samples using the dialysis buffer and this is often done gravimetrically.

3.2 Polydisperse and Interacting Macromolecular Systems

In a polydisperse system a molecular weight distribution of the form shown in *Figure 20* might be present. It is extremely difficult to devise approaches to define this distribution completely and is more common to obtain representative averages to characterise the distribution. The number average (M_n), weight average (M_w), and z-average (M_z), molecular weights are indicated where:

$$M_n = \frac{\Sigma n_i M_i}{\Sigma n_i}; \; M_w = \frac{\Sigma n_i M_i^2}{\Sigma n_i M_i}; \; M_z = \frac{\Sigma n_i M_i^3}{\Sigma n_i M_i^2}$$

Sedimentation equilibrium analysis gives values of the weight average molecular weight, but the other averages can be readily obtained. When graphs of ln$c$ against r^2 graphs slope upwards, this is an indication that the apparent weight average molecular weight is increasing and this offers an approach for analysing interacting systems.

3.2.1 *Self-associating Systems*

Great attention must be paid to the precision and accuracy of the experiments used to evaluate molecular weight and concentration data. Apparent weight average molecular weights are measured over a wide range of concentration and a composite graph constructed. From this data, various other molecular weight moments can be derived. A discrete model can then be assumed and the observed behaviour is matched with that predicted for different models of associating systems, the best fitting situation being selected. This process is carried out by computer analysis and a number of approaches are well

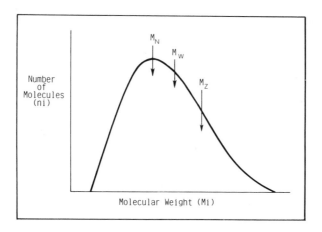

Figure 20. Molecular weight distribution for a polydisperse sample. The position of the number average (M_n), weight average (M_w) and z-average (M_z) molecular weights are indicated.

documented (24,25,26).

Example: test the self-association of lysozyme for a monomer-dimer fit.
Dialyse three samples of solutions containing 0.5, 5.0 and 10.0 mg/ml of lysozyme in 0.15 M NaCl, 5 mM phosphate buffer (pH 6.7) at 4°C for 24 h. Transfer the samples to a six-channel cell and centrifuge at 20 000 r.p.m. at 25°C for 24 h then scan the samples.
Repeat the experiment with three more samples at concentrations of 10.0, 15.0 and 20.0 mg/ml and centrifuge at 15 000 r.p.m. at 25°C for 24 h.
Construct a composite graph of apparent weight average molecular weight against concentration for all the samples, taking tangents to the curves of plots of $\ln c$ against r^2 graphs at selected intervals.
Derive a smoothed curve from this data. If the associating system is homogeneous, then the data of the weight average molecular weight and the c values will be superimposable. If they do not superimpose, then the system may be heterogeneous or there may be pressure effects on the equilibria involved. Tabulate values of $M_1/M_{w_{app}}$ and c and derive values of $M_1/M_{n_{app}}$ and $\ln f_1^{app}$ at the various concentrations where:

$$\frac{M_1}{M_{n_{app}}} = \frac{1}{c} \int_0^c \frac{M_1}{M_{w_{app}}} \quad \text{and} \quad \ln f_a = \int_0^c \left(\frac{M_1}{M_{w_{app}}} - 1 \right) \frac{dc}{c}$$

Use the equation:

$$\frac{2cM_1}{M_{n_{app}}} - c = \alpha \exp(-BM_1c) + BM_1c^2$$

where:

$$\alpha = \frac{f_c}{\exp(-BM_1c)} \quad \text{and} \quad \ln f = \ln f_a - BM_1c$$

Match the left hand side (experimental) of the equation with the right hand side by allowing the value of BM_1 to vary until the best fit is achieved. Calculate the monomer concentration, c_1, from:

$$c_1 = \alpha \exp(-BM_1c)$$

Derive the association constant, K_2, from the equation:

$$c = c_1 = K_2 c_1^2$$

It is not essential to assume a specific model to derive values for the association constants and virial coefficients and series expansion methods have been presented as an alternative approach (27).

3.3 Interaction of Small Molecules with Macromolecules

A single sedimentation equilibrium experiment is analogous to many dialysis equilibrium experiments with the macromolecule present at different concentrations. The equilibrium distribution of the small molecule (e.g., dyes) is determined at an appropriate wavelength with the photoelectric scanner and the distribution of the macromolecule is obtained with the Rayleigh optical system. Different ratios of small molecule to macromolecule are used and graphs of small molecule concentration against macromolecule concentration are constructed. Such graphs are linear with the relationship:

$$[A_0] = [A] + r[P_0]$$

where $[A_0]$ is the total concentration of small molecule, $[P_0]$ is the concentration of the macromolecule and where r is defined as:

$$r = \frac{nk[A]}{1 + k[A]}$$

Thus the experiment determines both r (the slope) and $[A]$ (the intercept) so that the number of binding sites n and the association constant k can be determined.

A classical experiment described by Steinberg and Schachman (28) to measure the binding of methyl orange to bovine plasma albumin is as follows. Transfer 0.025 ml of a mixture containing 0.4 g/100 ml bovine plasma albumin and methyl orange at 7.8×10^{-5} M in 0.1 M phosphate buffer (pH 5.66) to the sample sector of a 3 mm double-sector cell and buffer in the other. Centrifuge at 12 590 r.p.m. at 4°C for 18 h and scan at 440 nm with the photoelectric scanner to determine the distribution of dye.
Determine the protein distribution with the Rayleigh interferometer and determine c/c_0 against r.
Plot the absorbance at 440 nm against bovine plasma albumin concentration. This graph is linear and the slope and intercept can be found.
Repeat the experiment with different amounts of bovine plasma albumin and methyl orange, for example, 46.5 μM protein and 188.8 μM dye and 46.5 μM protein with 75.3 μM dye under the above conditions.
Substitute the slopes and intercepts to determine the binding constant.

3.4 **Analysis of Multicomponent Systems**

Systems of this kind, especially three-component systems, have received extensive experimental analysis in recent years. The treatment of oligomeric proteins with denaturing agents, such as urea and guanidine hydrochloride, or detergents, such as sodium dodecyl sulphate, to determine subunit composition is widespread. The added small molecules (component 3), and the solvent (component 1), interact preferentially with the protein (component 2). Sometimes it is relatively easy to determine the extent of binding, in other cases it is less certain. However, the theoretical basis of analysing three-component systems at sedimentation equilibrium is now well established. The other classical case to be considered here is the system H_2O-DNA-CsCl at buoyant equilibrium.

3.4.1 *Sedimentation Equilibrium Analysis of Proteins in Guanidine Hydrochloride*

Guanidine hydrochloride disrupts non-covalent interactions that maintain the native conformation of proteins and, in the presence of mild reducing agents to break disulphide bonds, the polypeptide is converted to a random coil conformation as revealed, for example, by hydrodynamic properties. These conditions are commonly employed to determine subunit molecular weights of proteins. In such a three component system, preferential interactions between solvent components and the macromolecules occur and virial effects (non-ideal effects) are an order of magnitude higher, since the effective hydrodynamic radius of the particles is larger and therefore any excluded volume effects are greater.

Experimental procedures are available that deal rigorously with the preferential interactions. The concentration distribution of a macromolecule (at low concentrations) in a three-component system (component 1 is H_2O, component 3 is the low molecular weight solute) is given (29) by the equation:

$$\frac{d\ln c_2}{dr_2} = \frac{\omega^2 M_2}{2RT} \left(\frac{\partial\varrho}{\partial c_2}\right)^0_\mu$$

Preferential interactions of proteins with a variety of salts is under active study (30).

The crucial feature is to determine $(\partial\varrho/\partial c_2)^0_\mu$ which accounts for preferential interactions. In the three-component system, samples are filtered through 0.8 μm pore Millipore filters into dialysis bags and dialysed at room temperature for $24-30$ h with two changes of guanidine hydrochloride. Dialysis is in a cylinder with rotation around the long axis and an air bubble inside the bag for mixing. Protein concentrations should be in the range $1-20$ mg/ml.

To fill the ultracentrifuge cell, the sample is withdrawn immediately into a syringe for transfer into the double-sector cell for short column equilibrium analysis either at high or low speed. Evaporation must be avoided at this stage by working quickly. For low speed work, protein concentration is $2-3$ mg/ml spun for $80-90$ h and high speed, 0.5 mg/ml for $60-70$ h.

In the low speed approach, graphs of ln$c$ against r^2 may be slightly curved downwards because of non-ideal effects, especially at concentrations used for Rayleigh interference work. The slope of the graph of ln$c$ against r^2 allows the calculation of $M_{w_{app}}$. Similarly, a graph of $(dc/dr)/r$ against c has a slope of $M_z(\partial\varrho/\partial c)\omega^2/RT$ and M_z can be found. Non-ideality has the effect to give low values for M_w and M_z (30).

One approach for obtaining an accurate estimate of molecular weights of a non-ideal, homogeneous sample in the three component system is to use the relationship:

$$\frac{1}{M_{app}} = \left(\frac{\partial\varrho}{\partial c}\right)(\omega^2cr/RT)\ (dc/dr)$$

and to plot $1/M_{app}$ against c. The intercept when c is zero is M_z/M_w^2. Both high speed and low speed experiments give similar extrapolated results. The value of M_w^2/M_z yields molecular weights with reasonable accuracy. The slope of the graph gives an estimate of the second virial coefficient, B.

3.4.2 *Sedimentation Equilibrium Analysis of Proteins in the Presence of Detergents*

When detergent-protein binding events are analysed by sedimentation equilibrium centrifugation a measure of $M(1 - \phi'\varrho)$ is obtained from the graph of ln$c$ against r^2. The effective specific volume (ϕ') reflects the preferential binding events. Binding of detergent has been resolved using the approximation:

$$M(1 - \phi'\varrho) = M_p(1 - \bar{\nu}\varrho) + nM_d(1 - \bar{\nu}_d\varrho)$$

where n is the number of moles of detergent bound per mole protein, and M_d and $\bar{\nu}_d$ are the molecular weight and partial specific volume of the detergent (31). The term $(1 - \phi'\varrho)$ is equivalent to $(\partial\varrho/\partial c)_\mu$ and, if c is the protein concentration, for example, by absorption at 275 nm, then M_p is the molecular weight of the polypeptide in the absence of detergent.

It is also possible to determine values of M and $\bar{\nu}$ for the entire complex by adopting the alternative approach of sedimentation equilibrium analysis using solvents of more than one density (Section 2.2.2).

3.5 **Banding of DNA in Self-generating Density Gradients**

The technique of equilibrium centrifugation in CsCl buoyant density gradients has played an important role in the development of molecular biology. Initially, the DNA is uniformly dispersed throughout the CsCl solution and when the centrifugal force is applied, the CsCl becomes distributed as a gradient and the DNA migrates to a point where the density of the DNA equals the buoyant density of the CsCl solution. At equilibrium the buoyant density of the DNA can be determined from the banding position and the molecular weight of a homogeneous sample can be calculated from the band width.

3.5.1 *Determination of the Buoyant Density of a DNA Sample in Relation to a Marker DNA of Known Buoyant Density*

When using the photoelectric scanner, $0.1 - 1.0 \, \mu g$ of each DNA is added to a solution of CsCl whose density is adjusted by addition of solid CsCl until the desired value is attained. The density of the solution can be calculated from the relationships:

CsCl

$$\varrho = 10.98 \eta_{25}^0 - 13.67$$

Cs_2SO_4

$$\varrho = 13.77 \eta_{25}^0 - 17.46$$

where η_{25}^0 is the refractive index (32).

To determine refractive index, place a drop $(15 - 20 \, \mu l)$ between the prisms of a refractometer, equilibrated to 25°C. Insert 0.3 ml samples into a double-sector cell assembly. If the DNA is of high molecular weight then it is subject to hydrodynamic shear. This can be minimised if the cell is assembled empty, the housing plugs screwed in, then the upper window assembly removed. The DNA solution can then be loaded into the partially assembled cell avoiding the shear that would accompany injection through the hole in the centrepiece. The buoyancy gradient defined as:

$$\left(\frac{\partial \varrho}{\partial r} \right)_B = \frac{\omega^2 r}{\beta_B}$$

is a composite of the salt gradient and the compression gradient, and is the gradient used to determine the density of an unknown DNA relative to a marker DNA. At 25°C, $1/\beta_B$ is equal to 9.35×10^{-10} gsec²/cm⁵ for CsCl and

$$\varrho = \varrho^0 + \left(\frac{\partial \varrho}{\partial r} \right)_B (r - r^0) = \varrho^0 + \frac{\omega^2 \bar{r}}{\beta_B} (r - r^0)$$

where ϱ^0 is the density of the marker DNA at radius r^0, ϱ is the density of the unknown DNA at radius r, and $\bar{r}$ is equal to $(r + r^0)/2$.

It is somewhat more complex and usually more inaccurate to determine buoyant densities without use of a marker DNA.

When additional components, for example, ethidium bromide or dimethyl sulphoxide, are added to the system, then the buoyant density of the macromolecule depends on its position in the cell. This is because interactions between the macromolecule and the added components influences the buoyant density and the concentration of these components varies throughout the cell. Such distributions have received a thorough theoretical analysis and are well understood (33,34).

Differential binding of added components to DNA of differing base composition, for example, $HgCl_2/Cs_2SO_4$ gradients or $AgNO_3/Cs_2SO_4$ gradients, has led to subfractionation of subclasses of DNA such as repetitive sequences in eukaryotes by preparative ultracentrifugation (35,36).

3.5.2 *Determination of the Molecular Weight of DNA from the Band Width*

The molecular weight of a homogeneous DNA preparation can be obtained without reference to markers by band width analysis at buoyant equilibrium. The shape of the band is influenced by hydration, pressure effects and concentration-dependent interactions. A typical protocol is as follows. Set up five 12 mm double-sector cells containing 0.1, 0.5, 1.0, 5.0 and 25 μg of SV40 DNA in CsCl density of 1.70 g/ml (56% w/w). Centrifuge at 30 000 r.p.m. for 72 h and scan at either 265 nm or 290 nm. The scanner data is integrated numerically to obtain values of $<m_2>$ and $<\delta^2>$ thus:

$$<m_2> = \frac{\int m^2_2\, d\delta}{\int m_2 d\delta} \text{ and } <\delta^2> = \frac{\int \delta^2 m_2 d\delta}{\int m_2 d\delta}$$

where δ is the distance (cm) from the band centre and m_2 is the DNA concentration. The apparent molecular weight M_{app} is derived from:

$$M_{app} = \frac{RT\beta_{eff}}{<\delta^2>\ \bar{v}_{s,0}\ r_0^2\ \omega^4}$$

where r_0 is the distance of the band centre from the centre of rotation, β_{eff} is related to the effective density gradient which accounts for hydration effects and pressure effects on the buoyant behaviour of DNA, and $\bar{v}_{s,0}$ the partial specific volume of the DNA. This is evaluated at each concentration and the molecular weight of the DNA obtained (37) from the equation:

$$\ln M_{app} = \ln M_{s,0} - B<m_2>$$

3.6 Analysis of Purified, Highly-charged Macromolecules

When using high or low speed equilibrium analysis, it is often difficult to obtain accurate values of M_{app} for Rayleigh analyses for charged macromolecules where aggregation readily occurs, and additionally the influence of the charge on the equilibrium distribution must be taken into account.

The variation of fringe displacement with radius follows an exponential equation of the form

$$y = y_0 + b \exp (qr^2/2)$$

where

$$q = M_2\left(\frac{\partial \varrho}{\partial c_2}\right)_\mu \omega^2/RT$$

The value of q is found by using the experimental values of y and searching by least squares for the best fitting value (38). The 'charge effects' are accommodated in the term $(\partial \varrho/\partial c_2)_\mu$.

4. CONCLUSION

Analytical ultracentrifugation is a powerful biophysical tool which has a valuable part to play in the study of biological macromolecules and their interactions, especially as cheap computer power is more readily available than ever before.

5. REFERENCES

5.1 General References

1. Schachman,H.K. (1959) *'Ultracentrifugation in Biochemistry'*, Academic Press, NY.
2. Many relevant articles in *'Methods in Enzymology'*, eds. Hirs,C.H.W. or Timasheff,S.N., vol. XXVII (1973) 3-140; vol. XLVIII (1978), 169-270.

5.2 References

1. Vinograd,J. and Bruner,R. (1966) *Biopolymers,* **4**, 131.
2. Bruner,R. and Vinograd,J. (1965) *Biochim. Biophys. Acta,* **108**, 18.
3. Freifelder,D. (1970) *J. Mol. Biol.,* **54**, 567.
4. Moller,W.J. (1964) *Proc. Natl. Acad. Sci. USA,* **51**, 501.
5. Edelstein,S.J. and Schachman,H.K. (1967) *J. Biol. Chem.,* **242**, 306.
6. Giri,L., Dijk,J., Labischinski,H. and Bradaczek,H. (1978) *Biochemistry (Wash.),* **17**, 745.
7. Gerhart,J.C. and Schachman,H.K. (1968) *Biochemistry (Wash.),* **7**, 538.
8. Schumaker,V. and Adams,P. (1968) *Biochemistry (Wash.),* **7**, 3422.
9. Wang,J.C. (1969) *J. Mol. Biol.,* **43**, 263.
10. Cohen,R. and Mire,M. (1971) *Eur. J. Biochem.,* **23**, 267.
11. Cohen,R., Giraud,B. and Messiah,A. (1967) *Biopolymers,* **5**, 203.
12. Cohen,R. and Claverie,J.-M. (1975) *Biopolymers,* **14**, 1701.
13. Pretorius,H.T., Nandi,P.K., Lippoldt,R.E., Johnson,M.L., Keen,J.H., Pastan,I. and Edelhoch,H. (1981) *Biochemistry (Wash.),* **20**, 2777.
14. Bloomfield,V.A. (1968) *Science (Wash.),* **161**, 1212.
15. Morel,J.E. and Garrigos,M. (1982) *Biochemistry (Wash.),* **21**, 2679.
16. Cox,D.J. (1969) *Arch. Biochem. Biophys.,* **129**, 106.
17. Cann,J.R. and Kegeles,G. (1974) *Biochemistry (Wash.),* **13**, 1868.
18. Claverie,J.-M., Dreux,H. and Cohen,R. (1975) *Biopolymers,* **14**, 1685.
19. Cox,D. (1969) *Arch. Biochem. Biophys.,* **129**, 106.
20. Harrington,W.F. (1975) *Fractions, Beckman,* **1**, 10.
21. Yphantis,D.A. (1964) *Biochemistry (Wash.),* **3**, 297.
22. Van Holde,K.E. (1967) *Fractions, Beckman,* **1**, 1.
23. Rowe,H.J. and Rowe,A.J. (1970) *Biochim. Biophys. Acta,* **222**, 647.
24. Johnson,M.L. and Yphantis,D.A. (1978) *Biochemistry (Wash.),* **17**, 1448.
25. Visser,J., Deonier,R.C., Adams,E.T. and Williams,J.W. (1972) *Biochemistry (Wash.),* **11**, 2634.
26. Adams,E.T. (1967) *Fractions, Beckman,* **3**.
27. Tung,M.S. and Steiner,R.F. (1974) *Eur. J. Biochem.,* **44**, 49.
28. Steinberg,I.Z. and Schachman,H.K. (1966) *Biochemistry,* **5**, 3728.
29. Reisler,E. and Eisenberg,H. (1969) *Biochemistry (Wash.),* **11**, 4572.
30. Munk,P. and Cox,D.J. (1972) *Biochemistry (Wash.),* **11**, 687.
31. Robinson,N.C. and Tanford,C. (1975) *Biochemistry (Wash.),* **14**, 369-378.
32. Szybalski,W. (1968) *Fractions, Beckman,* **1**, 1.
33. Bauer,W. and Vinograd,J. (1969) *Ann. N.Y. Acad. Sci.,* **164**, 192.
34. Williams,A.E. and Vinograd,J. (1971) *Biochim. Biophys. Acta,* **228**, 423.
35. Rinehart,F.P., Wilson,V.L. and Schmid,C.W. (1977) *Biochim. Biophys. Acta,* **474**, 69.
36. Filipski,J., Thiery,J.-P. and Bernardi,G. (1973) *J. Mol. Biol.,* **80**, 177.
37. Schmid,C.W. and Hearst,J.E. (1971) *Biopolymers,* **10**, 1901.
38. Ausio,J. and Subirana,J.A. (1982) *Biochemistry (Wash.),* **21**, 5910.
39. Goldberg,R.J. (1953) *J. Phys. Chem.,* **57**, 194.

Nomogram for Computing Relative Centrifugal Force

1. NOMOGRAM

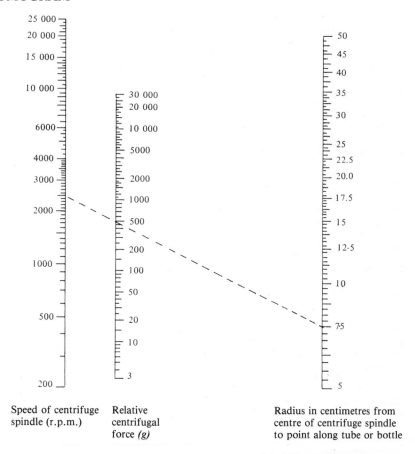

Speed of centrifuge spindle (r.p.m.)

Relative centrifugal force *(g)*

Radius in centimetres from centre of centrifuge spindle to point along tube or bottle

To calculate the RCF value at any point along the tube or bottle being used in the centrifuge, measure the radius in centimetres from the centre of the centrifuge spindle to the particular point. A straight line connecting the value of the radius on the right of the nomogram with the centrifuge speed on the left will enable the RCF value to be read off on the centre column.

2. FORMULA FOR THE CALCULATION OF RELATIVE CENTRIFUGAL FORCE

$$\text{RCF } (g) = 11.18 \text{ x } r \text{ x } \left(\frac{N}{1000}\right)^{2}$$

where r is the radius in centimetres from the centre of the rotor to the point at which the RCF value is required, and N is the speed of the rotor in revolutions per minute.

Chemical Resistance Chart for Tubes and Zonal Rotors

Key: S = Satisfactory, M = Marginal, test before using, U = Unsatisfactory, not recommended, − = Not tested.

Reagent	Polycarbonate	Polysulfone	Polypropylene	Polyethylene	Polyallomer	Cellulose nitrate	Cellulose acetate butyrate	Nylon	Kynar	Noryl	Stainless steel	Aluminium	Titanium
Acetaldehyde	U	−	M	M	M	U	U	−	−	−	S	S	S
Acetic acid (5%)	S	S	S	S	S	S	S	S	S	S	S	S	S
Acetic acid (60%)	U	S	S	M	S	U	U	M	S	S	S	S	S
Acetic acid (glacial)	U	M	M	M	S	U	U	−	S	−	S	S	S
Acetone	U	U	M	S	M	U	U	M	M	−	S	S	S
Allyl alcohol	S	−	S	S	S	−	U	M	−	−	−	−	S
Aluminium chloride	S	−	S	S	S	S	S	S	S	−	U	−	S
Aluminium fluoride	U	−	S	S	S	−	−	S	S	−	−	−	S
Ammonium acetate	S	−	S	S	S	−	−	−	−	−	−	−	S
Ammonium carbonate	U	S	S	S	S	S	S	S	S	−	S	S	S
Ammonium hydroxide (10%)	U	S	S	S	S	U	U	S	S	−	S	−	S
Ammonium hydroxide (conc.)	U	−	S	S	S	U	U	S	−	−	S	−	S
Ammonium sulphide	U	−	S	−	S	−	−	−	S	−	−	−	−
Amyl alcohol	S	−	M	S	M	U	U	S	S	S	−	S	S
Aniline	−	−	M	S	U	−	−	−	S	−	−	S	S
Aqua Regia	U	−	U	U	U	U	U	−	S	−	U	U	S
Benzene	U	U	U	U	M	S	S	S	S	−	S	S	S
Benzyl alcohol	U	−	U	U	U	S	U	M	S	−	−	−	S
Boric acid	S	S	S	S	S	S	S	S	S	S	−	−	S
N-Butyl alcohol	M	M	S	M	S	U	U	M	S	S	−	−	S
Caesium chloride	S	S	S	S	S	S	S	−	−	−	S	M	S
Caesium trifluoroacetate	S	S	S	S	S	U	M	−	−	−	M	M	−
Calcium chloride	M	S	S	S	S	S	S	S	S	S	S	M	S
Calcium hypochlorite	M	S	S	S	S	−	−	S	S	−	U	M	S
Carbon tetrachloride	U	S	M	U	U	S	S	S	S	−	M	M	S
Chlorine water	S	−	S	M	S	−	S	−	−	−	−	−	S
Chlorobenzene	U	−	U	U	U	U	U	−	S	−	−	−	S
Chloroform	U	U	M	U	M	S	M	M	S	−	S	−	S
Chromic acid (10%)	M	U	S	S	S	U	U	−	S	S	U	M	S
Chromic acid (50%)	U	U	S	S	S	S	U	−	S	−	U	U	M
Citric acid (10%)	S	S	S	S	S	−	S	S	S	S	S	S	S
Cresol	U	−	S	S	S	−	−	U	S	−	−	S	S
Cyclohexyl alcohol	M	−	S	S	S	−	U	S	−	−	−	S	S
Diacetone	−	−	S	S	S	−	U	−	−	−	−	S	S
Diethyl ketone	U	−	M	M	U	U	U	M	−	−	−	S	S

Reagent	Polycarbonate	Polysulfone	Polypropylene	Polyethylene	Polyallomer	Cellulose nitrate	Cellulose acetate butyrate	Nylon	Kynar	Noryl	Stainless steel	Aluminium	Titanium
Diethylpyrocarbonate	U	–	–	–	S	–	–	–	–	U	S	S	S
Dimethylformamide	U	–	S	S	S	–	–	–	–	–	–	S	S
Dimethylsulphoxide	U	–	S	–	S	–	–	–	–	–	S	S	S
Dioxane	U	–	M	M	M	–	U	–	S	–	–	S	S
Ether diethyl	U	–	M	M	M	U	U	–	–	–	–	S	S
Ethyl acetate	U	U	M	S	M	U	U	M	S	–	–	M	S
Ethyl alcohol (50%)	M	S	S	S	S	S	S	M	S	S	S	S	S
Ethyl alcohol (95%)	U	S	S	S	S	U	U	M	S	S	S	S	S
Ethylene dichloride	U	–	U	U	M	U	U	S	–	S	–	S	S
Ethylene glycol	S	S	S	S	S	S	S	M	S	–	–	S	S
Ferric chloride	–	–	S	S	S	–	–	S	S	S	U	U	S
Fluoboric acid	–	–	S	S	S	–	–	–	–	–	–	–	S
Formaldehyde (40%)	S	S	S	S	S	S	–	S	S	S	S	M	S
Formic acid (100%)	M	–	S	S	S	–	U	U	S	S	–	S	S
Gallic acid	–	–	S	S	S	–	–	S	S	–	–	–	S
Glycerol	S	S	S	S	S	S	–	–	S	S	S	S	S
2-Heptyl	–	–	S	S	S	–	U	–	–	–	–	–	S
Hydrochloric acid (10%)	S	S	S	S	S	S	S	S	S	S	U	U	–
Hydrochloric acid (50%)	M	–	M	S	M	U	U	–	S	S	U	U	–
Hydrofluoric acid (10%)	M	S	S	S	S	M	M	S	S	–	U	U	U
Hydrofluoric acid (100%)	U	–	S	S	S	U	U	–	S	–	U	U	U
Hydroformic acid (100%)	–	–	S	S	S	–	–	–	–	–	U	U	S
Hydrogen peroxide (100%)	S	S	S	S	S	S	S	–	–	S	S	S	U
Isobutyl alcohol	–	–	S	S	S	–	U	M	–	–	–	–	S
Isopropyl alcohol	U	M	S	S	S	U	U	M	–	S	–	U	S
Lactic acid (20%)	S	S	S	M	S	–	–	–	–	S	S	–	S
Lauryl alcohol	–	–	S	S	S	–	U	S	–	–	–	–	S
Lead acetate	–	–	S	S	S	–	S	–	S	S	–	M	S
Maleic acid	–	–	S	S	S	–	–	–	S	–	–	S	S
Magnesium hydroxide	U	–	S	S	S	–	U	–	S	S	–	U	S
Manganese salts	–	–	S	S	S	–	S	–	–	–	–	M	S
Methyl alcohol	U	S	S	S	S	U	U	M	S	–	S	S	S
Methyl ethyl ketone	U	U	S	S	S	U	U	M	M	–	–	S	S
Methylene chloride	U	U	M	M	M	U	U	M	S	–	S	S	S
Nitric acid (10%)	S	S	S	S	S	S	S	M	S	S	M	M	S
Nitric acid (50%)	M	–	M	M	S	M	M	M	S	S	M	M	S
Oleic acid	S	S	S	S	S	S	S	S	S	–	S	–	S
Oxalic acid	S	S	S	S	S	S	S	S	S	–	S	M	M
Perchloric acid (10%)	U	–	M	M	S	–	–	–	S	–	U	U	S
Phenol	U	U	U	S	M	–	–	U	S	–	S	–	U
Phenyl ethyl alcohol	–	–	S	S	S	–	U	S	–	–	–	–	S
Phosphoric acid (10%)	S	S	S	S	S	S	S	–	S	S	S	–	–
Phosphoric acid (conc.)	M	S	M	S	M	M	M	–	S	–	M	–	M
Phosphorus trichloride	U	–	S	S	S	–	–	–	S	–	–	–	–
Potassium acetate	M	–	S	S	S	–	–	–	–	–	S	M	S
Potassium carbonate	U	–	S	S	S	S	S	S	S	S	S	M	S
Potassium chlorate	S	–	S	S	S	S	S	S	S	S	S	M	S

Reagent	Polycarbonate	Polysulfone	Polypropylene	Polyethylene	Polyallomer	Cellulose nitrate	Cellulose acetate butyrate	Nylon	Kynar	Noryl	Stainless steel	Aluminium	Titanium
Potassium hydroxide	U	S	S	S	S	M	M	S	–	–	S	U	U
Potassium permanganate	–	–	S	S	S	–	–	S	S	–	–	–	–
Silver cyanide	–	–	S	S	S	–	–	–	S	–	–	–	S
Sodium bisulphate	S	–	S	S	S	S	S	S	S	S	S	M	S
Sodium borate	S	–	S	S	S	S	S	S	–	–	S	M	S
Sodium carbonate	U	–	S	S	S	S	S	S	S	S	S	M	S
Sodium chloride	S	S	S	S	S	S	S	–	–	–	S	M	S
Sodium dichromate	–	–	S	S	S	–	–	S	–	–	–	M	S
Sodium hydroxide (1%)	U	S	S	S	S	S	S	S	S	–	S	U	S
Sodium hydroxide (10%)	U	S	S	S	S	U	U	S	S	–	S	U	S
Sodium hydroxide (conc.)	U	–	M	S	M	U	U	S	–	–	S	U	M
Sodium hypochlorite	S	S	S	S	S	S	S	S	S	–	M	M	S
Sodium nitrate (10%)	U	–	S	S	S	–	–	S	S	S	S	–	S
Sodium peroxide	–	–	S	S	S	–	–	S	S	S	S	–	S
Sodium sulphide	U	–	S	S	S	–	S	S	S	–	S	S	M
Sodium thiosulphate	S	–	S	S	S	–	–	S	S	S	S	M	S
Sulphuric acid (10%)	U	S	S	S	S	S	S	S	S	S	U	M	S
Sulphuric acid (50%)	S	S	S	S	S	U	U	M	S	S	U	U	M
Sulphuric acid (conc.)	U	U	S	M	S	U	U	U	S	–	M	U	U
Tannic acid	–	–	S	S	S	–	–	S	S	–	S	–	S
Toluene	U	U	U	M	U	S	S	M	S	–	S	–	S
Trichloroacetic acid	M	–	S	S	S	–	–	–	–	–	–	–	S
Trichlorethylene	U	U	U	U	U	–	–	M	S	–	U	S	S
Trichloroethane	U	M	U	U	U	–	S	S	S	–	–	S	S
Turpentine	U	–	M	U	M	–	U	S	S	–	–	S	S
Urea	S	S	S	S	S	S	S	S	–	–	S	M	S
Xylene	U	U	U	U	U	–	S	U	S	–	S	S	S

291

Specifications of Ultracentrifuge Rotors

This appendix is an attempt to produce a comprehensive list of rotors manufactured for ultracentrifuges. This list includes some rotors which are no longer available and these are marked with an asterisk (*). These, now obsolete rotors, are included since they may still be in use and because their inclusion facilitates finding their modern equivalents when trying to reproduce earlier results carried out using these rotors. Because scientific journals often do not insist on authors giving full details of the rotors used in centrifugal separations, this appendix is also useful if one is trying to reproduce separations obtained by other workers using a make of centrifuge other than the type available to the reader.

A number of ultracentrifuge manufacturers have now standardised on a drive shaft and speed sensor control system identical to that manufactured by Beckman/Spinco. Hence, now it is possible to use the rotor of one manufacturer on the machine of another. Some manufacturers frown on this practice while others actively encourage it to the extent of selling rotors produced by other manufacturers. Hence, the reader will find that some rotors are listed under more than one supplier. In some cases manufacturers have changed their methods of designating rotors and in this case earlier designations are given in brackets. Ultracentrifuge rotors are usually manufactured from either titanium (Ti) or aluminium (Al). In addition, some rotors contain both, for example, some swing-out rotors have an aluminium central yoke, and titanium buckets. Carbon fibre (Cf) composite rotors have also been developed and are likely to become more widely available in the future.

In the case of the swing-out rotors, as described in Section 4.3.2 of Chapter 1, there are three basic methods of attaching the buckets. The buckets of three-bucket rotors are usually attached by the hinge pin (H/P) method while the buckets of six-position rotors either hook on (H/O) or are of the ball and socket type (B/S). Damon/IEC have modified the ball and socket system to allow top loading (T/L) of the buckets, hence making it easier to attach the buckets to the rotor.

The performance of rotors can be judged from the maximum centrifugal force that they generate. However, a more practical method is based on the calculation of k-factors since these can be used to calculate the time (t) in hours required to sediment a particle of known s-value (s) using the equation:

$$t = \frac{k}{s}$$

k-factors can be calculated using the following equation:

$$k = 2.53 \times 10^{11} \left[\frac{\ln\left(\frac{r_{max}}{r_{min}}\right)}{N^2} \right]$$

where r_{min} and r_{max} are the minimum and maximum radii of the rotor, respectively, and N is the maximum speed of the rotor. If the rotor is not used at its maximum speed (N) then the new k-factor (k_n) is calculated from the equation:

$$k_n = k\left(\frac{N^2}{N_n^2}\right)$$

The use of k-factors is only valid if the viscosity of the solution is not significantly different from that of water. Hence for swing-out rotors that are used for rate-zonal separations it is often more appropriate to compare the k*-values of rotors. The k*-value can be used to calculate the time (t) in hours taken by a particle of sedimentation rate, s, to pellet through a $5-20\%$ (w/w) sucrose gradient at 5°C using the equation:

$$t = \frac{k^*}{s}$$

For calculating the value of k*, one assumes a particle density of 1.3 g/cm^3 which corresponds to the density of both DNA and protein in sucrose solutions. RNA is slightly denser (~ 1.7 g/cm^3) and hence when sedimenting RNA the k*-value will slightly under-estimate the rate at which the RNA is pelleted.

Acknowledgements

The assistance of Jens Steensgaard (Aarhus) and individual centrifuge rotor manufacturers in the preparation of these tables is gratefully acknowledged.

SMALL VOLUME FIXED-ANGLE ROTORS

Supplier	Rotor designation	Tube number and volume (ml)	Rotor material	Tube angle	Maximum speed (r.p.m.)	R_{min} (cm)	R_{max} (cm)	RCF_{max} (g)	k-factor
Beckman/Spinco	TLA100	20 x 0.2	Ti	30	100 000	3.0	3.9	436 000	7
	TLA100.1	14 x 0.5	Ti	30	100 000	2.5	3.9	436 000	12
	TLA100.2	10 x 1.0	Ti	30	100 000	2.5	3.9	436 000	12
	TLA100.3	6 x 3.5	Ti	30	100 000	2.5	4.8	550 000	17
	Type 80Ti	8 x 13.5	Ti	26	80 000	4.1	8.4	601 000	28
	Type 75Ti	8 x 13.5	Ti	26	75 000	3.7	8.0	503 100	35
	Type 70.1Ti	12 x 13.5	Ti	24	70 000	4.1	8.2	449 200	36
	Type 65	8 x 13.5	Al	24	65 000	3.7	7.8	368 000	45
	Type 50Ti	12 x 13.5	Ti	26	50 000	3.7	8.1	226 400	79
	Type 50	10 x 10	Al	20	50 000	3.7	7.0	195 600	65
	Type 50.3Ti	18 x 6.5	Ti	20	50 000	4.9	8.0	223 600	50
	Type 42.2Ti	72 x 0.2	Ti	30	42 000	10.4	11.3	222 900	12
	Type 40	12 x 13.5	Al	26	40 000	3.7	8.1	144 900	124
	Type 40.2	12 x 6.5	Al	40	40 000	3.6	8.1	144 900	128
	Type 40.3	18 x 6.5	Al	20	40 000	4.9	8.0	143 000	78
	Type 30.2	20 x 10.5	Al	14	30 000	6.2	9.4	94 600	117
	Type 25	100 x 1	Al	25	25 000	8.1	10.0	69 900	85
						9.7	11.6	81 100	72
						11.3	13.2	92 200	62
Damon/IEC	494 (A321)	8 x 12	Al	35	60 000	2.7	7.9	318 000	75
	460 (A269)	10 x 12	Al	20	55 000	4.2	7.9	267 200	53
	468 (A168)	20 x 12	Al	14	40 000	6.5	9.4	168 100	58
Du Pont/Sorvall	T-875	8 x 12.5	Ti	24	75 000	4.6	8.7	547 100	29
	T-865.1	8 x 12.5	Ti	24	65 000	4.6	8.7	410 900	38
	TFT 65.13	12 x 13.5	Ti	26	65 000	3.8	8.1	382 600	45
	TFT 45.6	40 x 6.5	Ti	20	45 000	6.1	9.0	203 800	49
						7.6	10.6	240 000	42

295

SMALL VOLUME FIXED-ANGLE ROTORS (contd.)

Supplier	Rotor designation	Tube number and volume (ml)	Rotor material	Tube angle	Maximum speed (r.p.m.)	R_{min} (cm)	R_{max} (cm)	RCF_{max} (g)	k-factor
Heraeus Christ	W70/38Ti	8 x 12	Ti	25	70 000	3.6	7.6	416 300	39
	W65/40Ti	12 x 12	Ti	25	65 000	4.2	8.2	387 300	40
	W60/44	8 x 12	Al	20	60 000	4.0	7.5	301 800	44
	W40/77	20 x 12	Al	25	40 000	6.3	10.3	184 200	78
	W40/41	40 x 5	Al	20	40 000	6.8	9.2	164 500	48
						8.2	10.6	189 600	41
Kontron	TFT 80.4	10 x 4.4	Ti	25	80 000	3.7	6.6	468 600	23
	TFT 80.13	8 x 13.5	Ti	26	80 000	4.4	8.5	605 300	26
	TFT 75.13	8 x 13.5	Ti	26	75 000	4.0	8.2	514 700	32
	TFT 65.13	12 x 13.5	Ti	26	65 000	4.0	8.2	386 600	43
	TFT 50.13	12 x 13.5	Ti	26	50 000	4.0	8.2	228 700	73
	TFT 45.6	40 x 6.5	Ti	20	45 000	6.1	9.0	204 300	49
						7.6	10.6	238 900	42
	TFT 32.13	32 x 13.5	Ti	30	32 000	4.4	9.3	106 900	185
LKB/Hitachi	RP83T	8 x 12	Ti	26	83 000	3.8	8.0	616 100	27
	SRP70AT	12 x 12	Ti	24	70 000	4.1	8.3	454 600	36
	RP65AF	8 x 5	Cf	14	65 000	5.7	7.9	373 100	20
	RP65T/RPW65T	10 x 12	Ti	26	65 000	3.5	7.8	368 400	48
	RP65	10 x 12	Al	26	65 000	3.5	7.8	368 400	48
	RD55T	12 x 12	Ti	26	55 000	3.8	8.1	274 000	63
	SRP50AT	18 x 6.5	Ti	20	50 000	4.8	8.0	223 600	52
	RPW45	10 x 12	Al	26	45 000	3.5	7.8	176 600	100
	RPL42T	72 x 0.23	Ti	30	42 000	10.4	11.3	222 800	12
	RP40	12 x 12	Al	26	40 000	3.8	8.1	144 900	120
	RP40-2	12 x 6.5	Al	40	40 000	3.5	8.0	143 100	131
	RP40-3	18 x 6.5	Al	20	40 000	4.8	8.0	143 100	81
	RP30-3	20 x 10.5	Al	14	30 000	6.3	9.4	94 600	113

SMALL VOLUME FIXED-ANGLE ROTORS (contd.)

Supplier	Rotor designation	Tube number and volume (ml)	Rotor material	Tube angle	Maximum speed (r.p.m.)	R_{min} (cm)	R_{max} (cm)	RCF_{max} (g)	k-factor
MSE	MFT 75.14 (8 x 14 ml Ti)	8 x 14	Ti	29	75 000	3.7	8.0	503 100	35
	MFT 70.10 (10 x 10 ml Ti)	10 x 10	Ti	35	70 000	4.4	8.6	471 100	35
	MFA 65.14 (8 x 14 ml Al)	8 x 14	Al	26	65 000	3.5	7.8	368 400	48
	MFA 50.10 (10 x 10 ml Al)	10 x 10	Al	20	50 000	3.9	7.1	198 400	61
	MFT 50.6 (18 x 6.5 ml Ti)	18 x 6.5	Ti	20	50 000	4.9	8.0	223 600	50
	MFA 40.6 (18 x 6.5 ml Al)	18 x 6.5	Al	20	40 000	4.9	8.0	143 100	78
	8 x 10 ml Al*	8 x 10	Al	35	40 000	3.7	8.0	143 100	122

MEDIUM VOLUME FIXED-ANGLE ROTORS

Supplier	Rotor designation	Tube number and volume (ml)	Rotor material	Tube angle	Maximum speed (r.p.m.)	R_{min} (cm)	R_{max} (cm)	RCF_{max} (g)	k-factor
Beckman/Spinco	Type 70Ti	8 x 38.5	Ti	23	70 000	4.0	9.2	504 000	63
	Type 60Ti	8 x 38.5	Ti	24	60 000	3.7	9.0	362 200	63
	Type 55.2Ti	10 x 38.5	Ti	24	55 000	4.7	10.0	338 200	63
	Type 50.2Ti	12 x 38.5	Ti	24	50 000	5.4	10.8	301 900	70
	Type 42.1	8 x 38.5	Al	30	42 000	3.9	9.9	195 200	134
	Type 30	12 x 38.5	Al	26	30 000	4.9	10.5	105 700	214
Damon/IEC	404 (A237)	8 x 40	Al	23	50 000	3.4	8.4	234 800	92
	495 (A211)	8 x 40	Al	30	45 000	3.1	9.3	210 500	137
	410 (A192)	8 x 50	Al	33	40 000	3.7	10.6	189 600	166
	466 (A110)	12 x 40	Al	26	30 000	5.4	11.1	111 700	203
Du Pont/Sorvall	T-865	8 x 36	Ti	24	65 000	3.8	9.1	429 800	52
	TFT 50.38	12 x 38.5	Ti	24	50 000	5.3	10.8	301 800	72
	A841	8 x 36	Al	24	41 000	3.8	9.1	171 000	131
Heraeus Christ	W60/58	8 x 35	Ti	24	60 000	3.9	9.0	362 200	59
	W40/128*	8 x 35	Al	25	40 000	4.1	9.3	166 300	130
Kontron	TFT 70.38	8 x 38.5	Ti	20	70 000	4.3	9.2	504 000	39
	TFT 65.38	8 x 38.5	Ti	20	65 000	4.3	9.2	434 500	46
	TFT 50.38	12 x 38.5	Ti	24	50 000	5.8	10.8	301 800	63
LKB/Hitachi	RP70T	8 x 40	Ti	23.5	70 000	3.8	9.2	504 000	46
	RP50T-2	12 x 40	Ti	23.5	50 000	5.4	10.9	303 300	71
	RPW50T	8 x 40	Ti	23.5	50 000	3.6	9.0	251 500	93
	RP50-2	8 x 40	Al	30	50 000	3.9	9.9	276 700	94
	RP30-2	12 x 40	Al	26	30 000	4.9	10.6	106 700	217

MEDIUM VOLUME FIXED-ANGLE ROTORS (contd.)

Supplier	Rotor designation	Tube number and volume (ml)	Rotor material	Tube angle	Maximum speed (r.p.m.)	R_{min} (cm)	R_{max} (cm)	RCF_{max} (g)	k-factor
MSE	MFT 60.35 (8 x 35 ml Ti)	8 x 35	Ti	21	60 000	3.6	9.4	378 300	68
	MFT 60.25 (8 x 25 ml Ti)	8 x 25	Ti	30	60 000	4.4	9.4	378 300	53
	MFA 50.35 (8 x 35 ml Al)	8 x 35	Al	21	50 000	3.6	9.4	262 700	97
	MFA 50.25 (8 x 25 ml Al)	8 x 25	Al	30	50 000	4.4	9.4	262 700	77
	MFA 40.50 (8 x 50 ml Al)	8 x 50	Al	30	40 000	4.6	10.9	195 000	136

LARGE VOLUME FIXED-ANGLE ROTORS

Supplier	Rotor designation	Tube number and volume (ml)	Rotor material	Tube angle	Maximum speed (r.p.m.)	R_{min} (cm)	R_{max} (cm)	RCF_{max} (g)	k-factor
Beckman/Spinco	Type 45Ti	6 x 94	Ti	24	45 000	3.6	10.6	235 400	133
	Type 42*	6 x 94	Al	25	42 000	3.5	10.4	206 000	156
	Type 35	6 x 94	Al	25	35 000	3.5	10.4	142 400	225
	Type 21	10 x 94	Al	18	21 000	6.0	12.2	60 100	407
	Type 19	6 x 250	Al	25	19 000	4.4	13.3	53 600	776
	Type 15*	4 x 500	Al	15	15 000	4.1	14.2	35 700	1397
Damon/IEC	496 (A170)	6 x 75	Al	20	40 000	3.2	9.4	168 100	170
	486 (A54)	6 x 250	Al	20	20 000	3.7	12.3	55 000	760
	A28*	4 x 500	Al	20	14 000	1.5	13.2	28 900	2807
Du Pont/Sorvall	A641	6 x 98	Al	23	41 000	3.4	9.9	186 000	161
Heraeus Christ	W30/281	6 x 90	Al	20	30 000	3.5	9.6	96 500	284
	W23/534*	4 x 250	Al	20	23 000	3.7	12.0	71 000	563
	W15/1460*	4 x 500	Al	20	15 000	4.2	15.2	38 200	1446
Kontron	TFT 45.94	6 x 94	Ti	23	45 000	4.0	10.4	235 400	119
	TFA 20.250	6 x 250	Al	23	20 000	5.0	13.6	60 800	632
LKB/Hitachi	RP45T	6 x 94	Ti	23	45 000	3.8	10.5	237 700	127
	RP42	6 x 94	Al	26	42 000	3.6	10.4	205 100	152
	RPW35	6 x 94	Al	26	35 000	3.6	10.4	142 400	219
	RP21	10 x 94	Al	18	21 000	6.0	12.0	59 200	398
	RP19	6 x 230	Al	26	19 000	4.7	13.6	54 900	746
MSE	MFA 35.100 (6 x 100 ml Al)	6 x 100	Al	25	35 000	4.0	11.2	153 300	213
	MFA 25.100 (10 x 100 ml Al)	10 x 100	Al	18	25 000	6.3	12.7	88 700	284
	MFA 21.300 (6 x 300 ml Al)	6 x 300	Al	25	21 000	5.9	15.3	75 400	547
	4 x 500 ml*	4 x 500	Al	18	14 000	4.1	14.7	32 200	1648

VERTICAL ROTORS

Supplier	Rotor designation	Tube number and volume (ml)	Rotor material	Maximum speed (r.p.m.)	R_{min} (cm)	R_{max} (cm)	RCF_{max} (g)	k-factor	k*
Beckman/Spinco	TLV 100	8 x 2.0	Ti	100 000	2.5	3.6	402 500	10	18
	Type VTi80	8 x 5.1	Ti	80 000	5.8	7.1	508 000	8	14
	Type VTi65	8 x 5.1	Ti	65 000	7.2	8.5	401 500	10	18
	Type VTi65.2	16 x 5.1	Ti	65 000	7.5	8.8	415 600	10	17
	Type VTi50	8 x 39	Ti	50 000	6.1	8.7	243 000	36	65
	Type VA126	8 x 39	Al	26 000	6.7	9.3	70 300	123	219
Du Pont/Sorvall	TV 865	8 x 5	Ti	65 000	7.2	8.5	401 500	10	18
	TV 865B	8 x 17	Ti	65 000	5.9	8.5	401 500	22	39
	TV 850	8 x 36	Ti	50 000	5.9	8.5	237 500	37	66
Kontron	TV 865	8 x 5	Ti	65 000	7.2	8.5	401 500	10	18
	TV 865B	8 x 17	Ti	65 000	5.9	8.5	401 500	22	39
	TV 850	8 x 36	Ti	50 000	5.9	8.5	237 500	37	66
LKB/Hitachi	SRP83VT	8 x 5	Ti	83 000	5.8	7.2	549 100	8	19
	RP70VT	12 x 2	Ti	70 000	6.1	7.3	400 000	9	23
	RP67VF	8 x 5	Cf	67 000	6.8	8.2	410 000	11	26
	RPV65T	8 x 5	Ti	65 000	7.1	8.6	404 300	10	25
	RP55VF	12 x 12	Cf	55 000	6.3	8.9	299 300	29	72
	RP50VF	8 x 40	Cf	50 000	6.1	8.7	243 000	36	89
	RPV50T	8 x 40	Ti	50 000	6.1	8.7	243 000	36	89
	RPV45T	12 x 12	Ti	45 000	7.2	8.85	200 400	25	63
	RPV30	8 x 40	Al	30 000	6.1	8.7	87 500	100	247
MSE	MVT 50.35 (VWR 50)	8 x 35	Ti	50 000	6.0	8.6	240 300	36	65

SMALL VOLUME SWING-OUT ROTORS

Supplier	Rotor designation	Tube number and volume (ml)	Rotor material	Bucket system	Maximum speed (r.p.m.)	R_{min} (cm)	R_{max} (cm)	RCF$_{max}$ (g)	k-factor	k*
Beckman/Spinco	SW 65Ti	3 x 5	Ti	H/P	65 000	4.1	8.9	420 300	46	122
	SW 60Ti	6 x 4.4	Ti	H/O	60 000	6.3	12.0	482 900	45	120
	SW 56Ti*	6 x 4.4	Ti	H/O	56 000	5.6	11.6	407 700	59	156
	SW 55Ti	6 x 5	Ti	H/O	55 000	6.1	10.9	368 600	49	123
	TLS 55	4 x 2.2	Al	T/L	55 000	4.2	7.6	259 000	50	130
	SW 50*	3 x 5	Al	H/P	50 000	4.7	9.8	273 900	74	196
	SW 50.1	6 x 5	Al/Ti	H/O	50 000	6.0	10.7	299 000	59	156
	SW 39*	3 x 5	Al	H/P	39 000	4.7	9.8	126 600	161	425
	SW 30.1	6 x 8	Al/Ti	H/O	30 000	7.5	12.3	123 700	139	373
Damon/IEC	498 (SB405)	6 x 4	Ti	B/S	60 000	5.1	10.1	406 400	48	127
	648	6 x 4	Ti	T/L	60 000	5.1	10.1	406 400	48	127
Du Pont/Sorvall	TST 60.4	6 x 4.4	Ti	B/S	60 000	6.4	12.2	491 000	45	120
	AH 650	6 x 5	Al	H/O	50 000	6.0	10.7	299 000	59	156
Heraeus Christ	S52/61Ti	6 x 5	Ti	B/S	52 000	5.2	9.9	299 200	60	160
	S40/105*	3 x 5	Al	H/P	40 000	5.0	9.7	173 400	105	278
Kontron	TST 60.4	6 x 4.4	Ti	B/S	60 000	6.4	12.1	486 900	45	120
	TST 55.5	6 x 5	Ti	B/S	55 000	6.6	11.3	382 100	45	120
LKB/Hitachi	RPS65T	3 x 5	Ti	H/P	65 000	3.8	8.9	420 400	51	123
	RPS56T	6 x 4	Ti	H/O	56 000	5.9	11.6	406 700	55	131
	RPS55T-2	6 x 5	Ti	H/O	55 000	6.1	10.9	368 600	49	117
	RPS50	3 x 5	Al	H/P	50 000	4.7	9.8	273 900	74	180
	RPS50-2	6 x 5	Al	H/O	50 000	6.0	10.8	301 800	60	144

SMALL VOLUME SWING-OUT ROTORS (contd.)

Supplier	Rotor designation	Tube number and volume (ml)	Rotor material	Bucket system	Maximum speed (r.p.m.)	R_{min} (cm)	R_{max} (cm)	RCF_{max} (g)	k-factor	k*
MSE	MST 60.4 (6 x 4.2 ml Ti)	6 x 4.2	Ti	B/S	60 000	5.2	12.4	499 000	61	160
	MST 60.6 (3 x 6.5 ml Ti)	3 x 6.5	Ti	H/P	60 000	4.3	10.4	418 500	62	162
	3 x 5 ml*	3 x 5	Al	H/P	50 000	4.7	10.7	299 000	84	210
	MST 45.5 (6 x 5.5 ml Ti)	6 x 5.5	Ti	H/O	45 000	5.6	10.6	240 000	80	193

MEDIUM VOLUME SWING-OUT ROTORS

Supplier	Rotor designation	Tube number and volume (ml)	Rotor material	Bucket system	Maximum speed (r.p.m.)	R_{min} (cm)	R_{max} (cm)	RCF_{max} (g)	k-factor	k*
Beckman Spinco	SW 41Ti	6 x 13.2	Ti	H/O	41 000	6.7	15.3	287 500	136	355
	SW 40Ti	6 x 14	Ti	H/O	40 000	6.9	15.9	284 400	132	346
	SW 36*	4 x 13.5	Al	H/P	36 000	6.0	13.3	192 600	156	409
	SW 30*	6 x 20	Al	H/O	30 000	7.5	12.3	123 700	139	373
	SW 28.1	6 x 17	Al/Ti	H/O	28 000	7.3	17.1	149 900	275	720
	SW 27.1*	6 x 17	Al/Ti	H/O	27 000	6.8	16.6	135 200	310	810
Damon/IEC	488 (SB 283)	6 x 12	Ti	B/S	41 000	5.6	15.1	283 700	149	388
	628	6 x 12	Ti	T/L	41 000	5.6	15.1	283 100	149	388
	531	6 x 18	Al	B/S	28 000	5.9	15.4	135 400	309	809
Du Pont/Sorvall	TST 41.14	6 x 14	Ti	H/O	41 000	6.8	16.0	302 500	130	340
	AH 627	6 x 17	Al	H/O	27 000	6.8	16.6	135 200	313	810
Heraeus Christ	S40/135	6 x 15	Ti	B/S	40 000	6.8	16.0	286 100	135	355
Kontron	TST 41.14	6 x 14	Ti	H/O	41 000	6.8	16.1	302 500	130	340
	TST 28.17	6 x 17	Ti	H/O	28 000	7.4	17.2	150 700	273	714
LKB/Hitachi	RPS40T	6 x 13	Ti	H/O	40 000	6.3	15.9	284 400	147	352
	SRP28SAl	6 x 16	Al	H/O	28 000	6.5	16.6	145 600	303	719
MSE	MST 40.14 (6 x 14 ml Ti)	6 x 14	Ti	H/O	40 000	6.3	15.9	284 300	147	382
	MSA 30.16 (6 x 16.5 ml Al)	6 x 16.5	Al	B/S	30 000	6.1	15.8	159 900	268	697

LARGE VOLUME SWING-OUT ROTORS

Supplier	Rotor designation	Tube number and volume (ml)	Rotor material	Bucket system	Maximum speed (r.p.m.)	R_{min} (cm)	R_{max} (cm)	RCF_{max} (g)	k-factor	k*
Beckman/Spinco	SW 28	6 x 38.5	Al/Ti	H/O	28 000	7.5	16.1	141 100	247	650
	SW 27*	6 x 38.5	Al/Ti	H/O	27 000	7.5	16.1	131 200	265	699
	SW 25.1	3 x 34	Al	H/P	25 000	5.6	12.9	90 100	338	886
	SW 25.2*	3 x 60	Al	H/P	25 000	6.7	15.2	106 200	332	871
Damon/IEC	402 (SB110)	6 x 40	Al	B/S	25 000	6.3	15.8	110 400	348	910
	603	6 x 40	Ti	T/L	25 000	6.3	15.8	110 400	348	910
Du Pont/Sorvall	AH 627	6 x 36	Al	H/O	27 000	7.7	16.1	131 200	265	699
Heraeus Christ	S30/230*	3 x 30	Al	H/P	30 000	5.7	12.9	129 800	230	603
	S27/245	6 x 38	Al	H/O	27 500	7.6	15.9	134 200	245	645
	S20/584*	3 x 75	Al	H/P	20 000	6.3	15.9	71 100	586	1527
Kontron	TST 28.38	6 x 38.5	Ti	H/O	28 000	7.8	16.1	141 100	234	618
LKB/Hitachi	SRP28SA	6 x 40	Al	H/O	28 000	7.1	16.1	141 100	265	632
	RPS25-2	3 x 60	Al	H/P	25 000	6.1	15.3	106 900	373	886
MSE	MSA 30.25 (3 x 25 ml)	3 x 25	Al	H/P	30 000	5.9	12.9	129 800	220	579
	MSA 25.38 (6 x 38 ml)	6 x 38	Al	H/O	25 000	8.1	16.5	115 300	288	762
	MSA 23.70	3 x 70	Al	H/P	23 500	6.4	16.3	100 600	429	1117

ZONAL ROTORS

Supplier	Rotor designation	Rotor volume (ml)	Rotor material	Maximum speed (r.p.m.)	Maximum pathlength (cm)	Depth of rotor chamber (cm)	RCF$_{max}$ (g)
Beckman/Spinco	Z-60	330	Ti	60 000	5.2	3.1	256 000
	Ti-14	665	Ti	48 000	5.4	5.4	171 800
	Al-14	665	Al	35 000	5.4	5.4	91 300
	Ti-15	1675	Ti	32 000	7.6	7.6	102 000
	Al-15	1675	Al	22 000	7.6	7.6	48 100
Damon/IEC	Z-15*	780	Al	8000	9.5	2.5	9100
	1073 (B30)	659	Ti	50 000	5.2	5.4	186 500
	1071 (B29)	1674	Ti	35 000	7.5	7.6	121 800
Du Pont/Sorvall	TZ28	1330	Ti	28 000	5.9	9.8	83 500
Heraeus Christ	B-14*	650	Ti	45 000	5.2	5.4	151 000
	B-15*	1650	Ti	30 000	7.4	7.6	89 500
	B-29*	1400	Ti	30 000	7.1	7.6	86 000
Kontron	TZT 48.650	650	Ti	48 000	5.2	5.4	124 500
	TZT 40.325	325	Ti	40 000	5.2	5.4	86 500
	TZT 40.160	160	Ti	40 000	5.2	5.4	86 500
	TZT 40.140	140	Ti	40 000	4.0	5.4	82 200
		60			1.0	5.4	82 200
	TZT 32.1650	1650	Ti	32 000	7.4	7.6	75 200
LKB/Hitachi	RPZ48T	660	Ti	48 000	6.7	5.5	172 000
	RPZ35T	1690	Ti	35 000	7.4	7.6	121 800
	RPC35T	430	Ti	35 000	(2.4)	7.6	121 800
MSE	B-14	650	Ti	47 000	5.4	5.4	165 000
	B-14	650	Al	35 000	5.4	5.4	91 000
	B-15	1670	Ti	35 000	7.6	7.6	121 800
	B-15	1670	Al	25 000	7.6	7.6	62 000

Equations Relating the Refractive Index to the Density of Solutions

When solutes are dissolved in water the refractive index of the resulting solution differs from that of water. The increase in the refractive index is proportional to the concentration of solute. In the case of gradient solutes the density of the solution is directly related to the solute concentration as well as the refractive index. This relationship can be expressed in terms of the following equation:

$$\varrho = a\eta - b$$

The tables below list the coefficients a and b for a number of ionic and nonionic gradient media. However, before applying the equations it is essential that allowance is made for the presence of other solutes (e.g., EDTA buffers). The following equation should be used:

$$\eta_{corrected} = \eta_{observed} - (\eta_{buffer} - \eta_{water})$$

1. Ionic Gradient Media

Gradient solute	Temperature (°C) η	for ϱ	Coefficients a	b	Valid density range (g/cm³)
CsCl	20	20	10.9276	13.593	1.2 – 1.9
	25	25	10.8601	13.497	1.3 – 1.9
Cs$_2$SO$_4$	25	25	12.1200	15.166	1.1 – 1.4
	25	25	13.6986	17.323	1.4 – 1.8
Cs(HCOO)	25	25	13.7363	17.429	1.7 – 1.8
	25	20	12.8760	16.209	1.8 – 2.3
CsTCA			7.6232	9.1612	1.1 – 1.7
CsTFA	20	20	23.041	29.759	1.2 – 1.8
NaBr	25	25	5.8880	6.852	1.0 – 1.5
NaI	20	20	5.3330	6.118	1.1 – 1.8
KBr	25	25	6.4786	7.643	1.0 – 1.4
KI	20	20	5.7317	6.645	1.0 – 1.4
	25	25	5.8356	6.786	1.1 – 1.7
RbBr	25	25	9.1750	11.241	1.1 – 1.7
RbCl	25	25	9.3282	11.456	1.0 – 1.4
RbTCA			6.5869	7.7805	1.1 – 1.6

2. **Nonionic Gradient Media**

Gradient	*Temperature (°C)*		*Coefficients*	
solute	η *for*	ϱ	*a*	*b*
Sucrose	20	0	2.7329	2.6425
Ficoll	20	20	2.381	2.175
Metrizamide	20	5	3.453	3.601
Metrizamide/D_2O	25	25	3.0534	2.9541
Nycodenz	20	20	3.242	3.323
Metrizoate	25	5	3.839	4.117
Renografin	24	4	3.5419	3.7198
Iothalamate	25	25	3.904	4.201
Chloral hydrate	4	4	3.6765	3.9066
Bovine serum albumin	24	5	1.4129	0.8814

Marker Enzymes and Chemical Assays for the Analysis of Subcellular Fractions

J. GRAHAM AND T.C. FORD

1. ASSAYS OF MARKER ENZYMES

Aminopeptidase
Aryl sulphatase
Catalase
Galactosyl transferase
Glucose-6-phosphatase
NADPH-cytochrome c reductase and NADH-cytochrome c reductase
5'-Nucleotidase
Ouabain-sensitive Na^+/K^+-ATPase
Phosphatase assays
Succinate-cytochrome c reductase

2. ASSAYS FOR NUCLEIC ACIDS

DAPI fluorescent assay for DNA
Diphenylamine assay for DNA
Ethidium bromide assay for DNA and RNA
Methyl green assay for DNA
Orcinol assay for RNA

3. ASSAYS FOR PROTEINS

Amido-black filter assay
Coomassie blue filter assay
Coomassie blue solution assay
Fluorescamine assay
Folin-Ciocalteu assay
Microbiuret assay

4. ASSAYS FOR POLYSACCHARIDES AND SUGARS

Anthrone assay
Phenol-H_2SO_4 assay

1. ASSAYS OF MARKER ENZYMES

1.1 **Aminopeptidase**

Wachsmuth, E. D., Fritze, E. and Pfleiderer, G. (1966) *Biochemistry,* **5**, 169-174.

Solutions

25 mM leucine *p*-nitroanilide;
50 mM sodium phosphate (pH 7.2).

Assay method

Use a 1.0 ml cuvette and add 1.0 ml of phosphate buffer, 0.1 ml leucine *p*-nitroanilide and 0.1 ml of the sample. Monitor the absorbance at 405 nm continuously using a chart-recorder with a full scale deflection equivalent to 0.2 absorbance units. The molar extinction coefficient of *p*-nitrophenol at 405 nm is 9620.

Notes

1.2 **Aryl Sulphatase**

Chang, P. L., Rosa, N. E. and Davidson, R. G. (1981) *Anal. Biochem.,* **117**, 382-389.

Solutions

0.1 M sodium acetate buffer (pH 5.6);
30 mM lead acetate;
5.0 mM 4-methyl umbelliferyl sulphate (4-MUS);
0.2 M glycine-carbonate buffer (pH 10.3) containing 1.0 M EDTA.

Assay method

Incubate a mixture of 100 μl acetate buffer, 20 μl lead acetate, 20 μl 4-MUS, 50 μl water and 10 μl of sample (containing 10 μg protein) at 37°C for 15 min. Stop the reaction by the addition of 1.0 ml of 0.2 M glycine-carbonate buffer. Measure the fluorescence of the solution using an excitation wavelength of 387 nm and an emission wavelength of 470 nm. Use 4-methyl umbelliferone (4-MU) in glycine-carbonate buffer as a standard to calibrate the instrument. The range of standard concentrations used depends entirely on the sensitivity of the fluorimeter.

Notes

The method used is based upon that of Chang *et al.* (1981): it is a very sensitive fluorimetric assay which measures the production of 4-methyl umbelliferone from 4-methyl umbelliferyl sulphate.

1.3 **Catalase**

Cohen, G., Dembiec, D. and Marcus, J. (1970) *Anal. Biochem.,* **34**, 30-38.

Solutions

0.01 M sodium phosphate buffer (pH 7.0);
6 mM H_2O_2 in sodium phosphate buffer;
3 M H_2SO_4;
2 mM $KMnO_4$.

Assay method

Set up the following tubes at 4°C:
(1) Test; 50 μl sample, 0.5 ml H_2O_2, vortex and after 3 min add 0.1 ml H_2SO_4 and vortex.
(2) Sample blank; 50 μl sample, 0.1 ml H_2SO_4, 0.5 ml H_2O_2, in that order.
(3) Standard; 0.55 ml of buffer, 0.1 ml H_2SO_4.
(4) Reagent blank; 0.1 ml H_2SO_4, 0.5 ml H_2O_2, 0.5 ml buffer.
(5) Spectrophotometer blank; 0.7 ml H_2O, 0.1 ml H_2SO_4, 0.55 ml buffer.
To tubes 1 − 4 add 0.7 ml $KMnO_4$ and read the optical density at 480 nm against tube 5 within 1 min.

Notes

1.4 **Galactosyl Transferase**

Fleisher, B., Fleisher, S. and Ozawa, H. (1969) *J. Cell Biol.,* **43**, 59-79.

Solutions

40 mM Hepes-NaOH (pH 7.0);
300 mM $MnCl_2$;
30 mM mercaptoethanol;
1.0 mg/ml uridine diphospho-[6-^{3}H]-galactose, 2 μCi/ml (74 kBq/ml);
400 mM N-acetylglucosamine;
300 mM EDTA (pH 7.4).

Assay method

Set up assay tubes containing 20 μl Hepes-NaOH buffer, 10 μl of each of the $MnCl_2$, mercaptoethanol and uridine diphosphogalactose solutions and 10 μl of sample (20 − 50 μg protein), in the presence or absence of 10 μl of 400 mM N-acetylglucosamine as an acceptor. At the same time, set up blanks with no sample added. Incubate the mixtures at 37°C for 60 min and stop the reaction by the addition of 20 μl of 300 mM EDTA (pH 7.4). Cool the tubes to 4°C. Prepare 2 cm Dowex-1 (chloride form) columns in Pasteur pipettes and wash them through with three volumes of water. Apply each incubation mixture to separate Dowex-1 columns and wash each twice with 0.5 ml water. Collect the eluant in scintillation vials, add a suitable water-soluble scintillation fluid (e.g., Beckman Ready Solv. HP) and measure the radioactivity in each case.

Notes

Appendix V

1.5 Glucose-6-phosphatase

Aronson, N. N. and Touster, O. (1974) *Methods Enzymol.,* **31**, 90-102.

Solutions

Prepare the assay solution by mixing 0.1 M glucose-6-phosphate (sodium salt), 35 mM histidine buffer (pH 6.5) and 10 mM EDTA in the ratio 2:5:1:1; 8%(w/v) trichloroacetic acid (TCA); 2.5% ammonium molybdate in 2.5 M H_2SO_4; reducing solution: 0.5 g 1-amino-2-naphthol-4-sulphonic acid, 1.0 g anhydrous Na_2SO_3, in 200 ml of 15% $NaHSO_3$.

Assay method

Add 50 μl of sample ($\sim$1 mg/ml protein) to 0.45 ml of the assay solution and incubate at 37°C for 30 min. For every sample prepare a blank containing no substrate, together with a reagent blank to determine the background hydrolysis of the substrate. Stop the reaction by adding 2.5 ml of cold 8% (w/v) TCA and keep the acidified sample at 0°C for 20 min. Centrifuge at 700 g for 20 min, take 1.0 ml of the supernatant and add 1.15 ml distilled water. Add 0.25 ml of the ammonium molybdate solution and 0.1 ml of the $NaHSO_3$ reducing solution. Measure the absorbance after 10 min at 820 nm.

Notes

The reducing solution should be freshly prepared each time.

1.6 **NADPH-cytochrome c Reductase and NADH-cytochrome c Reductase**

Williams, C. H. and Kamin, H. (1962) *J. Biol. Chem.,* **237**, 587-595.

Solutions

For NADPH-cytochrome c reductase:
  NADPH (2 mg/ml);
  cytochrome c (25 mg/ml);
  10 mM EDTA;
  all made up in 50 mM sodium phosphate (pH 7.7).
For NADH-cytochrome c reductase:
  NADH (1 mg/ml);
  cytochrome c (25 mg/ml);
  both made up in 50 mM sodium phosphate (pH 7.2).

Assay method

Use a 1 ml cuvette and add 0.9 ml of the appropriate buffer, 50 μl cytochrome c, 10 μl EDTA (for the NADPH-cytochrome c reductase only) and $10-20$ μl of sample. Using a chart-recorder (full scale deflection equivalent to 0.2 absorbance units) record the absorbance at 552 nm until steady, then add 0.1 ml of NADPH or NADH as appropriate and measure the increase in absorbance due to the reduction of cytochrome c. The molar extinction coefficient of reduced cytochrome c is 27 000 at 552 nm.

Notes

NADPH and NADH solutions must be freshly prepared, protected from light and kept at 0°C until used.

1.7 5'-Nucleotidase

Avruch, J. and Wallach, D. F. H. (1971) *Biochim. Biophys. Acta,* **233**, 334-347.

Solutions

10 mM [U-^{14}C]AMP, 0.5 μCi/ml (18.5 kBq/ml);
1.8 mM $MgCl_2$;
500 mM Tris-HCl (pH 8.0);
0.3 N $ZnSO_4$;
0.3 N $Ba(OH)_2$.
The last two solutions are best obtained as solutions through Sigma Chemical Co. (Poole, Dorset) as preparation of the solutions from solids has been found to be generally unsatisfactory.

Assay method

Set up assay tubes containing 0.1 ml of each of the AMP, $MgCl_2$ and Tris-HCl (pH 8.0) solutions. Add 50 μl H_2O and 50 μl of membrane suspension (0.1 – 1.0 mg/ml protein). Prepare a blank as above, but without the sample suspension, to measure the background rate of AMP hydrolysis.

Incubate the reaction mixtures at 37°C for 60 min and stop the reaction by addition of 0.3 ml of each of the $ZnSO_4$ and $Ba(OH)_2$ solutions (unhydrolysed AMP is precipitated by $BaSO_4$). Stand the tubes at 4°C for 30 min with occasional shaking and centrifuge at 700 *g* for 20 min. Transfer 0.5 ml of the supernatant to a scintillation vial, add a suitable water-soluble scintillator (e.g., Beckman Ready-Solv. HP) and measure the [U-^{14}C]adenosine using a liquid-scintillation counter.

Notes

1.8 Ouabain-sensitive Na$^+$/K$^+$-ATPase

Avruch, J. and Wallach, D. F. H. (1971) *Biochim. Biophys. Acta,* **233**, 334-347.

Solutions

10 mM [γ-^{32}P]ATP, 1 μCi/ml (37 kBq/ml);
1.5 M NaCl;
10 mM MgCl$_2$;
0.3 M KCl;
200 mM Tris-HCl (pH 7.4);
10 mM ouabain;
0.01 M HCl, 1 mM sodium phosphate;
activated charcoal suspension (Norit): contains 4% (v/v) Norit A, 0.1 M HCl, 0.2 mg/ml bovine serum albumin, 1 mM sodium phosphate, 1 mM sodium pyrophosphate.

Assay method

Set up assay tubes containing 0.1 ml of each of the ATP, NaCl, MgCl$_2$, KCl and Tris-HCl (pH 7.4) solutions and 0.45 ml H$_2$O. Add 50 μl of sample suspension (0.2 − 1.0 mg/ml protein) to each tube in the presence or absence of 10 μl of ouabain solution. Set up a blank assay as above, but without the sample suspension, to measure the background rate of ATP hydrolysis. Incubate the assay mixture for 60 min at 37°C and stop the reaction by the addition of 3.0 ml of the charcoal suspension. Unhydrolysed ATP is adsorbed onto the activated charcoal. Stand the tubes at 4°C for 30 min with occasional shaking and then filter the solution through a 2.5 cm Whatman glass-fibre filter disc (GF/C), in a Millipore micro-analysis filter-holder, directly into a scintillation vial. Wash the residue twice with 3 ml of 0.01 M HCl containing 1 mM phosphate. Measure the non-adsorbed [^{32}P]phosphate by Cerenkov counting in a liquid-scintillation counter.

Notes

1.9 **Phosphatase Assays**

Engstrom, L. (1961) *Biochim. Biophys. Acta,* **52**, 36-48.

1.9.1 *Alkaline Phosphatase*

Solutions
16 mM *p*-nitrophenyl phosphate;
50 mM sodium borate buffer (pH 9.8);
1.0 M $MgCl_2$;
0.25 M NaOH.

Assay method
Make up a stock assay mixture of 5.0 ml substrate, 5.0 ml sodium borate buffer and 20 μl $MgCl_2$. Take 0.2 ml of the assay mixture and add 50 μl of sample (0.1 – 1.0 mg/ml protein) and incubate at 37°C for 60 min. To stop the reaction add 0.6 ml of 0.25 M NaOH and centrifuge at 700 *g* for 20 min. Measure the absorbance of the supernatant at 410 nm.

1.9.2 *Acid Phosphatase*

Carry out the assay as for alkaline phosphatase (1.9.1) substituting 180 mM sodium acetate buffer (pH 5.0) instead of the borate buffer and omit the $MgCl_2$.

Notes

1.10 Succinate-cytochrome c Reductase

1.10.1 *Method 1*

Mackler, B., Collip, P. J., Duncan, H. M., Rao, N. A. and Heunnekens, F. M. (1962) *J. Biol. Chem.*, **237**, 2968-2974.

Solutions

50 mM sodium phosphate buffer (pH 7.4);
100 mM potassium cyanide (*CAUTION*, very poisonous);
660 mM sodium succinate;
cytochrome c (12.5 mg/ml).
Make up all reagents in the phosphate buffer.

Assay method

In a 1 ml cuvette, make up an assay mixture containing, 1.0 ml phosphate buffer, 20 μl potassium cyanide, 50 μl cytochrome c and 10 − 50 μl of the sample. Record the absorbance at 552 nm, using a chart-recorder with a full scale deflection equivalent to 0.2 absorbance units. When the trace is steady, add 0.1 ml of sodium succinate solution and record the increase in absorbance due to the reduction of cytochrome c. The molar extinction coefficient of reduced cytochrome c at 552 nm is 27 000.

1.10.2 *Method 2*

Modification of the method of Butcher, R. G. (1970) *Exp. Cell Res.,* **60**, 54-58.

Solutions

20 mM Tris-HCl (pH 7.5), 0.1 mM EDTA;
0.2 M sodium succinate (pH 7.5);
10 mM 2-*p*-iodophenyl-3-*p*-nitrophenyl tetrazolium chloride (INT) in dimethyl formamide.
2% sodium dodecyl sulfate (SDS)

Assay method

Incubate the following incubation mixture 1.5 ml microcentrifuge tubes: 0.5 ml Tris-HCl, EDTA, 0.1 ml sodium succinate, 0.05 ml INT and 0.1 ml of the sample (added last to start the reaction).
Incubate the mixture for 5 min at room temperature (20 − 23°C) and stop the reaction by the addition of 0.025 ml of 2% SDS. Centrifuge the tubes for 2 min at top speed in the microcentrifuge and remove the supernatant. Measure the absorbance of the supernatant at 500 nm.

Notes

The sodium succinate solution may be stored indefinitely at −20°C. The potassium cyanide and cytochrome c solutions must be freshly prepared. **CAUTION; potassium cyanide is extremely poisonous.**

2. CHEMICAL ASSAYS FOR NUCLEIC ACIDS

2.1 DAPI Fluorescent Assay for DNA

Brunk, C. F., Jones, K. C. and James, T. W. (1979) *Anal. Biochem.*, **92**, 497-500.

Solutions

DAPI reagent: 100 mM NaCl, 10 mM EDTA, 10 mM Tris-HCl (pH 7.0) containing 100 ng/ml 4′,6-diamidino-2-phenylindole (DAPI);
DNA standard solution (20 μg/ml).

Assay method

The fluorescence of 3.0 ml of the DAPI solution is determined in the absence of DNA (excitation wavelength 360 nm and emission wavelength 450 nm). Add 15 μl of the DNA standard solution and remeasure the fluorescence; add a further 15 μl DNA solution and again measure the fluorescence, repeat twice more to obtain four points for a standard curve. Take four further fluorescence measurements after each addition of four 15 μl aliquots of the DNA sample solution. Calculate the DNA content of the sample material by comparing the fluorescence yield of the standards and sample material.

Notes

When DAPI is complexed with DNA, the fluorescence of the complex is enhanced about 20-fold as compared with the fluorescence of the dye alone. The fluorscence yield is unaffected by pH between pH 5 and pH 10. Cations, especially divalent or heavy metal ions, cause significant quenching of the fluorescence and quenching is also observed at low ionic strengths (hence the composition of the buffer used). Quenching is found to increase with decreasing temperature and therefore it is necessary that all measurements are made at a uniform temperature. The binding of DAPI is highly specific for A – T base pairs hence it is essential that the DNA of the standard and sample have the same base composition.

2.2 **Diphenylamine Assay for DNA**

Schneider, W. C. (1957) *Methods Enzymol.,* **3**, 680-684.

Solutions

Diphenylamine reagent: dissolve 1.0 g diphenylamine in 100 ml glacial acetic acid and add 2.75 ml of concentrated H_2SO_4;
DNA standard solution, 500 μg DNA/ml (A_{260} = 11);
20% (w/v) trichloroacetic acid (TCA).

Assay method

Dilute the DNA stock solution to give sample volumes of 1.0 ml in 5% TCA containing 0 – 200 μg DNA. In the case of sucrose, Ficoll or metrizamide gradients it is necessary to remove the gradient medium by precipitating the DNA; to do this add an equal volume of cold 20% (w/v) TCA. Pellet the DNA by centrifugation at 1000 *g* for 20 min and wash the pellet twice with cold 10% TCA to remove all of gradient medium which otherwise interferes with this assay. Hot-acid digest the standards and gradient fractions in 5% TCA in a water bath at 90°C for 20 min to solubilise the DNA. To 1.0 ml samples add 2.0 ml of the diphenylamine reagent and incubate in a water bath at 100°C for 10 min. Cool to room temperature and measure the optical density at 595 nm using a mixture of 2.0 ml of diphenylamine reagent and 1.0 ml 5% TCA as the blank.

Notes

This assay cannot be used for Percoll or metrizoate gradient fractions. Do not store the diphenylamine reagent in the cold since it solidifies.

2.3 Ethidium Bromide Assay for DNA and RNA

Karsten, U. and Wollenberger, A. (1972) *Anal. Biochem.*, **46**, 135-148.
Karsten, U. and Wollenberger, A. (1977) *Anal. Biochem.*, **77**, 464-469.

Solutions

Phosphate buffered saline (PBS): contains 0.1 g $CaCl_2$, 0.2 g KCl, 0.2 g KH_2PO_4, 0.1 g $MgCl_2.6H_2O$, 8.0 g NaCl, 1.15 g Na_2HPO_4, adjusted to pH 7.5 and made up to 1 litre;
ethidium bromide (25 μg/ml) in PBS;
heparin (25 μg/ml) in PBS;
RNase (50 μg/ml) in PBS (heat to 100°C for 10 min to destroy any DNase activity);
DNA standard solution (25 μg/ml DNA) in PBS. Stock DNA solutions should be stored frozen. Before use dilute the solution by addition of four volumes of PBS.
The sample material should be either in PBS or another appropriate buffer.

Assay method

Make up the following assay mixtures in 3 ml cuvettes:
(1) Standard: 0.5 ml DNA standard solution, 0.5 ml heparin solution, 1.0 ml PBS.
(2) Blank I: 0.5 ml heparin solution, 1.5 ml PBS.
(3) Blank II: 2.5 ml PBS.
(4) Sample (DNA + RNA): 0.5 ml homogenate, 0.5 ml heparin solution, 1.0 ml PBS.
(5) Sample (DNA only): 0.5 ml homogenate, 0.5 ml heparin solution, 0.5 ml RNase solution, 0.5 ml PBS.
(6) Sample (background correction): 0.5 ml homogenate, 2.0 ml PBS.
Incubate mixtures (1 − 6) in a waterbath at 37°C for 20 min then add 0.5 ml of ethidium bromide solution to tubes (1), (2), (4) and (5). Full intensity of fluorescence is attained within 60 sec of the addition of ethidium bromide and remains constant for at least 60 min. Briefly stir the reaction mixture before taking any measurements. The fluorescence is measured using an excitation wavelength of 360 nm and an emission wavelength of 580 nm. The fluorescence of the standard is set at 100. Temperature during the measurements is critical and should be kept constant to within ±0.5°C.

Calculation of nucleic acid content
DNA

$$A_{dna} = \frac{A_{std} (F_{(5)} - F_{(2)} - F_{(6)} + F_{(3)})}{F_{(1)} - F_{(2)}}$$

A_{dna} = amount of DNA/mixture (μg). A_{std} = amount of standard DNA/mixture. F = fluorescence intensity (units).

RNA

$$A_{rna} = \frac{A_{std} (F_{(4)} - F_5)}{0.46 (F_{(1)} - F_{(2)})}$$

The factor 0.46 is empirically derived from the ratio of the fluorescence yield of RNA and DNA with ethidium bromide. Alternatively, an RNA standard solution can be used.

Notes

Up to 5 μg/ml DNA or RNA fluorescence increases linearly with concentration. Check the RNase solution for fluorescence and subtract from the reading of mixture (5).

2.4 **The Methyl Green Assay for DNA**

Peters, D. L. and Dahmus, M. E. (1979) *Anal. Biochem.*, **93**, 306-311.

Solutions

Methyl green reagent: dissolve methyl green (Kodak-Eastman, Rochester, NY, USA) in 100 mM Tris-HCl (pH 7.9) to a concentration of 0.01% (w/v). To remove the contaminating crystal violet present in the solution, extract the solution with an equal volume of water-saturated chloroform. Allow the mixture to separate in a separating funnel and draw off the chloroform in the lower part of the funnel together with the dissolved crystal violet. Repeat if necessary, using fresh chloroform, until the chloroform layer is clear.
DNA standard solution (500 μg/ml);
proteinase K solution, 1 mg/ml (pre-digested at 37°C for 30 min).

Assay method

Dilute the DNA stock solution to give 0.2 ml samples containing $0-100$ μg DNA, or make up gradient fractions to 0.2 ml. Add 5 μg (5 μl) of proteinase K to each fraction and incubate at about room temperature for 60 min. Add 1.0 ml of the methyl green reagent to each fraction and incubate either overnight at room temperature (20°C) or for $2-3$ h at $40-45$°C. The optical density of the solutions is measured at 640 nm. Prepare a blank using 0.2 ml of distilled water (treated in the same way as the sample fractions) and 1.0 ml of the methyl green reagent.

Notes

In the absence of DNA, the blue colour of the reagent fades, but in the presence of DNA, the colour is retained in proportion to the amount of DNA present. The assay is compatible with samples containing Nycodenz, metrizamide or metrizoate. Methyl green from some sources may not give consistent results.

2.5 **Orcinol Assay for RNA**

Schneider, W. C. (1957) *Methods Enzymol., ***3**, 680-684.

Solutions

Orcinol reagent: 0.5 g orcinol and 0.25 g $FeCl_3.6H_2O$ in 50 ml of concentrated HCl (the reagent should be freshly prepared and kept at $4°C$ until required); RNA standard solution (250 $\mu g/ml$) in distilled water; 20% (w/v) trichloroacetic acid (TCA) solution.

Assay method

Dilute the RNA stock solution to give 0.5 ml volumes containing $0-250$ μg RNA for the preparation of a standard curve, or dilute gradient fractions to 0.5 ml. Add TCA to a final concentration of 5% (w/v) and incubate each fraction for 20 min at $90°C$. Centrifuge for 2 min in a microcentrifuge, remove the supernatant and add to it an equal volume of the orcinol reagent. Incubate for 20 min in a boiling water-bath, cool the solutions and measure the optical densities at 660 nm. Use 1.0 ml of reagent and 1.0 ml of 5% TCA as a blank.

Notes

The yellow colour of the reagent becomes green in the presence of RNA. This assay is incompatible with the presence of Percoll and metrizoates, also the orcinol reagent reacts with sucrose, Ficoll and the glucosamide group of the metrizamide molecule. Metrizamide can be removed by acid precipitation (see 2.2). Proteins, and formaldehyde also interfere with the orcinol assay.

3. ASSAYS FOR PROTEINS

3.1 **Amido-black Filter Assay**

Schaffner, W. and Weissman, C. (1973) *Anal. Biochem.*, **56**, 502-514.

Solutions

Stain: dissolve 0.1 g Amido-black 10B in 100 ml of methanol/glacial acetic acid water, 45/10/45 by vol;
destain: methanol/glacial acetic acid/water, 90/2/8 by vol;
eluant solution: (25 mM NaOH and 0.05 mM EDTA in 50% aqueous ethanol);
standard protein solution (150 μg/ml);
stock solution of 60% (w/v) TCA;
1% SDS in 1 M Tris-HCl (pH 7.5).

Assay method

Dilute the standard protein solution to a final volume of 0.27 ml containing $0-150$ μg protein for the standard curve, or dilute gradient fractions to 0.27 ml. Add 0.03 ml of the Tris-SDS solution and then 0.6 ml of 60% TCA to each fraction. Filter each sample through a Millipore filter (0.45 μ pore size) rinse the tube with 0.3 ml 6% TCA and then wash the filter with 2 ml of 6% TCA. Place each filter in stain for $2-3$ min with gentle agitation and remove to a water rinse for about 30 sec. Pass each filter through three changes of destain, about 1 min per change, rinse again in water and blot with a tissue. Place the filter in a test-tube with 0.6 ml of eluant solution for 10 min, vortexing three times for $2-3$ sec, add a further 0.9 ml eluant and mix by vortexing. The optical density of the solutions is measured at 630 nm using the eluant solution as a blank.

Notes

Using this assay, no interference is experienced from the presence of gradient solutes unless, like Percoll or metrizoates, they are not acid soluble. Sucrose, metrizamide and Nycodenz are acid soluble and thus such gradients may be analysed using this assay. Potassium ions must be excluded from the sample solutions since they precipitate the dodecyl sulphate ions.

3.2 Coomassie Blue Filter Assay

McKnight, G. S. (1977) *Anal. Biochem.,* **78**, 86-92.

Solutions

Stain: 0.25% (w/v) Coomassie blue G-250, 7.5% acetic acid, 5% methanol in distilled water;
destain: 7.5% acetic acid, 5% methanol;
eluant solution: 0.12 M NaOH, 20% H_2O, 80% methanol;
3.0 M HCl;
20% (w/v) TCA.

Assay method

Cut 25 mm discs from Whatman glass-fibre filters (GF/C). Number the filters along the edge, taking care not to touch the filters with bare fingers. Protein samples are pipetted onto the centre of the discs and allowed to absorb completely (5 – 30 sec). The sample volume to be used depends upon the protein (see notes). Prepare blank filters as described, using distilled water instead of the protein sample. After the absorption period, place the filters in 20% TCA at 4°C and swirl gently for several minutes. Transfer the filters individually to the stain solution at 4°C for 20 min. Transfer the filter to destain and swirl for several minutes, then decant the solution and repeat twice with fresh destain. Place the filters on a filtering apparatus and wash several times with destain until the blank filters are completely white. Cut out the stained spots using a 1 cm cork borer and resting the filters on several sheets of filter paper. Put the cut out spots into 1.5 ml microcentrifuge tubes and add 0.6 ml of eluant solution, vortex and stand at room temperature until the colour disappears from the filters (~ 5 min). Acidify the eluant with 0.03 ml 3 M HCl (Coomassie blue is colourless in basic solutions), vortex and centrifuge for 2 min in a microcentrifuge to remove any pieces of glass fibre from the solution. Carefully remove the supernatant and measure its absorbance at 590 nm using acidified eluant as a blank.

Notes

This assay is not compatible with Percoll or metrizoates. The sample volume that can be pipetted onto the filter depends on the salt concentration used and on the particular protein. Some proteins, such as RNase and histones, do not spread and volumes of up to 100 μl can be used to give concentrated spots, but BSA and ovalbumin, for example, diffuse into the filter and in such cases volumes need to be less than 50 μl. In high salt (e.g., 1 M NaCl), all proteins bind tightly to the filter and produce a concentrated spot.
The sensitivity of this assay can be increased 3-fold by using 200 μl of eluant solution and 10 μl of 3 M HCl.

3.3 Coomassie Blue Solution Assay

Bradford, M. (1976) *Anal. Biochem.,* **72,** 248-254.

Solutions

Coomassie blue reagent: dissolve 100 mg Coomassie blue G-250 in 50 ml 90% ethanol and add 100 ml 85% (w/v) phosphoric acid, make up to 200 ml with distilled water. This concentrated solution can be stored for at least a month at 5°C. Dilute by the addition of four volumes of distilled water before use. Standard protein solution (100 μg/ml).

Assay method

Dilute the protein standard solution to final volume of 0.1 ml containing $0-100$ μg protein, dilute the sample fractions to 0.1 ml. Add 5.0 ml of the reagent to each fraction and measure the optical density at 595 nm. Measurements should be carried out after 2 min but before 60 min has elapsed after the addition of the reagent. Use 5.0 ml of reagent and 0.1 ml protein buffer as a blank.

Notes

This assay provides a swift and easy determination of protein distribution in a gradient and is compatible with sucrose, metrizamide and Nycodenz but not metrizoates.

This assay can be used for Percoll gradient fractions if the Percoll is first removed by centrifugation at 12 000 g for 15 min after precipitation with 0.025% Triton X-100 in 0.25 M NaOH, as described by Vincent, R. and Nadeau D. in *Anal. Biochem.,* **135,** 355.

3.4 **Fluorescamine Assay**

Bohlen, P., Stein, S., Dairman, W. and Udenfriend, S. (1973) *Arch. Biochem. Biophys.,* **155**, 213-220.

Solutions

Fluorescamine reagent: 30 mg of fluorescamine in 100 ml dioxane (histological grade);
0.05 M sodium phosphate buffer (pH 8.0).

Assay method

Place the sample, containing $0.5-50$ μg protein, into a tube and add phosphate buffer to bring the total volume up to 1.5 ml. Add 0.5 ml of the fluorescamine solution with vortexing to ensure vigorous mixing. Rapid mixing is essential for reproducible results since the reagent is hydrolysed rapidly. The pH must be maintained at pH $8-9$. The reaction takes place at room temperature.

Measure the fluorescence using a fluorimeter with an excitation wavelength of 390 nm and an emission wavelength of 475 nm. More reproducible results are obtained by carrying out direct fluorescence measurements in the reaction tubes, using a cell adaptor, instead of transferring the mixtures to a cuvette. Apparent protein concentrations are calculated by comparison to bovine serum albumin standard solutions. All values should be corrected using a blank, prepared using 1.5 ml of sample buffer.

Notes

Fluorescamine reacts with amino groups (primary amines), thus Tris and other primary amine buffers cannot be used. Large amounts of secondary and tertiary amine buffers also interfere with the assay. Hence inorganic buffers, such as phosphate or borate, are most successful. A modification of this procedure uses gel filtration to remove contaminating low molecular weight amines in the sample which interfere with the assay (Bohlen *et al., Anal. Biochem.,* **58**, $559-562$). Using this modification of the assay as described, it should be noted that contaminants with molecular weights in excess of 300 do interfere due to incomplete separation from the proteins on the Sephadex G-25 columns.

3.5 **Folin-Ciocalteu Assay**

Lowry, O. H., Rosebrough, N. J., Farr, A. L. and Randall, R. J. (1951) *J. Biol. Chem.,* **193**, 265-267.

Peterson, G. L. (1979) *Anal. Biochem.,* **100**, 201-220.

Solutions

Folin-ciocalteu reagent, is available commercially;
solution A: 20 g Na_2CO_3, 4.0 g NaOH, 0.2 g sodium potassium tartrate, made up in one litre water;
solution B: 0.5% (w/v) $CuSO_4.5H_2O$;
solution C: 50 ml solution A and 1 ml solution B mixed immediately before use.

Assay method

Dilute samples to 1.0 ml, add 4.0 ml of solution C, mix and allow to stand for 10 min. Add 1.5 ml of Folin-Ciocalteu reagent (diluted 1:9 immediately before use), mix again and stand for 30 min in the dark. Measure the optical density at 660 nm. Use 1.0 ml water instead of sample solution as blank.

Notes

This is the most widely used method for estimating proteins but over the years it has been extensively modified. Most modifications are aimed at improving sensitivity and in avoiding the many sources of interference in this assay. Gradient media that interfere with this assay include: sucrose, glycerol, metrizamide, Nycodenz and Percoll.

3.6 **Microbiuret assay**

Itzhaki, R. F. and Gill, D. M. (1964) *Anal. Biochem.,* **9**, 401-410.

Solutions

Benedict's reagent: (i) 17.3 g trisodium citrate and 10.0 g Na_2CO_3 dissolved in warm water. (ii) 1.73 g $CuSO_4$ dissolved in 10.0 ml water. Mix solutions (i) and (ii) and make up to 100 ml with distilled water;
1.0 M NaOH;
protein standard solution: 250 μg/ml protein.

Assay method

Make up sample fractions to 1.0 ml, add 3.0 ml 1.0 M NaOH and 0.2 ml of Benedict's reagents. Mix well and allow to stand at room temperature for 15 min. Measure the optical density at 330 nm. Use water in place of sample solution as a blank.

Generate a standard curve using the standard protein solution diluted to give 1.0 ml samples containing 0 − 250 μg protein.

Notes

The presence of amide bonds in compounds (e.g., metrizamide) will cause interference with this assay.

4. ASSAYS FOR POLYSACCHARIDES AND SUGARS

4.1 **Anthrone Assay**

Seifter, S., Dayton, S., Norvic, B. and Muntwyler, E. (1950) *Arch. Biochem.,* **25**, 191.

Solutions

Anthrone reagent; 0.2% anthrone in 95% H_2SO_4;
Glucose standard solution: 50 μg/ml glucose.

Assay method

Place 2.5 ml of sample solution in a boiling tube. Cool the tube in ice-water and add 5.0 ml of anthrone reagent from a fast-flowing burette and mix thoroughly, still in the ice-water. Cover the tube with a marble and heat for 10 min in a boiling water bath. Cool quickly and measure the optical density at 620 nm. Use 2.5 ml distilled water as a blank. Prepare a standard curve of $0-125$ μg of glucose using the standard glucose solution.
CARE, when boiling the H_2SO_4 solution.

Notes

Gradient media containing sugars, such as sucrose, Ficoll and metrizamide, will interfere with this assay. Other media may be unstable in the hot acid conditions of the assay.

4.2 **Phenol-H$_2$SO$_4$ Assay**

Dubois, M., Gilles, K. A., Hamilton, J. K., Rebers, A. and Smith, F. (1956) *Anal. Chem.*, **28**, 350-356.

Solutions

80% (w/w) phenol;
concentrated H$_2$SO$_4$ (95.5%);
standard sugar solution (100 μg/ml).

Assay method

Use tubes of at least 10 mm diameter to allow good mixing and to minimise heat dissipation. To 1.0 ml of standard or sample sugar solution add 50 μl 80% phenol; add 5.0 ml H$_2$SO$_4$ allowing the acid to leave the pipette quickly and fall directly onto the liquid surface, so ensuring good mixing. Let the tubes stand for 10 min, shake and place into a water-bath at $25-30°$C for $10-15$ min. Blanks are prepared by substituting water for the sample solution. Measure the optical density at 490 nm in the case of hexoses and at 480 nm in the case of pentoses and uronic acids.

Notes

5% (w/w) phenol may be used in place of 80% phenol, in which case 1.0 ml of phenol solution is added, other volumes as given above. Most compounds that interfere with the anthrone assay also interfere with this assay.

Names and Addresses of Suppliers of Centrifuges and Ancillary Equipment

Anderman & Co. Ltd., Central Ave., East Molesey, Surrey KT8 0QZ, UK
Anton Paar: see Paar K.G.
Artisan Industries, Waltham, Mass., USA
Astra, Södertalje, Sweden
Baird & Tatlock Ltd., Freshwater Rd., Chadwell Heath, Romford, Essex, UK
Baskerville & Lindsay Ltd., Barlow Moor Rd., Manchester M21 2AX, UK
BDH Chemicals Ltd., Broom Rd., Poole, Dorset BH12 4NN, UK
Beckman Instruments Spinco Division, 1117 California Ave., Palo Alto, CA 94304, USA
Beckman Instruments International SA, 17, Rue des Pierres du Niton, P.O. Box 308, CH-1207 Geneva, Switzerland
Beckman-RIIC Ltd., Turnpike Rd., Cressex Industrial Estate, High Wycombe, Bucks HP12 3NR, UK
Bio-rad Laboratories, 2200 Wright Ave., Richmond, CA 94804, USA
Biorad Laboratories Ltd., Caxton Way, Watford, Herts, UK
Braun Melsungen International GmbH, Schwarzenbergerweg, D-3508 Melsungen, FRG
Buchler Instruments Inc., 1327 Sixteenth Street, Fort Lee, NJ 07024, USA
Burkhard, P.O. Box 55, Uxbridge, Middx UB8 1LA, UK
Christ: see Heraeus Christ
Clandon Scientific Ltd., Lysons Ave., Ash Vale, Aldershort, Hants GU12 5QR, UK
Corex: see Dupont-Sorvall and Dupont (UK) Ltd.
Corning Ltd., Stone, Staffs. ST15 0BG, UK
Damon/IEC, 300, 2nd Av., Needham Heights, MA 02194, USA
Damon/IEC (UK) Ltd., Unit 1, Lawrence Way, Brewers Hill Rd., Dunstable, Beds LU6 1BD, UK
Denley, Daux Road, Billingshurst, Sussex RH14 9SJ, UK
Du Pont-de Nemours & Co. Inc., Wilmington, DE 19898, USA
Du Pont-Sorvall, Pecks Lane, Newtown, CT 06470, USA
Du Pont (UK) Ltd., Wedgewood Way, Stevenage, Herts SG1 4QN, UK
Electro Nucleonics Inc., 368 Passaic Ave., Fairfield, NY 07006, USA
Esco (Rubber) Ltd., 14-16 Great Portland Street, London W1N 5AB, UK
Finnegans Speciality Paints Ltd., Eltringham Works, Prudhoe, Northumberland, UK

Fisons Scientific Ltd., Bishop Meadow Rd., Loughborough, Leics LE11 0RG, UK

F.T. Scientific Ltd., Station Industrial Estate, Bredon, Tewkesbury, Glos GL20 7HH, UK

Gallenkamp & Co. Ltd., P.O. Box 290, Technico House, Christopher Street, London, EC2P 2ER, UK

Grant Instruments Ltd., Barrington, Cambridge CB2 5QZ, UK

Heraeus-Christ, D-3360 Osterode am Marz, Postfach 1220, FRG

Hermle Laborgeräte GmbH, Walldorfer Strasse 14, Postfach 1150, D-6837 St. Leon-Rot 2, FRG

Hettich, Postfach 4255, D-7200 Tuttlingen, FRG

A.R. Horwell Ltd., 2 Grangeway, Kilburn High Road, London NW6 NB, UK

V.A. Howe & Co. Ltd., 88 Peterborough Road, London SW6 3EP, UK

I.E.C. (International Equipment Co.): see Damon/I.E.C.

ISCO (Instrument Specialities), Box 5347, Lincoln, NE 68505, USA

Janetzki, Leipzig, GDR

Jouan SA, Box 403, F-44608 St. Nazaire Cédex, France

Kontron, Analytical Division, Bernerstrasse Süd 169, CH-8048 Zürich, Switzerland

LKB Instruments Inc., 12221 Parklawn Dr., Rockville, MD 20852, USA

LKB Instruments Ltd., 232 Addington Road, South Croydon, Surrey CR2 8YD, UK

LKB Produkter AB, Box 305, S-16126 Bromma, Sweden

Merck, Postfach 4119, D-6100 Darmstadt 1, FRG

Monsanto Chemicals Limited, 10-18 Victoria Street, London SW1 UK

MSE Scientific Instruments Ltd., Manor Royal, Crawley, Sussex, UK

Nyacol Inc., Megunco Road, Ashland, MA 01721, USA

Nyegaard & Co., Postbox 4220, Torshov, Oslo 4, Norway

Paar KG, Anton, Graz, Austria

Paar Scientific Ltd., 594, Kingston Road, Raynes Park, London SW20 8DW, UK

Pennwalt Ltd., Doman Rd., Camberley, Surrey GU15 3DN, UK

Pharmacia Fine Chemicals A.B., Box 175, S-75104 Uppsala 1, Sweden

Pharmacia (UK) Ltd., Pharmacia House, Midsummer Boulevard, Milton Keynes MK9 3HP, UK

Schering, A.G., 1000 Berlin 65, Mullerstrasse 170-172, FRG

Schering Chemicals Ltd., Burgess Hill, Sussex, UK

Scientific Supplies Co. Ltd., Scientific House, Vine Hill, London EC1 5EB, UK

Searle Instrument Division, West Rd., Temple Fields Industrial Estate, Harlow, Essex, UK

Shandon Southern Ltd., Frimley Road, Camberley, Surrey, UK

Sharples: see Pennwalt

Sorvall: see Du Pont-Sorvall and Du Pont (UK)

E.R. Squibb & Sons, Regal House, Twickenham, Middlesex, UK

Townson & Mercer Ltd., Chadwick Rd., Astmoor, Runcorn, Cheshire WA7 1PR, UK

Whyteleafe Scientific Ltd., Exbridge Works, Exbridge, Dulverton, Somerset, UK

Wifug AB, Box 147, S-34300 Elmhult, Sweden

Wifug Ltd., Lustra Works, Parry Lane, Bradford, Yorks BD4 8TQ, UK

Winthrop Laboratories, Winthrop House, Surbiton, Surrey, UK

Glossary of Terms

Active band centrifugation. Sedimentation velocity analysis of an enzyme preparation in the analytical ultracentrifuge in which a small volume of the enzyme solution is layered on top of a supporting solution that contains the substrate for the enzyme. The movement of the active form of the enzyme is usually followed spectrophotometrically, for example at 340 nm when NAD is converted to NADH.

Analytical ultracentrifuge. A centrifuge in which it is possible to monitor the sedimentation of particles during centrifugation.

Angular velocity (ω). Time rate of angular motion about an axis usually expressed in terms of radians/sec.

Band (or zone). A discrete region containing particles sedimenting together.

Band centrifugation. Sedimentation velocity analysis of a macromolecule in the analytical ultracentrifuge in which a small volume of the solution of the macromolecule is layered on top of a solution which forms a supporting density gradient in the course of the centrifuge run.

Boundary analysis. A sedimentation velocity technique in the analytical ultracentrifuge where the ultracentrifuge cell is filled with the solution of the macromolecule. When the centrifugal force is applied, a boundary is formed between the macromolecule solution and the buffer in which it is dissolved and the rate of movement of the boundary down the cell is measured in order to determine the sedimentation coefficient of the macromolecule.

Batch-type zonal rotors. These rotors (e.g., BXIV, BXV, AXII, BXXIX) can only accommodate sample volumes less than their own internal capacities. They do not possess the facility for continuous-flow operation.

Bottom unloading. Recovery of gradients by piercing the tube bottom and collecting fractions as the gradient flows out.

Boundary. The region in a solution during a sedimentation velocity experiment in an analytical centrifuge where the concentration of particles changes abruptly.

Bubble trap. This is a T-shaped device, inserted into the fluid line, whose free end is closed off.

Buoyancy gradient. In buoyant equilibrium analysis, a number of factors influence the distribution of the various components in the ultracentrifuge cell. One factor is the gradient formed by the low molecular weight salt and another factor is the effect of hydrostatic pressure on the compressibility of the components. The buoyancy gradient is the net gradient comprising both the salt gradient and the commmpression gradient.

Buoyant density. The effective density of particles in a gradient medium; it differs from one medium to another.

Buoyant equilibrium analysis. Sedimentation equilibrium analysis using a self-generating density gradient in which macromolecules form bands at positions in the gradient that correspond to their buoyant densities.

Capacity of a band. The maximum concentration of particles which will sediment through a particular gradient without becoming unstable.

Centrepiece. This is a component of an analytical ultracentrifuge cell assembly. It may be double or single sector or of a special type. It specifies the pathlength of the solution to be analysed.

Centrifugal elutriation. A technique for separating cells mainly according to size based on differences in the equilibrium between two opposing forces (centrifugal force and fluid velocity) within a separation chamber inside an elutriator rotor.

Centripetal movement. Movement towards the axis of rotation.

Coaxial seal. Seal which gives access to a spinning rotor and yet ensures that the fluid in the channels leading to the centre and edge of a zonal rotor are kept separate. Composed of a stationary seal and a rotating seal which are kept in firm contact by spring tension.

Continuous density gradient. Gradient with gradual changes in solution density.

Continuous-flow zonal rotors. These rotors (e.g., CF32Ti and TZ 28) possess a fluid seal which allows the continuous passage of sample into the rotor while it is spinning at high speed: they can therefore process volumes of sample far in excess of their own capacity.

Core. Central part of a zonal rotor.

Coriolis's forces. These forces are generated during acceleration and deceleration of centrifuge rotors and they tend to swirl the contents within the tube. The forces are most significant at speeds less than 1000 r.p.m. and when using solutions of low viscosity in wide tubes or bottles. Care must be taken to minimise these forces which otherwise disturb sample zones within the solution.

Counterstreaming centrifugation. See centrifugal elutriation.

Cross-leakage. Leakage of liquid being fed into the zonal rotor across the face of the seal.

Cushion. Dense solution placed at the bottom of gradients to prevent particles pelleting.

Cytochemical staining. Staining to reveal the presence of a certain enzymatic activity within cells.

Denaturing gradients. Gradients which contain a solvent, usually dimethyl sulphoxide or formamide, which has the property of maintaining nucleic acids in their denatured form during centrifugation.

Density gradient. The solution in the centrifuge tube through which the sample will sediment is chosen such that the density of the solution increases with increasing distance down the centrifuge tube.

Differential centrifugation. Sedimentation of an initially homogeneous suspension to produce a pellet of the faster sedimenting particles leaving the slower moving components unpelleted. Further centrifugation for longer times or at higher speeds pellets the remaining particles in order of their sedimentation rates.

Discontinuous gradient. Gradient formed by introducing 'blocks' of sucrose solution of increasing density into the rotor or tube.

Droplet formation. Sample instability resulting from the diffusion of gradient medium into the sample zone.

Dynamic loading/unloading. Loading or unloading of a spinning zonal rotor. Made possible by the use of a feed head.

Edge of rotor. The wall of the zonal rotor chamber which is equivalent to the bottom of a centrifuge tube.

Edge unloading. Unloading of the gradient and the separated bands from the edge of the rotor and not from the centre, which is the more normal method.

Elutriation. This is a method of fractionating delicate particles, notably cells, with minimum disruption, and it requires the use of a special continuous-flow rotor (elutriator rotor) which allows one to balance sedimentation against a centripetal flow of liquid. This technique separates primarily on the basis of size.

Equivolumetric gradients. Specialised gradients for use with zonal rotors designed such that the volume through which a particle moves is proportional to its sedimentation constant.

Feed head. A device incorporating the rotating and stationary seals which fits over the core of a zonal rotor.

Fixed-angle rotor. The centrifuge tubes are inserted into holes in the rotor body which are set at a defined angle to the axis of rotation.

Flotation gradients. These are gradients that can be used to separate particles on the basis of their flotation rate. For these gradients the sample is loaded in a dense solution into the bottom of the gradient. This technique is especially useful for sub-fractionating membrane fractions.

Flow equations. The thermodynamics of irreversible processes provide a rigorous framework to understand the processes that occur during a sedimentation velocity experiment. The net flow of all the components can be defined by equations that incorporate their sedimentation and diffusion characteristics.

Fluid seal. See feed head.

Freeze-thaw gradients. Gradients produced *in situ* by freezing and thawing a uniform solution.

Gradient capacity. The maximum amount of sample that can be sedimented on a gradient without apparent deviation from the ideal sedimentation pattern.

Gradient maker. Mixing chamber(s) arranged to allow the formation of gradients.

Gradient profile. See gradient shape.

Gradient reorientation. A phenomenon observed in fixed-angle and vertical tube rotors where the centrifuge tube does not move during the centrifugation but the density gradient reorientates itself to lie parallel to the centrifugal force field during centrifugation and then reorientates back to its original position as the rotor comes to rest.

Gradient shape. This is the concentration profile of that particular gradient, i.e., the variation of the gradient solute concentration along the centrifuge tube.

Guard tray. A Perspex tray which fits over the rotor, secured to the centrifuge chamber, and supporting the feed head.

Hypertonic solutions. Solutions whose osmolarity is greater than the osmolarity of the cytoplasm of cells.

Hypotonic solutions. Solutions whose osmolarity is less than the osmolarity of the cytoplasm of cells.

Ideal solution. The behaviour of solutions of macromolecules is influenced by the concentration of the macromolecule. Thermodynamic treatments take account of concentration-dependent behaviour by the introduction of, for example, virial coefficients into the equations that describe the behaviour of the macromolecules in a centrifugal field. When the concentration of the macromolecule becomes vanishingly small, the virial effects disappear and the behaviour of the solution approximates to that of a thermodynamically 'ideal' solution.

Immunochemical staining. Staining to reveal the presence of a certain antigenic property.

Integrator. An electronic arrangement which automatically integrates the angular velocity and the time of the run.

Isokinetic gradients. Gradients which permit sedimenting particles to move at a constant speed throughout the length of the gradient.

Iso-osmotic gradients. Gradients in which the osmolarity is constant throughout. Often these gradients are isotonic in nature.

Isopycnic centrifugation. Sample particles sediment through a density gradient which eventually exceeds the density of the particles to be separated such that the particles sediment only until they reach a density equal to the particle density and then band at that position.

Isotonic solutions. Solutions whose osmolarity is equal to that of the cytoplasm of cells.

Isovolumetric gradients. See equivolumetric gradients.

Lines. Term used in conjunction with zonal rotors for the channels of liquid entering and leaving the rotor.

Magnification factor. In analytical ultracentrifuge experiments there are many ways of recording or photographing the distributions in the cell. Most of these records or photographs have dimensions that differ from the dimensions of the cell itself so that the factor by which the image has been magnified must be determined and distances can be referred back to actual distances in the ultracentrifuge rotor.

Marker enzymes. Enzymatic activities associated with specific subcellular organelles; some of these may differ between different cell types.

Monodisperse solutions. Solutions of macromolecules that contain only one species of the macromolecule.

Non-denaturing gradients. Aqueous gradients which maintain the native form of the sedimenting particles.

Overlayering. A method for performing density gradients which involves manually layering less dense gradient solutions over more dense gradient solutions in a centrifuge tube.

Partial density increment. This is a thermodynamic term that is extremely useful for the study of multicomponent macromolecular systems. When a macromolecule is present in a solution with a number of low molecular weight solutes, it will interact with them differentially according to the particular solution conditions. It can be difficult to measure exactly the amount of each component that is bound to the macromolecule so that the exact nature of the 'macromolecule' can be hard to define. The difference in density between the solution of the macromolecule and the dialysate at dialysis equilibrium divided by the concentration of the macromolecule is the partial density increment. This term takes account of all the interactions that occur and greatly assists unambiguous interpretation of ultracentrifuge data.

Partial specific volume. The reciprocal of the non-hydrated density of a particle, that is the volume occupied by 1.0 g of the particles and it is denoted by $\bar{v}$.

Plateau region. In sedimentation velocity the plateau region refers to the region between the boundary and the bottom of the cell (or tube).

Polydisperse. A solution containing more than one macromolecular species.

Polysomes (polyribosomes). The complex of mRNA with associated ribosomes.

Preformed gradients. Density gradients which are formed in the centrifuge tube before centrifugation begins.

Radial dilution. In a sector-shaped cell the concentration of particles in the plateau region will constantly decrease during a sedimentation velocity run by an amount specified by the radial dilution square law.

Rate controller. A device available for most modern high-speed and ultra-centrifuges which allows extremely slow acceleration and deceleration during centrifugation, thus minimising disturbance to the sample and gradient.

Rate-zonal centrifugation. The sedimentation of sample particles through a solution which is less dense that the particles themselves causing particles with different sedimentation rates to separate into distinct sedimenting zones.

Rayleigh optical system. An optical system present on most analytical ultra-centrifuges that uses narrow slits in the windows at the base of a double-sector cell assembly as part of an interferometer so that interference fringes are produced. In areas of the cell in which changes in refractive index occur fringes are deflected and changes in composition can be detected in the ultracentrifuge cell.

Reorientation of gradients. See gradient reorientation.

Reorienting (Reograd) rotor. A zonal rotor in which the gradient is loaded while the rotor is stationary. When the rotor accelerates the gradient reorients to the vertical position, on deceleration the gradient reorients to the horizontal plane.

Resolution. The separation of different particles into zones on the basis of their size.

Rotating seal. Cylindrical block of Tribolon (graphite-filled polyimide plastic) or Rulon (a filled fluorocarbon) situated on top of the rotor core which carries separated fluid lines of the inlet and outlet of the rotor.

Rotor core. Central pillar which carries (a) fluid lines to the centre and edge of the zonal rotor and (b) the septum assembly.

Schlieren optical system. An arrangement of optical lenses in the analytical ultracentrifuge that detects changes of refractive index in solutions. The lines seen in the schlieren image indicate changes of the concentration gradient at each position in the cell.

Sedim (values) function. Used in the calculation of sedimentation coefficients; these values are dependent on temperature, density of the particle and concentration of the gradient.

Sedimentation coefficient. This is defined as the velocity of a particle per unit centrifugal field. It is usually corrected to standard conditions (20°C in water) and extrapolated to infinite dilution (see standard sedimentation coefficient).

Sedimentation equilibrium analysis. Ultracentrifuge experiments carried out for a sufficient time so that the forces of sedimentation and diffusion become balanced and no further net transport of any component occurs.

Sedimentation velocity analysis. Ultracentrifuge experiments in which net transport of components occurs at rates dictated by their size and shape and from which their sedimentation characteristics can be determined.

Self-forming density gradients. A solution of a density gradient solute and water in a centrifugal field will form a gradient of concentration and therefore density. The gradient produced at equilibrium is predictable from the chemical and physical properties of the solute. Many solutes can form gradients in which a large difference in density at the top and bottom of the cell is present and can be used to separate macromolecules on the basis of differences in buoyant density.

Septum (vane) assembly. Noryl or Perspex cylinder supporting four vanes which fit over the zonal rotor core. It carries liquid to the edge of the zonal rotor.

Standard sedimentation coefficient. The sedimentation coefficient of a macromolecule measured in an ultracentrifuge experiment depends on many variables such as concentration, temperature and low molecular weight solutes present. In order to compare sedimentation coefficients measured under different experimental conditions, it is necessary to correct for these factors and calculate a value corresponding to standard conditions, usually in water at $20°C$ ($s_{20,w}$).

Stationary (static) seal. Hollow stainless steel block whose lower face, which carries central and edge fluid lines, is in contact with the rotating seal.

Swing-out (swinging bucket) rotor. The centrifuge tubes sit in buckets attached to the rotor via pivots such that the buckets are vertical whilst the rotor is at rest but swivel through 90° to lie horizontally (perpendicular to the rotor's axis of rotation) during centrifugation.

Svedberg (S). The unit of sedimentation rate, equivalent to 10^{-13} sec.

Three component system. A solution of a macromolecule in water to which has been added a third component such as a detergent or a large amount of salt. In such systems, preferential binding events will occur with the macromolecule.

Top unloading. Recovery of gradients by pumping the gradient out of the centrifuge tube via a narrow tube passed through the gradient to the tube bottom; not generally recommended.

Transport process. This is a situation in which net movement of the macromolecules in a solution takes place under the influence of an external driving force such as a centrifugal force. The principles of the thermodynamics of irreversible processes offer a unifying approach when applied to different techniques in which transport occurs.

Two-component system. A solution of a pure solute, possibly a macromolecule, in water. Water is conventionlly taken to be component one and the solute is component two. A solution of a macromolecule in dilute buffer is often taken to be effectively a two component system.

Underlayering. A method for preforming density gradients which involves the use of a syringe to layer more dense gradient solutions underneath less dense gradient solutions already in the centrifuge tube.

Unit gravity velocity sedimentation. Separation of cells on the basis of sedimentation rate without applying a centrifugal force.

Upward displacement fractionation. Recovery of gradients by displacing the gradient upwards and out of the centrifuge tube by the introduction of a denser displacing medium into the bottom of the centrifuge tube.

Vanes. See septum assembly.

Velocity sedimentation. See rate-zonal centrifugation.

Vertical (tube) rotor. The centrifuge tubes are inserted into holes set vertically into the rotor body.

Wall effects. Deviation from the ideal sedimentation pattern of bands of particles as the result of the axial centrifugal forces sedimenting particles onto the tube wall.

Weight average molecular weight. A polydisperse sample can be fully characterised by the appropriate molecular weight distribution, but it is often difficult to determine this directly. Analytical ultracentrifugation is used to measure representative average molecular weights from which it may be possible to deduce the corresponding molecular weight distribution. The molecular weight directly measured in ultracentrifugation is the weight average molecular weight.

Zonal rotor. A rotor which is not divided into separate centrifuge tube holding compartments; rather, the rotor itself is hollow and is filled with one sample only.

Zone. See band.

Zone-sharpening. A phenomenon observed during rate-zonal sedimentation in density gradients where there is also a concomitant gradient in viscosity along the length of the centrifuge tube; particles at the leading edge of the zone travel slower than those in the lower viscosity medium in the trailing edge of the zone causing a narrowing or sharpening of the zone.

INDEX